Lecture Notes in Computer Science 2673

Edited by G. Goos, J. Hartmanis, and J. van Leeuwen

W0114710

Springer
Berlin
Heidelberg
New York
Hong Kong
London
Milan
Paris
Tokyo

Nicholas Ayache   Hervé Delingette (Eds.)

Surgery Simulation and Soft Tissue Modeling

International Symposium, IS4TM 2003
Juan-Les-Pins, France, June 12-13, 2003
Proceedings

 Springer

Series Editors

Gerhard Goos, Karlsruhe University, Germany
Juris Hartmanis, Cornell University, NY, USA
Jan van Leeuwen, Utrecht University, The Netherlands

Volume Editors

Nicholas Ayache
Hervé Delingette
INRIA, Projet Epidaure
2004 route des Lucioles, 06902 Sophia-Antipolis, France
E-mail:{Nicholas.Ayache/Herve.Delingette}@sophia.inria.fr

Cataloging-in-Publication Data applied for

A catalog record for this book is available from the Library of Congress

Bibliographic information published by Die Deutsche Bibliothek
Die Deutsche Bibliothek lists this publication in the Deutsche Nationalbibliografie;
detailed bibliographic data is available in the Internet at <http://dnb.ddb.de>.

CR Subject Classification (1998): I.6.,I.4, J.3, I.2.9-10, I.3, I.5

ISSN 0302-9743
ISBN 3-540-40439-2 Springer-Verlag Berlin Heidelberg New York

Springer-Verlag Berlin Heidelberg New York
a member of BertelsmannSpringer Science+Business Media GmbH

http://www.springer.de

© Springer-Verlag Berlin Heidelberg 2003

Printed on acid-free paper   SPIN 10933374   06/3142   5 4 3 2 1 0

Preface

This book contains the written contributions to the International Symposium on Surgery Simulation and Soft Tissue Modeling (IS4TM 2003) which was held in Juan-Les-Pins, France, during June 12–13, 2003.

The articles are organized around four thematic sections which cover some of the major aspects of this rapidly growing field: soft tissue modeling, haptic rendering, cardiac modeling, and patient-based simulators. This book also includes the written contributions of two important poster sessions and, last but not least, the contributions of three invited speakers: Prof. David Hawkes of King's College (London), Prof. Simon Warfield of Harvard Medical School (Boston), and Prof. Naoki Suzuki of Jikei University (Tokyo).

The objective of the symposium was to bring together the researchers of this emerging field to present their most innovative and promising research work, to highlight research trends and foster dialogue and debates among participants.

This event was decided on after two preliminary successful meetings organized, respectively, by J. Duncan and G. Szekely in Boston in 1998, and by E. Keeve in Bonn in 2001. The organization by INRIA (French Research Institute in Computer Science and Automatic Control) was decided on to celebrate the 20th anniversary of the INRIA research unit based in Sophia-Antipolis, very close to the conference location.

We received 45 submitted full papers; each of them was evaluated by three members of the scientific committee. Based on the written reviews, we selected 16 papers for oral presentation and 17 papers for poster presentation. All accepted articles were allowed a written contribution of equal maximum length in these proceedings. All contributions were presented in a single track, leaving time for two poster sessions, and a panel session.

The geographical breakdown of the contributions is: 33 from European countries (excluding France), 7 from France, 6 from North America, and 2 from Asia.

The quality of the various contributions will make this conference an important milestone of this new but rapidly growing field at the confluence between several disciplines including medical image analysis and biomechanics.

We enjoyed welcoming all participants to what proved to be an intense and stimulating scientific event.

March 2003

Nicholas Ayache
Hervé Delingette

Acknowledgments

We would like to thank the members of the program committee for reviewing the submitted papers in a timely manner and also for supporting the idea and the organization of the symposium. Additional reviewers included François Faure (Université Joseph Fourier, France).

We thank the members of the Epidaure team at INRIA Sophia-Antipolis for their involvement and support in the organization of the conference. In particular, we thank Rupert Colchester and Guillaume Flandin for managing the conference website and Olivier Clatz for his help in preparing the camera-ready version of the proceedings.

We also thank Monique Simonetti, Isabelle Strobant and the "Bureau des Missions et Colloques" of INRIA Sophia-Antipolis for the overall organization of the conference.

Sponsoring Institutions

Institut National de Recherche en Informatique et en Automatique (INRIA, France)
Centre National de la Recherche Scientifique (CNRS, France)
Conseil Régional de la Région Provence Alpes Côte d'Azur (France)
Conseil Général des Alpes Maritimes (France)

Organization

IS4TM 2003 was organized by the Epidaure Laboratory and the Service des Relations Extérieures at INRIA Sophia Antipolis, France.

Conference Chairs

Nicholas Ayache   Epidaure, INRIA, Sophia-Antipolis, France
Hervé Delingette   Epidaure, INRIA, Sophia-Antipolis, France

Program Committee

M. Brady	Oxford University, UK
M.-P. Cani	INPG, France
C. Chaillou	INRIA, France
E. Coste-Manière	INRIA, France
S. Cotin	CIMIT, USA
C. Davatzikos	University of Pennsylvania, USA
S. Dawson	CIMIT, USA
T. Dohi	Tokyo University, Japan
J. Duncan	Yale University, USA
H. Fuchs	University of North Carolina, USA
D. Hawkes	King's College London, UK
D. Hill	King's College London, UK
K.-H. Hoehne	Hamburg University, Germany
R. Howe	Harvard University, USA
E. Keeve	CAESAR, Germany
R. Kikinis	Harvard Medical School, USA
F. Kruggel	Max Planck Leipzig, Germany
U. Kühnapfel	Karlsruhe University, Germany
J.-C. Latombe	Stanford University, USA
C. Laugier	INRIA, France
D. Metaxas	Rutgers University, USA
M. Miga	Vanderbilt University, USA
K. Montgomery	Stanford University, USA
W. Niessen	Utrecht University, The Netherlands
D. Pai	Rutgers University, USA
J. Prince	Johns Hopkins University, USA
R. Robb	Mayo Clinic, USA
F. Sachse	Karlsruhe University, Germany
L. Soler	IRCAD, France
P. Suetens	K.U. Leuven, Belgium
N. Suzuki	Jikei University, Japan
G. Szekely	ETH Zurich, Switzerland
R. Taylor	Johns Hopkins University, USA
D. Terzopoulos	New York University, USA
D. Thalmann	EPF Lausanne, Switzerland
M. Thiriet	INRIA, France
J. Troccaz	TIMC, France
M. Viergever	Utrecht University, The Netherlands
S. Warfield	Harvard Medical School, USA

Table of Contents

Invited Speaker

Session 1: Soft Tissue Models

Poster Session 1

Invited Speaker

Session 2: Haptic Rendering

Invited Speaker

Session 3: Cardiac Modeling

Poster Session 2

Session 4: Patient Specific Simulators

Measuring and Modeling Soft Tissue Deformation for Image Guided Interventions

David J. Hawkes, P.J. Edwards, D. Barratt, J.M. Blackall,
G.P. Penney, and C. Tanner

Imaging Sciences, 5th Floor Thomas Guy House, Guy's Hospital, London Bridge
London SE1 9RT, United Kingdom
david.hawkes@kcl.ac.uk
http://www-ipg.umds.ac.uk

Abstract. This paper outlines the limitations of the rigid body assumption in image guided interventions and describes how intra-operative imaging provides a rich source of information on spatial location of key structures allowing a pre-operative plan to be updated during an intervention. Soft tissue deformation and variation from an atlas to a particular individual can both be determined using non-rigid registration. Classic methods using free-form deformations have a very large number of degrees of freedom. Three examples – motion models, biomechanical models and statistical shape models – are used to illustrate how prior information can be used to restrict the number of degrees of freedom of the registration algorithm to produce solutions that could plausibly be used to guide interventions. We provide preliminary results from applications in each.

1 Introduction: Image Guided Surgery Assuming the Validity of the Rigid Body Transformation

Image guidance is now well established in neuro, otolaryngology, maxillo-facial and orthopaedic surgery. A 3D pre-operative plan, based on images acquired before the procedure, is registered to the physical space of the patient in the operating room. This is usually done by establishing correspondence between a set of point landmarks or fiducials fixed to the patient and their corresponding location in the pre-operative images. These point locations are then used to derive the rigid body transformation that minimizes the sum of squared displacements between corresponding points. Alternatively we can use corresponding sets of points acquired on an accessible, visible surface in the patient, such as the skin surface or exposed bone, together with the corresponding surface in the pre-operative images. Palpation of points on the surface using a tracked pointer or by triangulation using a laser beam captures the point locations in physical space registration of these points to the surface determines the transformation. Once registration has been established navigation proceeds by tracking a pointer together with a tracker attached rigidly to the patient and computing the corresponding location of the pointer tip in the pre-operative plan. Trackers can take the form of bone screws attached directly to bone for maximum accuracy (and invasiveness), other restraining devices fixed to the skull such as the Mayfield

N. Ayache and H. Delingette (Eds.): IS4TM 2003, LNCS 2673, pp. 1–14, 2003.

Clamp, rigidly fixed head bands, or dental stents or moulds rigidly attached to the teeth or palate. A variety of graphical user interfaces (GUIs) have been devised for display but by far the most widespread is the near real-time display of three orthogonal views intersecting at the tracked pointer tip (figure 1). Surface rendered displays are also used and various specialised GUIs have been devised to guide the use of specific tools and instruments such as biopsy needles, rigid endoscopes and treatment delivery systems. These navigation systems are now widely available as commercial products, but their application is restricted to bone or structures near to or encased by bone, where the rigid body assumption remains reasonably valid throughout the procedure.

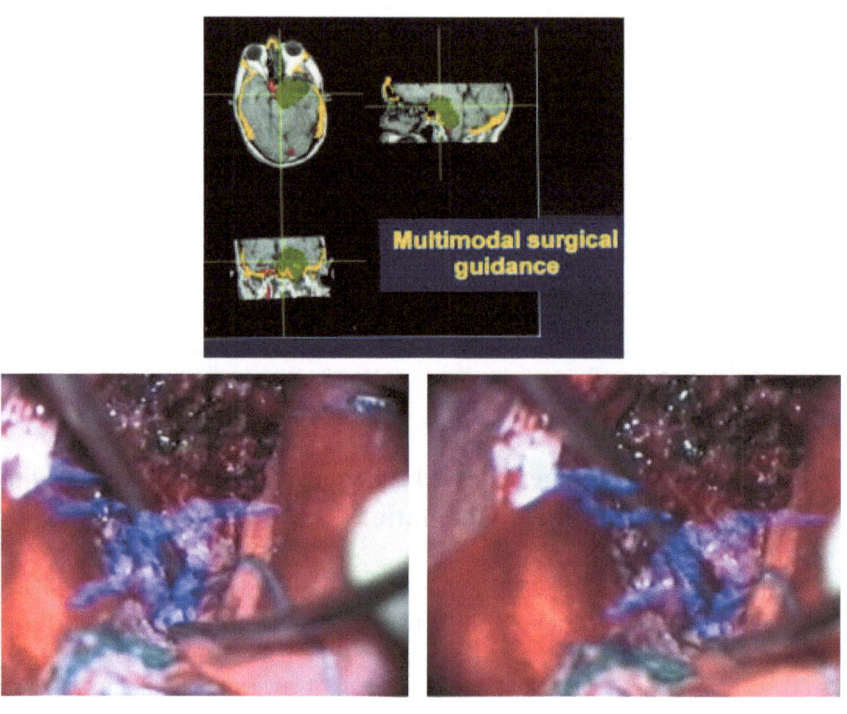

Fig. 1. Displays for image guided surgery. (*Left*) Orthogonal cuts through a fused 3D volume obtained from MR, bone delineated from CT, vascular structures from MRA and tumour from Gd-DTPA enhanced MR. (*Right*) A stereo pair (relaxed eye) of vascular structures derived from pre-operative MRA taken towards the end of a deep excision to remove a meningioma as seen through the MAGI system - an augmented reality system based on stereo projection into an operating microscope [2].

A recent development is the use of augmented reality (AR), in which the real surgical scene and information derived from the pre-operative plan are combined. This can be done in one of two ways. In "see through" AR appropriately calibrated and rendered views of the pre-operative plan are projected into a view of the real surgical scene. Examples of this include the Varioscope see-through head mounted system [1] and the MAGI system in which images are projected view beam splitters into the binocular optical paths of a stereo operating microscope (figure 1) [2]. Alternatively video images can be acquired of the surgical scene and fused with the pre-operative plan prior to display. The advantage of the former is that the integrity of the real surgical scene is largely preserved and remains in view in real-time, while the advantage of the latter is that images are fused prior to display stopping the disconcerting effect of the AR displays "swimming" above the real scene, albeit with a penalty of a small time delay and some inevitable degradation of perceived quality of the real scene. Development of appropriate AR displays remains a research issue, although these technologies have the potential to provide navigation where it is needed and can provide an immediate feedback that navigation accuracy has been lost.

All these systems rely on the rigid body transformation remaining valid throughout the operation. While some checking of registration against identifiable landmarks or structures is provided in most surgical protocols, there is no immediate compensation for any patient movement or non-rigid deformation relative to the tracker fixed to the patient. Re-registration in the event of such motion may be feasible during certain procedures but usually is impractical once the intervention is underway. This is the most common cause of failed navigation in image-guided surgery.

2 Intra-operative Imaging for Guidance

Imaging taken during an intervention provides real-time feedback of any motion of the patient and allows tracking of flexible instruments and catheters. Interventional radiology has been available for about 50 years using X-ray fluoroscopy, which provides real-time tracking of catheters and needles relative to X-ray visible structures such as bone and vascular structures, bile duct or ureters after the administration of iodine containing contrast material. Ultrasound and CT fluoroscopy are used to guide placement of biopsy needles and therapy delivery devices. Multislice CT and interventional MR can provide 3D support for interventions.

Intra-operative imaging, as long as the image sensor is carefully calibrated, provides a rich source of accurate spatial information that can be used to aid image registration during image-guided interventions. There is much more spatial information in an image than can be gathered with a hand held pointer. Intra-operative imaging provides information on invisible, deep structures and, potentially, provides a route to tracking motion and updating the pre-operative plan to compensate for soft tissue deformation.

A wide range of imaging modalities can, in principle, be used in the operating room including X-ray, ultrasound, interventional MR, CT and video.

2.1 X-ray

X-ray imaging has been available in the operating room for many years in orthopaedics, maxillo-facial surgery and orthopaedics. Recent advances in X-ray generator technology, solid state detectors and digital image manipulation have improved image quality, increased versatility and greatly simplified the process of integration of X-ray imaging into guidance systems.

X-ray images can be used to track the location of X-ray visible markers. A pair of suitably calibrated X-ray views can be used top locate structures in physical space to sub-millimetre accuracy. These locations can be used to register pre-operative images and plan as described above.

X-ray images can also be used to locate bone structures such as the vertebrae, skull or pelvis. Lavallee [3] has shown how aligning projections of the silhouettes of X-ray images and tangent planes of surfaces in CT derived 3D models can achieve registration. Using the concept of the digitally reconstructed radiograph (DRR), pseudo-X-ray projections can be generated from a pre-operative CT scan in any arbitrary direction. Registration proceeds by testing different DRR projections until there is a match between the X-ray image and the DRR. A range of similarity measures have been proposed for this match and we have shown that two of these - pattern intensity and gradient difference – provide a robust and accurate method for registering vertebrae in image guidance procedures [4].

Post operative imaging can also be used to verify surgical procedures. For example X-ray imaging is usually used to verify prosthesis location in total hip and total knee replacement surgery. These images can be registered in an identical way to provide measures of the accuracy of placement of prostheses. In total hip replacement, for example, the relative pose of the acetabular cup and the pelvis are separately determined from a pair of X-ray views and compared with the surgical plan, figure 2. [5].

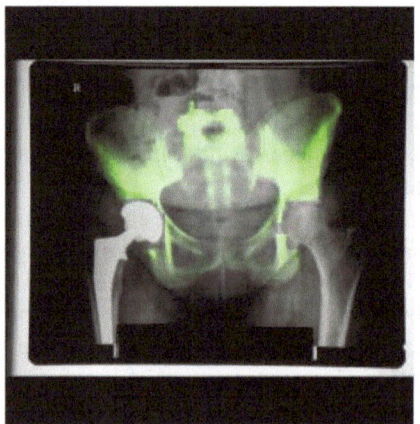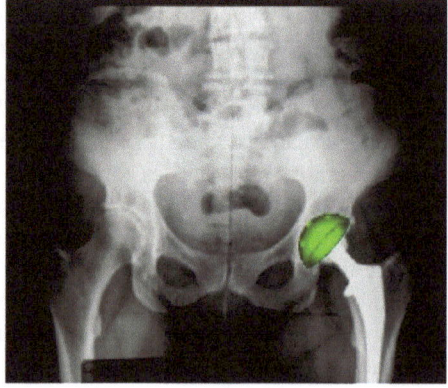

Fig. 2. Overlay of a DRR from CT (*left*) and CAD model of acetabulum (*right*) illustrating how 2D-3D registration can be used to verify prosthesis placement with respect to a surgical plan [5].

The DRR concept can also be extended to other X-ray visible structures and we have recently shown how 3D vascular networks derived from MR angiography, CT angiography or 3D rotating X-ray angiography can be aligned automatically with the 2D X-ray fluoroscopy images to provide 3D support in complex neurovascular interventions such as insertion of coils in aneurysms or embolisation of feeding vessels in arterio-venous malformations [6].

2.2 Ultrasound

Ultrasound provides a rich source of information on the structure and location of anatomy. The technology is relatively cheap, widely available and convenient to use in the operating room environment, but the images are notoriously difficult for the untrained observer to interpret although image quality has improved significantly in the last few years. The convenient size and ease of manipulation of an ultrasound probe make it an ideal device for intra-operative guidance. The standard 2D or B-mode probe can be tracked accurately in 3D in the same way as a tracked pointer and careful probe calibration means that visible structures in the ultrasound image can be located in physical space with an accuracy of about 1mm [7], [8]. The ultrasound effectively becomes a tracked pointer that can identify multiple anatomical features deep within the patient.

In the context of rigid body registration bone surfaces provide a very strong ultra-sound signal and many bone surfaces are easily accessible by conventional ultrasound from the skin surface. A-mode ultrasound has been used to locate the surface of the skull for image guided surgery [9] and B-mode ultrasound has been used to locate the surface of the skull, vertebrae [10], pelvis and femur. Registration accuracy is ap-proximately 1mm, which is comparable to bone implanted markers and more accurate than skin markers or palpation of the skin surface.

As ultrasound can also provide information on the boundaries of soft tissue struc-tures its role extends to compensating for soft tissue motion and deformation as de-scribed below.

The velocity of sound is not the same through all tissues, varying by 10-15% be-tween adipose tissue (fat) and lean tissue. This will lead to in-plane distortions and out-of-plane diffraction effects that may compromise spatial integrity of the images. Ultrasound is also prone to signal drop out due to sudden changes in acoustic imped-ance that might occur at tissue-bone or tissue air interfaces, resulting in acoustic shadows cast from these surfaces away from the probe.

2.3 Optical Imaging

Optical imaging is the method of choice for tracking markers in image guided surgery and most commercial systems are based on optical tracking devices with an intrinsic accuracy of locating infra-red LEDs or reflecting spheres of 0.1 to 1mm. Surfaces can also be reconstructed using optical methods usually with two calibrated cameras and a projected light pattern [11]. These surfaces are then registered to the corresponding surface in the preoperative plan as described above using ICP or its variants.

Optical images are available from a variety of devices used in surgery including the endoscopes and surgical microscopes. With appropriate calibration the images can provide information on the location of structures visible during an intervention (e.g.

the ventricles within the brain and structures visible in bronchoscopy, colonoscopy or functional endoscopic sinus surgery (FESS)).

Recently we have proposed an alternative registration method based on the concept of photoconsistency that establishes the pose of a known surface in relation to a pair of optical images based on the fact that a Lambertian reflecting surface will have similar brightness independent of viewpoint. The pose of the surface "most consistent with" the two optical images is found by an optimization process [12].

2.4 Interventional Magnetic Resonance Imaging (iMRI)

Interventional MRI has the potential to provide 3D imaging during an interventional procedure. In principle the 3D information acquired will be sufficient to guide the procedure. In practice certain information such as contrast enhancement characteristics, neuro-activation using functional MR, diffusion tensors, tissue perfusion and accurate segmentation using atlases may be very difficult to obtain during an intervention. Guidance can however be augmented by accurate alignment of pre-operative images with intra-operative imaging. Registration is 3D to 3D and robust and accurate registration methods based on voxel similarity measures such as mutual information or its variants are available [13].

Some interventional MRI scanners are more prone to geometric distortion and although the assumption is that the anatomical structures have not changed shape nonrigid registration may be required to bring pre- and inter-operative image into alignment.

3   Modeling Soft Tissue Deformation in Image Guided Interventions

All the methods described above assume that there is a rigid body transformation between the physical anatomy of the patient at the time of pre-operative imaging and during the intervention. This restricts application to interventions on bony structures (ortopaedics, maxillofacial surgery) or soft tissue structures that are close to or encased in bone (skull-base, neuro and sinus surgery). This excludes the vast majority of surgical procedures that could, in principle, benefit from image guidance. Intraoperative imaging provides a source of information on soft tissue structures. How can this information be used to guide interventions?

In the last few years there has been significant progress in non-rigid registration, i.e. automating the non-rigid alignment of 3D datasets. These algorithms require a very large number of degrees of freedom in describing the transformation between pre and intra-operative image space in order to be able to compensate both for local deformation and large-scale tissue motion. For example an algorithm that we have devised and found to be useful in a wide range of non-rigid 3D to 3D registration tasks is based on free form deformation defined by an array of node points defining a set of approximating B-splines. For a typical node spacing of 10mm and a field of view of 250mm by 250mm by 200mm this corresponds to over 14,000 degrees of freedom. Many 3D images provide sufficient information to drive such a registration. The algorithm has been used successfully to align serially acquired time sequences of volumes of the breast [14] and liver in the study of the dynamics of contrast enhancement following administration of Gd-DTPA.

The algorithm has also been used to align pre and post-operative images of the brain acquired in an interventional MR system in the study of brain deformation during surgery, so-called brain-shift [15]. Unfortunately this and similar algorithms are extremely computationally demanding. On a single machine registration can take many hours and only become a practical proposition for interventional work on very large multi-processor supercomputers that are unlikely to be available for routine image guidance work for the foreseeable future.

In addition intra-operative information from images rarely comprises a completely sampled 3D volume. X-ray imaging provides a 2D perspective projection, standard tracked B-mode ultrasound provides a set of discrete slices whose relative position is known but is likely to sample only a small proportion of the volume of interest, optical images only record visible surfaces.

In order to use this information to update the pre-operative plan to compensate for soft tissue motion and deformation we need additional information. In effect we need to decrease the number of degrees of freedom of the non-rigid transformation of the plan to the intra-operative image. We propose a number of ways this might be done using recent advances in statistical shape modeling, biomechanical modeling and motion modeling of structures that move cyclically due to respiratory or cardiac motion.

The rest of the paper will describe methods and applications, all work in progress, to illustrate our approaches.

3.1 Motion Modeling

Respiratory and cardiac motion are cyclic and to some extent predictable. It is their predictable nature that allows gated image acquisition protocols to produce good quality images although image acquisition times may be long compared with respiratory or cardiac cycles. However, the accuracy of image guided interventions using pre-operative imaging in the heart, lung, liver and other abdominal organs are severely limited by organ motion and deformation.

In recent work we have shown how non-rigid registration technology can be used to align series of images taken over the breathing cycle to generate a 4D motion model of the heart, lung and liver from CT or MR [16] [17]. There are a number of strategies available to compensate for respiratory motion. Images can be taken very quickly, for example using echo planar imaging in MR. They can be taken at a series of individual breath holds spread over the breathing cycle. The acquisition can be gated to particular stages of the breathing cycle using external bellows attached to the patient or, in the case of MR, using navigator echoes placed over the diaphragm. These images are aligned and the motion of landmarks from the resulting deformation field is used to build the motion model. Landmarks can either be placed on segmented structures such as surfaces and the location propagated using the deformation field determined by non-rigid registration [18] or the node points used in the registration can provide the motion model [19].

Figure 3 shows an example of this process for the liver over the breathing cycle. These models are 4 dimensional although the 4th dimension is only a parametrisation of true time, i.e. a parametrisation of a trajectory through shape space over the breathing cycle. The time "scale" is neither fixed nor uniform as the breathing rate and amplitude can vary significantly.

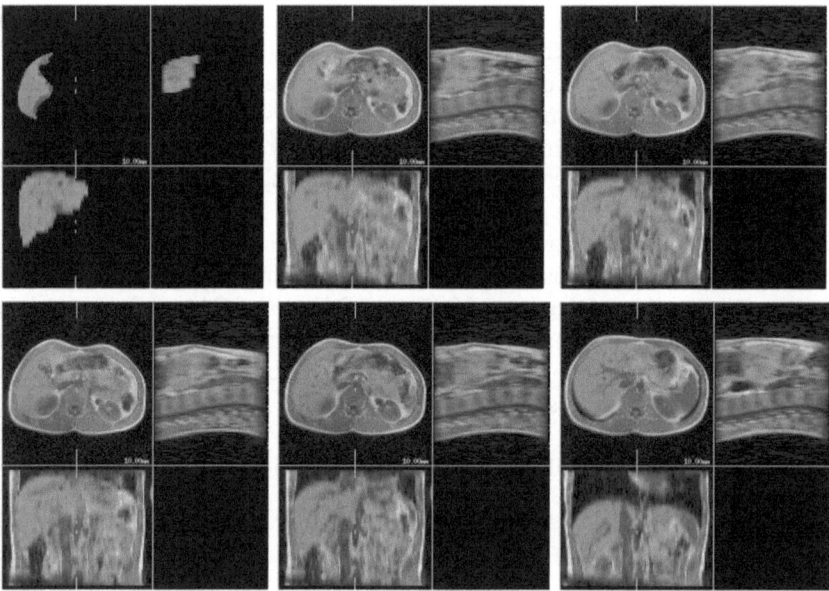

Fig. 3. Template at maximum exhale (*top left*) is registered non-rigidly to 5 volumes acquired over the breathing cycle to generate a breathing motion model of the liver.

We are using this technology to develop applications in improving the accuracy of radiotherapy treatment of the lung and in improving the accuracy of ultrasound guidance of ablation of liver metastases.

Image Guided Ultrasound Ablation of Metastatic Disease in the Liver

Radio frequency ablation of liver metastases provides an alternative to liver resection in the treatment of primary and secondary liver disease, with metastatic spread from colon and bowel cancer a significant source of fatal liver disease. Lesions are visible in contrast enhanced CT and most interventionists use CT to guided ablation by inserting needles that carry laser or radio frequency energy for thermo-ablation or refrigerants in cryoablation. We propose an alternative scheme in which CT is used to plan a procedure and ultrasound to guide ablation.

First we collected breath-hold MR volumes of 7 volunteers at 6 different stages of the breathing cycle from maximum inhale to maximum exhale. The 3D T1 weighted MR volumes encompassed the whole liver volume with a voxel size of 1.33mm x 1.33mm x 10.0mm and each took approximately 40 seconds to acquire. Analysis of these images showed displacements of the liver of up to 28mm over the breathing cycle with non-rigid deformations of 4.3mm (RMS) and 14.1mm (max). We generated patient-specific statistical shape models from the coordinates of surface points generated on one image which were propagated to the five other images using non-rigid registration. We generated a point distribution model (PDM) [20] from these point coordinates and fitted the weights of each mode of variation of the PDM to a polynomial function of position in the breathing cycle. These models were used as priors in a Bayesian formulation and the registration algorithm was extended from the 6 parameters of rigid body registration to 7 parameters where the 7th is the position in the breathing cycle [21][22]. Initial results were promising and showed TREs of less than 8mm as judged by displacement between hand-picked points on the liver surface in the ultrasound images and the nearest surface points from MR. This was the first time that a model of breathing motion had been used to constrain non-rigid registration from many thousands of degrees of freedom to just one extra degree of freedom, representing position in the breathing cycle. We also demonstrated how the registration process could be split into an initial rigid body component, using several ultrasound images, followed by a tracking component consisting of a single parameter search using just one ultrasound image.

This algorithm was still based on alignment of the surface of the liver and, although displacements of surface points after registration were small, closer examination of true TREs showed significant errors. We therefore developed a volume-based rather than surface-based registration method. We compared two motion models derived directly from the displacement of the control point grid generated by the non-rigid registration algorithm. One of these was a PDM as described above and the other was a direct polynomial fit to the displacement of the control points. We tested the resulting algorithms using data from four additional MR volumes spanning the shallow breathing cycle for each of five volunteers. These additional MR volumes were not used to generate the motion model. Slices were extracted from these MR datasets to simulate ultrasound at orientations corresponding to typical ultrasound slices. Voxel-based registration was then performed using cross correlation as the similarity measure. While both motion models performed well, the single-parameter model was more robust when reduced data was used. Using just a single US slice, registration succeeded in all but one of twenty cases, with RMS TRE values over the whole liver ranging from 2.9mm to 12.5mm.

Encouraged by these results we have recently incorporated an automatic alignment tool based on maps of vessel probability into the non-rigid registration algorithm constrained by the single-parameter motion model. This has produced excellent results on three volunteers who have had tracked ultrasound images acquired of their liver. No gold standard exists for these datasets but results showed that only two of thirty registrations failed to produce a good visual alignment of MR and US features.

Figure 4 shows corresponding MR and US slices before and after compensation for soft tissue deformation using the motion model.

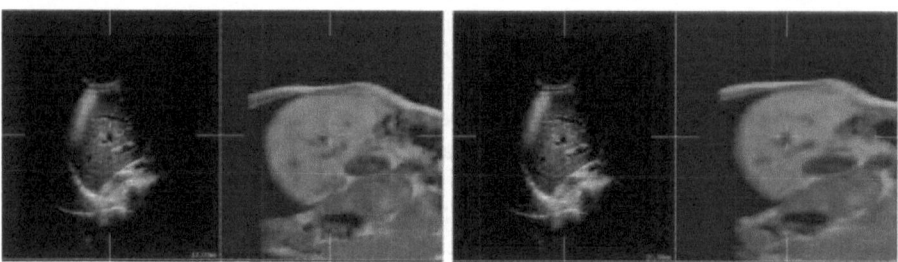

Fig.4. A single ultrasound slice (*left of each pair*) of a volunteer aligned automatically to the corresponding slice of an MR volume (*right of each pair*) assuming rigid-body motion only (*left*) and after alignment using the non-rigid breathing model (*right*). Note that the vascular structures and liver outline are much more accurately aligned using the breathing motion model.

Breathing Motion Correction of Radiotherapy Treatment Plan and Delivery

Person-specific motion models of the lungs were constructed using a voxel-based image registration technique to co-register a number of images acquired throughout the breathing cycle. Anatomic features were identified from the apex to the dome of the diaphragm on high-quality MRI volumes acquired using respiratory bellows to gate image acquisition. The reference image was automatically aligned to each of a sequence of rapidly acquired free-breathing MRI scans. The resulting affine transformations were used to transfer the landmarks from the reference image to the corresponding positions in the images throughout the rest of the cycle. Lung motion models were formed for four volunteers. Displacement of landmarks at the lung apex was in the range 2.5 - 6.2mm from inhale to exhale during shallow breathing. In all volunteers the magnitude of displacement was greater at the lung base measuring 7.7 - 16.0mm [17]. Work is now in progress to generate motion models during free breathing using EPI MR.

3.2 Biomechanical Modeling

We have shown how biomechanical models based on finite element methods can predict deformation in a highly mobile and deformable structure such as the female breast. We have shown previously that our non-rigid registration (NRR) algorithm [14] can produce visually plausible compensation for soft tissue deformations that occur during contrast enhanced dynamic MR volume acquisitions. To validate this method we constructed finite element models (FEMs) on 5 patients, imposed plausible artificial deformations and used the NRR algorithm to retrieve these artificial deformations. The average registration error over the whole breast was less than 0.5mm [23]. We have validated our FEM against tracking of visual landmarks on 3 volunteers and have found that the FEM, given appropriate boundary conditions, could predict the deformation of natural structures to within about 2mm [24]. Interestingly the accuracy of the FEM was remarkably independent of the precise form and biomechanical parameters of the model.

This work suggests that we could use biomechanical models to constrain NRR. Work is in progress to explore how this might be done with applications in the alignment of images of the liver, breast and brain.

We have shown how we can use non-rigid registration to propagate an accurately defined mesh from a brain atlas to an individual's MR scan and hence use this model to propagate deformations that occur during placement of cortical electrodes during surgery for epilepsy [25].

3.3 Statistical Shape Modeling

In certain circumstances an anatomical shape can be parametrised by a relatively small number of parameters with sufficient accuracy for image-guided surgery. One way to do this is to form Principle Component Analysis of the locations of corresponding points on a training set. Such a Point Distribution Model [20] can provide a compact description of a shape and its variation across a population. We can use this technology to both reduce the number of degrees of freedom in non-rigid registration to account for soft tissue deformation (as described above for the liver example) but also to reconstruct a reasonably accurate 3D model of the shape. In certain circumstances this might be sufficiently accurate for image-guided interventions. This concept has been exploited recently in orthopaedic surgery [26] [27] whereby X-ray imaging is used to instantiate 3D models of the vertebrae and pelvis for image guided orthopaedic surgery.

We have developed a system to use ultrasound to instantiate and register a statistical shape model of the femur, again for orthopaedic surgery [28]. A statistical shape model is created by registering, using NRR, a set of 10 CT scans of femurs and using the resulting deformation field from the NRR to propagate a set of surface points. Principle component analysis is undertaken on this set of points. The intersection of ultrasound images with the bone surface of a cadaver femur immersed in water was extracted from a set of calibrated and tracked B-mode ultrasound images. These data is used to instantiate the shape model for this particular instance. Using 5 modes of variation the residual RMS error over the points used in the model was 1.9mm. RMS error after registration of this model to a CT scan of the femur yielded a residual error of 4.0mm. Although currently too high for clinical use these results are very promising and point the way to the use of ultrasound in orthopaedics to both instantiate a 3D model with recourse to CT scanning and registration of that model to guide the intervention.

Important problems remain to be resolved, not least the effect of abnormalities in bony anatomy that may well be the reason for the surgical intervention in the first place. The surgeon is usually attempting to reconstruct bony anatomy to be as close as possible to the normal pre-disease configuration. So this may be less of an issue than might first be thought. Another important issue is how specific the model should be and how many examples should be used to generate it. Almost all work to date has used of the order of 10 examples. Should separate models be generated for left and right, male and female, different ethnic groups or even different age groups or can all variation be usefully contained within one model.

Work is in progress to extend these concepts to other bony structures and other organs such as the liver, lungs and brain.

4 Conclusions and Discussion

In this paper we have discussed the limitations of the rigid body transformation in image-guided interventions. We have shown how intra-operative imaging using a wide range of technologies can provide a rich source of intra-operative spatial information. If image guidance technology is to find application outside the brain and away from bone then methods must be found to compensate for soft tissue deformation before and during an intervention. Classic non-rigid registration methodologies are too slow and poorly constrained for this task but we have shown above how we can constrain registration by using additional information. We have presented three examples of how this might be done using constraints derived from models of breathing motion, knowledge of the biomechanics of the tissue and knowledge of variation in shape across the population.

With this technology and efficient implementations we predict that there will be a dramatic increase in the range of interventions in which guidance incorporating pre-operative information can be brought to bear. As we use different types of information in the guidance process and information processing becomes more complex, checks of integrity of the guidance information provided become more difficult as the intervention proceeds. As this technology is introduced we must have effective validation strategies in place to confirm the robustness and accuracy of our technology. In addition we must devise effective quality assurance procedures to give the interventionist or surgeon faith that the spatial information provided is indeed valid at all the different stages of an intervention at which guidance is required.

Acknowledgements

The work described in this paper was supported by grants from the UK EPSRC, Philips Medical Systems, Depuy and Brainlab.

References

1. Birkfellner, W., Figl, M., Matula, C., Hummel, J., Hanel, R., Imhof,H., Wanshitz, F., Wanger, A., Watzinger, F., Bergmann, H.: Computer endhanced stereoscopic vision in a head-mounted operating binocular. Phys Med Biol 48 (2003) N1-N9
2. Edwards, P.J., King, A.P., Maurer C., Jr., DeCunha, D.A., Hawkes, D.J., Hill, D.L.G., Gaston, R.P., Clarkson, M.J., Pike, M.R., Fenlon, M.R., Chandra, S., Strong, A.J., Chandler, C.L., Gleeson, M.J.: Design and evaluation of a system for microscope-assisted guided interventions (MAGI). IEEE trans. Med. Imag. 19 (2000) 1082-1093

3. Lavallee, S., Szeliske, R.: Recovering the position and orientation of free-form objects from image contours using 3D distance maps. IEEE Trans Patt Anal and Machine Intell, 17 (1995) 378-390

4. Penney, G.P., Weese, J., Little, J.A., Desmedt, P., Hill, DL.G., Hawkes, D.J.: A comparison of similarity measures for use in 2D-3D medical image registration. IEEE Trans. Med. Imag 17 (1998) 586-595

5. Edwards, P.J., Penney, G.P., Slomczykowski, M., Hawkes, D.J.: Accurate measurement of hip and knee prosthesis placement from postoperative X-rays. Proc Computer Assisted Orthopaedic Surgery (CAOS), Santa Fe, USA, (2002).

6. Hipwell, J.H., Penney G.P., Cox, T.C.S., Byrne, J., Hawkes, D.J. "2D-3D intensity based registration of DSA and MRA – a comparison of similarity measures". In Proc. Medical Image Computing and Computer-Assisted Interventions (MICCAI 2002), Tokyo, Japan, September 2002, Lecture Notes in Computer Science, Vol. 2489, Springer Verlag, (2002) 501-508

7. Prager, R.W., Rohling R.N., Gee, A.H., Berman, L.: Rapid Calibration for 3D free-hand ultrasound. Ultrasound in Med and Biol 24 (1998) 855-869

8. Blackall, J.M., Rueckert, D., Maurer, C.R., Penney G.P., Hill, D.L.G., Hawkes, D.J.: An Image registration approach to automated calibration for freehand 3D ultrasound. Proc. MICCAI'00 Springer Lecture Notes in Computer Science, Vol. 1935, Springer Verlag, (2000) 462-471

9. Maurer, C.R., Jr, Gaston, R.P., Hill, D.L.G., Gleeson, M.J., Taylor, M.G., Fenlon, M.R., Edwards, P.J., Hawkes, D.J.: AcouStick: A tracked A-mode ultrasonography system for resgiatrtion in image guided surgery. Proc MICCAI 1999, Vol. 1679, LNCS, Springer, (1999) 953-962

10. Lavallee, S., Troccaz, J., Sautot, P., Mazzier, B., Cinquin, P., Merloz, P., Chirossel, J.P.: Computer assisted spinal surgery using anatomy based registration. In Registration for Computer Integrated Surgery: Methodology, Sate of the Art, eds. Taylor, R.H., Lavallee, S., Burdea, G.C., Mosges, R.W., Cambridge Mass (1995) 425-449

11. Henri, C.J., Colchester, A.C.F., Zhao, J., Hawkes, D.J., Hill, D.L.G., Evans, R.L.: Registration of 3-D surface data for intra-operative guidance and visualisation in frameless stereotactic neurosurgery. Computer Vision, Virtual Reality and Robotics in Medicine (CVRMed '95). Ayache N (ed), Springer,Berlin, (1995) 47-69.

12. Clarkson, M.J., Rueckert, D., Hawkes, D.J.: Using Photo-Consistency to Register 2D Optical Images of the Human Face to a 3D Surface Model. Transactions in Pattern Analysis and Machine Intelligence (T-PAMI) 23 (2001) 1266-1280

13. Studholme, C., Hill, D.L.G., Hawkes, D.J.: An overlap invariant entropy measure of 3D medical image alignment. Pattern Recognition 32 (1999) 71-86

14. Rueckert, D., Sonoda, L.I., Hayes, C., Hill, D.L.G., Leach, M.O., Hawkes, D.J.: Non-rigid Registration using Free-Form Deformations: Application to Breast MR Images. IEEE Trans. Medical Imaging 18 (1999) 712-721

15. Maurer, C.R. Jr, Hill, D.L.G., Martin, A.J., Liu, H., McCue, M., Rueckert, D., Lloret, D., Hall, W.A., Maxwell, R.E., Hawkes, D.J., Truwit, C.L.: Investigation of intraoperative brain deformation using a 1.5T interventional MR system: preliminary results IEEE Trans. Med. Imag. 17 (1998) 817-825

16. Blackall, J.M., Landau, D., Crum, W.R., McLeish, K., Hawkes, D.J.: MRI Based modeling of respiratory motion for optimization of lung cancer radiotherapy. Proc UKRO 2003

17. Blackall, J.M., King, A.P., Penney, G.P., Adam, A., Hawkes, D.J.: A statistical model of respiratory motion and deformation of the liver. Proc MICCAI 2001, Vol. 2208 LNCS Springer, eds W. Niessen and M. Viergever (2001) 1338-1340

18. Rueckert, D., Frangi, A.F., Schnabel, J.A.: Automatic construction of 3D statistical deformation models of the brain using non-rigid registration. Proc MICCAI 2001, Springer (2001) 77-84
19. Frangi, A.F., Rueckert, D., Schnabel, J.A., Niessen, W.J.: Automatic construction of multiple-object three-dimensional shape models: Application to cardiac modeling. IEEE Transactions on Medical Imaging, 21 (2002). In press.
20. Cootes, T.F., Taylor, C.J., Cooper, D.H., Graham, J.: Active shape models – their training and application. Computer Vision and Image Understanding, 61 (1995) 38-59.
21. King, A.P., Blackall, J.M., Penney, G.P., Hawkes, D.J.: Tracking liver motion using 3-D ultrasound and a surface based statistical shape model. Proc Mathematical Methods in Biomedical Image Analysis MMBIA 2001, IEEE Computer Society, (2001) 145-152
22. King, A.P., Batchelor, P.G., Penney, G.P., Blackall, J.M., Hill, D.L.G., Hawkes, D.J.: Estimating Sparse Deformation Fields Using Multiscale Bayesian Priors and 3-D Ultrasound. Proc. Information Processing in Medical Imaging, (2001).
23. Schnabel, J.A., Tanner, C., Castellano-Smith, A.D., Degenhard, A., Leach, M.O., Hose, D.R., Hill, D.L.G., Hawkes, D.J. Validation of Non-Rigid Image Registration using Finite Element Methods: Application to Breast MR Images IEEE Transactions on Medical Imaging (2003) in press
24. Tanner, C., Degenhard, A., Schnabel, J.A., Smith, A.D., Hayes, C., Sonoda, L.I., Leach, M.O., Hose, D.R., Hill, D.L.G., Hawkes, D.J.: A method for the comparison of biomechanical breast models. Proc Mathematical Methods in Biomedical Image Analysis MMBIA 2001, IEEE Computer Society, (2001) 11-18
25. Smith, A.D., Hartkens, T., Schnabel, J., Hose, D.R., Liu, H., Hall, W.A.,Truwitt, C.L., Hawkes, D.J., Hill, D.L.G.: Constructing patient specific models for correcting intraoperative brain deformation. Proc MICCAI 2001, Vol. 2208 LNCS, eds W Niessen and M Viergever Springer (2001) 1091-1098
26. Fleute, M., Lavallee,S.: Non-rigid 3D/2D registration of images using statistical models. Proc MICCAI 1999, Cambridge, Lecture Notes in Computer Science, (2001) 365-372
27. Jianhua, Y., Taylor, R.: A multiple layer flexible mesh template matching method for non-rigid registration between a pelvis model and CT images. Proc SPIE Medical Imaging, 2003.
28. Chan, C., Edwards, P.J., Hawkes, D.J.: Integration of ultrasound based registration with statistical shape models for computer assisted orthopaedic surgery. Proc SPIE Medical Imaging, 2003.

Real-Time Simulation of Self-collisions for Virtual Intestinal Surgery

Laks Raghupathi, Vincent Cantin, François Faure, and Marie-Paule Cani

GRAVIR/IMAG, joint lab of CNRS, INPG, INRIA and UJF
INRIA RA, 655 avenue de l'Europe, Montbonnot
38334 Saint Ismier Cedex, France
{Francois.Faure,Marie-Paule.Cani}@imag.fr

Abstract. The context of this research is the development of a peda-gogical surgery simulator for colon cancer removal. More precisely, we would like to simulate the gesture which consists of moving the small intestine folds away from the cancerous tissues of the colon. This paper presents a method for animating the small intestine and the mesentery (the tissue that connects it to the main vessels) in real-time, thus enabling user-interaction through virtual surgical tools during the simulation. The main issue that we solve here is the real-time processing of multiple col-lisions and self-collisions that occur between the intestine and mesentery folds.

Keywords: Surgical simulators, physically-based animation, soft tissue modeling, collision detection and response

1 Introduction

Enabling surgeons to train on virtual organs rather than on a real patient has recently raised a major interest for the development of pedagogical surgery sim-ulators. Such simulators would be particularly useful in the context of minimally invasive surgery, where learning the right gestures while observing results on a screen causes major difficulties. The long term aim of this research is the devel-opment of a virtual-reality based simulator for minimally invasive colon cancer removal. Here, as the patient is resting on his back (Fig. 1), the small intestine is positioned just above the colon region, thus hiding the colon beneath. This requires the surgeon to interact with the intestine (by pulling and folding it) so that he can operate on the colon without any constraint. Hence, our aim is to simulate the behavior of the intestine when the surgeon is practicing in the virtual surgical environment. Note that the current scope of this research work does not include the simulation of the removal of the cancer itself.

The intestinal region of a human body is characterized by a very complex anatomy. The small intestine is a tubular structure, about 4 meters long, con-strained within a small space of the abdominal cavity, resulting in the creation of numerous *intestinal folds*. This is further complicated by a tissue known as the *mesentery* which connects the small intestine to the blood vessels. The mesen-tery suspends the small intestine within the abdominal cavity, at a maximal

N. Ayache and H. Delingette (Eds.): IS4TM 2003, LNCS 2673, pp. 15–26, 2003.

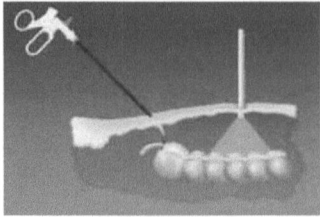

Fig. 1. Position of the small intestine when the patient is lying on his back.

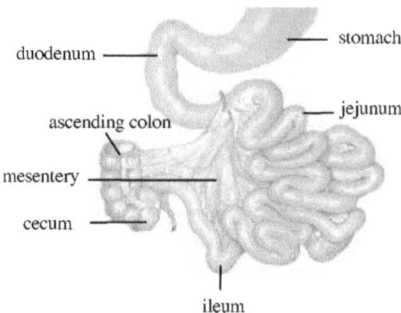

Fig. 2. Anatomy showing the intestine (*duodenum, jejunum* and *ileum*) and mesentery.

distance of 15 cm from the main vessels [1] (Fig. 2). Our challenge is to detect the collisions and self-collisions occurring in the intestinal region and to provide a realistic response at interactive frame rates.

Section 2 describes the earlier works in the area, focussing on the main techniques for efficient collision detection and response. Section 3 describes our geometrical and mechanical models of the small intestine and the mesentery. We then describe our collision detection method and our novel approach for providing response in Sect. 4. This is followed by results in Sect. 5 and conclusions in Sect. 6.

2   Related Work

Recently, numerous researchers have focussed on the efficient simulation of deformable models [4,5,7,9,15,16,17,23]. Several of them relied on adaptive, multi-resolution techniques for reaching real-time performances for complex volumetric bodies [4,5,9,15]. In particular, some of these techniques were successfully applied to surgery simulators [8,9,16,17,22,25]. In all these works, volumetric deformable bodies were simulated either in isolation, or were interacting with a single rigid tool, enabling the use of very specific techniques for collision detection and response, such as methods based on graphics hardware [19].

The problem we have to solve here is different: as will be shown in section 3, no volumetric deformable model will be needed since the intestine and the mesentery can be represented as a 1D and 2D structure respectively. Accordingly, a simple chain of masses and springs were used by France [12,13] for simulating the intestine. France used a grid-based approach for detecting self-collisions of the intestine and collisions with its environment. All objects were first approximated by bounding spheres, whose positions were stored, at each time step, in the 3D grid. Each time a sphere was inserted into a non-empty voxel, new colliding pairs were checked within this voxel. Though this method achieved real-time performances when the intestine alone was used, it failed when a mesentery surface was added.

A well-known technique for accelerating collision detection consists of approximating the objects by a hierarchy of bounding volumes [3,6,14,24,28]. It enables to quickly get rid-off most not-intersecting cases. In particular, thanks to the tight-fitting volumes used, the OBB-trees [14] are known as the best representation for detecting collisions between volumetric rigid bodies. The hierarchies can be recursively updated when the objects undergo small deformations. However, this is not suitable for intestine-mesentery interaction where, even a small local deformation can cause a large movement of the folds. This creates a *global deformation* at large scale, which prevents the hierarchy from being efficiently updated. An alternate multi-resolution method, based on layered shells, was recently presented by Debunne [10]. It is well-suited for collision detection between deformable objects since the shells themselves are deformable structures extracted from a multi-resolution representation of these objects. Though suitable for volumetric deformable bodies, this method will not be appropriate for intestine and mesentery, since the time-varying folds cannot easily be approximated at a coarse scale.

Finally, Lin and Canny [18] exploited temporal coherence by detecting collisions between convex polyhedra by tracking pairs of closest vertices. These pairs were very efficiently updated at each time-step by propagating closest distance tests from a vertex to its neighbors. Debunne [10] adapted this technique for detecting collisions between his volumetric layered shells very efficiently. Since these shells were neither convex nor rigid, a stochastic approach was used at each time step to generate new pairs of points anywhere on the two approaching objects. These pairs were made to converge to local minima of the distance, disappearing when they reached an already detected minimum. Our work inspires from this idea of stochastic collision detection exploiting temporal coherence. It has been adapted, in our case, to the specific processing of multiple collisions and contacts between the intestine and the mesentery folds.

3   Modeling of the Intestinal System

3.1   Geometric Model

As shown in Fig. 2, the mesentery is a folded surface membrane, approximately 15 cm thick, which links the small intestine, a long tubular structure 4 m in length, to the main vessels of 10 cm length. Since the mesentery cannot be developed onto a plane, setting up its initial geometry free of self-intersections, is

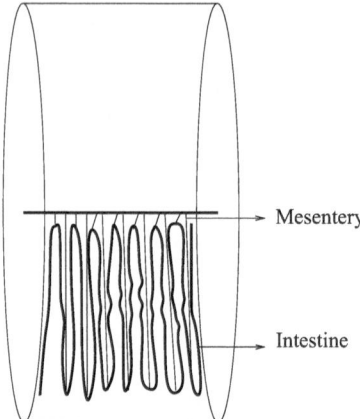

Fig. 3. Initialization of the geometric model of the intestine and the mesentery.

quite difficult. We solved the problem by approximating a possible rest position for the intestine as a folded curve lying at the surface of a cylinder of radius 15 cm. The axis of the cylinder, 10 cm in length, represents the main vessels. The folds are drawn on the cylinder such that their total length is 4 m (Fig. 3). Then the mesentery can be defined as the surface generated by a set of non-intersecting line segments linking the cylinder axis to the curve. Though this initial geometry is too symmetric to be realistic, it gives adequate local geometric properties to the mesentery membrane. This will enable the system to take correct arbitrary positions when animated under the effect of gravity. The geometry of the intestine is defined by creating tubular surface of radius 2 cm along its skeleton curve. The thickness of the mesentery membrane, which can be parameterized based on patient-specific data, was set to 1 cm.

3.2   Mechanical Model

The mechanical model representing the mesentery and its bordering curve, the intestine, should allow large displacements with local, elastic deformations. For animation, we used a simple mass-spring system since most of the computational time will be required for self-collision detection. Since the mesentery has a much larger length (4 m near the intestine) than thickness (15 cm near the vessel), we sampled it by four sets of 100 masses each, connected by damped springs. The last set of masses requires no computation since they are attached to the main vessels, requiring only 300 masses to be integrated at each time step. No specific model is needed for the intestine since it can be simulated by adjusting the masses and stiffness values along the first bordering curve of the mesentery surface (depicted as a darker curve in Fig. 4). To increase robustness and efficiency, we relied on the integration method recently proposed by Lyard [20].

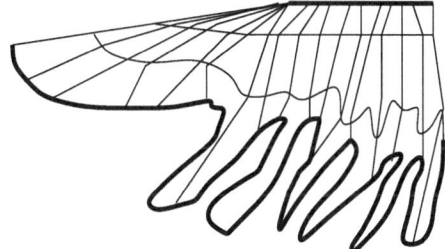

Fig. 4. Network of masses and springs used for the mechanical model.

4   Real-Time Collision Processing

4.1   Collision Detection

Our method for real-time collision detection exploits temporal coherence as in [10,18], i.e., to track the pairs of closest points between the colliding bodies. The main differences here are: (1) the interacting objects have a tubular (intestine) and a membrane structure (mesentery), and (2) most collisions will be self-collisions between different folds of the same body. We first explain the collision detection method for the intestine alone, and then explain the mesentery case.

Collision detection between cylinders can be processed by computing the closest distance between their axes [11], and comparing it to the sum of their radii. For intestine, computing the distance between two segments is done by considering the distance between their principal axes. Then, we store the normalized abscissa (s, t) $(0 < s < 1, 0 < t < 1)$ of the closest points within the segments, and the corresponding distance d_{min}.

Adapting the notion of "closest elements pairs" to this skeleton curve means that we are willing to track the local minima of the distance between non-neighboring segments along the curve (Fig. 5). Of course, only the local minima satisfying a given distance threshold are of interest to us. We call these pairs of segments as "active pairs". Each active pair is locally updated at each time step, in order to track the local minima, when the intestine folds move. This is done by checking whether it is the current segment pair or a pair formed using one of their neighbors which now corresponds to the smallest distance. This update requires nine distance tests (Fig. 6), and the pair of segments associated to the closest distance becomes the new active pair. When two initially distant active pairs converge to the same local minimum, one of them is suppressed. The pair is also suppressed if the associated distance is greater than a given threshold.

The above process tracks the existing regions of interest but does not the detect new ones. Since the animation of the intestine may create new folds nearby, a method for creating new active pairs of segments is needed. Our approach is inspired from the stochastic approach of [10]. At each time step, in addition to the update of the currently active pairs, n additional random pairs of segments,

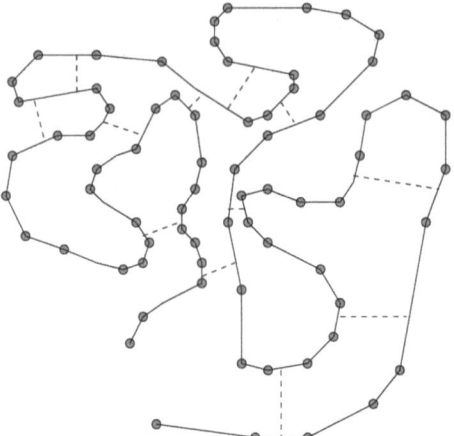

Fig. 5. Tracking of local minima of the distance between non-neighboring segments.

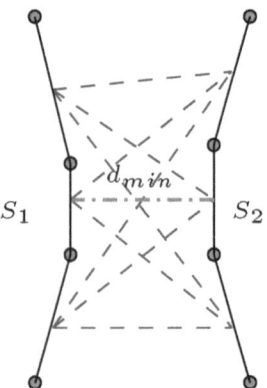

Fig. 6. Update of closest segment pairs of two adjoining intestinal folds.

uniformly distributed between the end-points but under the distance threshold, are generated. The update of these extra active pairs is similar to the update of the existing local minima, i.e., they are made to converge to a local distance minimum, the pair elements moving from a segment to one of its neighbors, and disappearing when an already detected minimum is reached. The complexity of the detection process thus linearly varies with user-defined parameter n. At each time step, collision detection consists in selecting, among the currently active pairs, the pairs of segments which are closer that the sum of their radii. Reaction forces, described in the collision response section, will then be generated between these segments.

For the mesentery, the total number of segments to be considered during each time-step is very large for real-time computation. Hence, we use the following approximation to reduce the complexity of the problem. First, since the mesentery

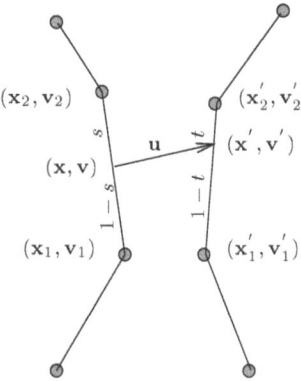

Fig. 7. Collision response by displacement-velocity correction.

is very thin and soft compared to the intestine, self-collisions of the membrane will almost have no effect on the overall behavior of the system. Hence, we neglect the testing of these collisions and only consider the pairs of segments from the intestine or pairs with one intestine segment and one non-neighboring mesentery segment.

Secondly, we use adaptive convergence to reduce the large number of distance computation required in this case. We first replace the first segment S_1 of the pair (S_1, S_2) by its closest neighbor S to S_2 (S is S_1 if all neighbors are farther than S_2). We then update S_2 by replacing it, if needed, by its neighbor which is the closest to S. This update requires 12 distance computations at most (i.e., when one segment belongs to the intestine, and the other to the inside of the mesentery). When a collision is detected, a recursive search starts across the neighbors to find all the colliding pairs in the area.

4.2 Collision Response

We initiate the response whenever the distance between the two segments is less than the sum of their radii. The earlier approaches such as penalty method [2,27] and reaction constraint method [21,26] implemented collision response by altering the force matrix in the mass-spring method. In our simulations, we observed that the stability of the system was reduced when we applied penalty and constraint methods.

Our new method alters the displacements and velocities of the two colliding segments in such a way so as to avoid interpenetration. Let the end-point velocities of segment S_1 be $\mathbf{v}_1$ and $\mathbf{v}_2$ and that of segment S_2 be $\mathbf{v}_1'$ and $\mathbf{v}_2'$ respectively. Let $\mathbf{x}_1$, $\mathbf{x}_2$, $\mathbf{x}_1'$ and $\mathbf{x}_2'$ be the corresponding positions. Let $\mathbf{v}$ and $\mathbf{v}'$ be the velocities of the closest approaching point within each segment (already stored in the neighborhood data structure) and $\mathbf{x}$ and $\mathbf{x}'$ be the positions of the closest points (Fig. 7).

If s and t are the normalized abscissa of the closest points on the two segments we have:

$$\mathbf{v} = (1-s)\mathbf{v}_1 + s\mathbf{v}_2 \qquad \mathbf{v}' = (1-t)\mathbf{v}_1' + t\mathbf{v}_2' \tag{1}$$

Let two forces per time-step, f and $f'(=-f)$, be applied along the direction of collision $\mathbf{u}$ to cause a change in the velocities such that the relative velocities along the direction of collision is zero. These forces should set the new velocities $\mathbf{v}_{new}$ and $\mathbf{v}_{new}'$ to values satisfying the condition:

$$(\mathbf{v}_{new} - \mathbf{v}_{new}').\mathbf{u} = 0 \tag{2}$$

The force f acting on the point of collision can be split between the end-points according to their barycentric coordinates. Expressing the new velocities in terms of the force and old velocities at the segment end-points yields:

$$\begin{aligned} \mathbf{v}_{new1} &= \mathbf{v}_1 + (1-s)f\mathbf{u} & \mathbf{v}_{new2} &= \mathbf{v}_2 + sf\mathbf{u} \\ \mathbf{v}_{new1}' &= \mathbf{v}_1' + (1-t)f\mathbf{u} & \mathbf{v}_{new2}' &= \mathbf{v}_2' + tf\mathbf{u} \end{aligned} \tag{3}$$

Again, expressing the new velocity of the colliding point $\mathbf{v}_{new}$ in terms of the end-point velocities $\mathbf{v}_{new1}$ and $\mathbf{v}_{new2}$:

$$\begin{aligned} \mathbf{v}_{new} &= (1-s)\mathbf{v}_{new1} + s\mathbf{v}_{new2} \\ &= \mathbf{v} + ((1-s)^2 + s^2)f\mathbf{u} \end{aligned} \tag{4}$$

Similarly for segment S_2:

$$\mathbf{v}_{new}' = \mathbf{v}' - ((1-t)^2 + t^2)f\mathbf{u} \tag{5}$$

Substituting the new velocity values from (4) and (5) into (2) and solving for f, we have:

$$f = \frac{(\mathbf{v}' - \mathbf{v}).\mathbf{u}}{(1-s)^2 + s^2 + (1-t)^2 + t^2} \tag{6}$$

Using this value of f, we compute the new velocities of the end-points from (3). We use a similar formulation for correcting the positions of colliding segments. The only difference is in the condition for avoiding interpenetration, which takes the segment's radii r and r' into account:

$$(\mathbf{x}_{new} - \mathbf{x}_{new}').\mathbf{u} = r + r' \tag{7}$$

The force value g to change the positions in order to enforce the above condition, is then:

$$g = \frac{(\mathbf{x} - \mathbf{x}').\mathbf{u} + r + r'}{(1-s)^2 + s^2 + (1-t)^2 + t^2} \tag{8}$$

g is used for modify the positions $\mathbf{x}_{new1}$, $\mathbf{x}_{new2}$, $\mathbf{x}'_{new1}$ and $\mathbf{x}'_{new2}$ of the segments end points using similar expressions as in (3).

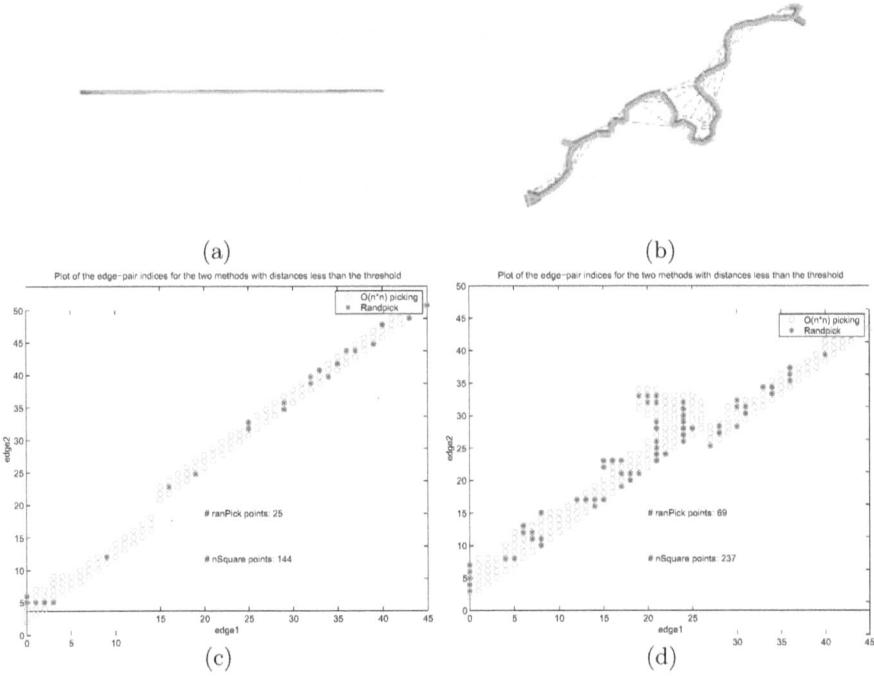

Fig. 8. (a) and (b) Snapshots of the simulation of an isolated intestine. (c) and (d) Plot of the active pairs of segments compared with the pairs from the $O(n^*)$ method.

5   Results

5.1   Validation

In order to compare the effectiveness of this method, we developed a simple testing methodology for case of the intestine in isolation. We would like to know if our algorithm detects all the regions of collisions. To do so, we compared our method with a naive $O(n^2)$ approach (i.e., do a check of all possible pairs to detect all the active collision regions). So, we let the simulation run and took snapshots of the model during different time intervals (Fig. 8a and 8b). At the same time, we also collected data on the segment pairs (index values) stored in the neighborhood data structure. We carried out this procedure for both the methods. During the simulation, the mass points were resting on a plane. Figures 8b and 8d plot the values of the active segment-pair indices for the two configurations (all segments under the distance threshold in the case of the $O(n^2)$ detection). Results show that since our method only tracks local minima of the distance, it considers much fewer segments (Table 1). They also show that all regions of interest are adequately detected, i.e., there are no colliding folds with no detected local minima, as depicted in Fig. 8b. The corresponding plot in Fig. 8d shows an increased density of points in the local minima region, thereby indicating that it has been detected by the algorithm. The resulting animations

(a) The system in its initial state with the intestine represented by the curve along the cylindrical surface edge and the mesentery by segments connecting it to the cylinder axis.

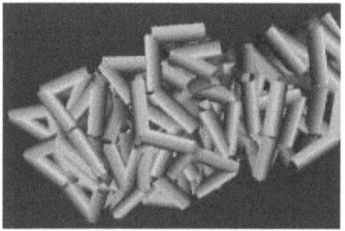

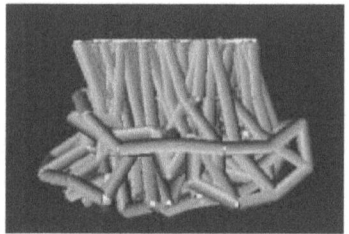

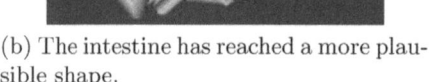

(b) The intestine has reached a more plausible shape.

(c) The intestine makes a loop around the mesentery (medically unrealistic).

Fig. 9. Snapshots from our real-time simulator of the intestinal system.

Table 1. Running time (in ms) for both the methods.

Number of segments	Time (ms)	
	Our method $O(n^{\cdot})$	
50	10	30
100	17	120
200	27	473

also showed no interpenetration with realistic collision response. The figure only shows the collisions detected by the converging pairs. Once a pair converges to a collision, the recursive propagation ensures that all the collisions of a collision region are detected. Our method can detect all the regions of collisions at 30 Hz on a PC. Comparing with the $O(n^2)$ method, this method uses a relatively small number segment-pairs and is hence a lot faster.

5.2 Qualitative Results

Snapshots from the real-time animation of the intestinal system including both the small intestine and the mesentery are depicted in Fig. 9. Note that realistic rendering is not our concern here. Our displacement-velocity method for collision response produces highly stable simulations.

As a companion to this paper, a dynamic real-time demonstration of our results is available at: http://www-imagis.imag.fr/Membres/Francois.Faure/papers /intestine/index.html.

6   Conclusion

We have developed a model which can accurately determine all the active regions of self-collisions in the intestine and the mesentery. We have also developed a method for providing realistic and stable collision response. Our models run at interactive frame rates which can be used in a virtual surgical environment.

Future work will include the incorporation of these algorithms in the intestinal surgery simulator developed by our collaborators in Lille [12,13], thus enabling the use of convincing geometric coating, texturing and rendering of the organs, in addition to the use of a force-feedback device for user-interaction.

Acknowledgments

This work is supported by INRIA (French National Institute for Research in Computer Science and Control) as part of the ARC SCI (research action for Intestine Surgery Simulator). The authors would like to thank Luc Soler (IRCAD) for his insights and suggestions for creating the geometry of the mesentery and for providing anatomical information and Laure France (LIFL) for fruitful discussions and for providing data from her earlier research.

References

1. L. Augusten, R. A. Bowen, and M. Rouge. *Pathophysiology of the Digestive System - Hypertexts for Biological Sciences.* Colorado State University, Fort Collins, CO, http://arbl.cvmbs.colostate.edu/hbooks/pathphys/digestion/index.html, 2002.
2. D. Baraff and A. Witkin. Large steps in cloth simulation. In *Proc. SIGGRAPH '98*, pages 43–54. ACM Press, July 1998.
3. G. Bradshaw and C. O'Sullivan. Sphere-tree construction using dynamic medial axis approximation. In *ACM SIGGRAPH Symposium on Computer Animation*, pages 33–40. ACM Press, July 2002.
4. S. Capell, S. Green, B. Curless, T. Duchamp, and Z. Popović. Interactive skeleton-driven dynamic deformations. In *Proc. SIGGRAPH '02*, pages 586–593. ACM Press, July 2002.
5. S. Capell, S. Green, B. Curless, T. Duchamp, and Z. Popović. A multiresolution framework for dynamic deformations. In *ACM SIGGRAPH Symposium on Computer Animation*, pages 41–48. ACM Press, July 2002.
6. J. D. Cohen, M. C. Lin, D. Manocha, and M. K. Ponamgi. I-COLLIDE: An Interactive and Exact Collision Detection System for Large-Scale Environments. In *Symposium on Interactive 3D Graphics*, pages 189–96, 1995.
7. S. Cotin, H. Delingette, and N. Ayache. Real-time elastic deformations of soft tissues for surgery simulation. *IEEE TVCG*, 5(1):62–73, March 1999.
8. S. Cotin, H. Delingette, and N. Ayache. A hybrid elastic model for real-time cutting, deformations, and force feedback for surgery training and simulation. *The Visual Computer*, 16(8):437–452, 2000.
9. G. Debunne, M. Desbrun, M. P. Cani, and A. H. Barr. Dynamic real-time deformations using space and time adaptive sampling. In *Proc. SIGGRAPH '01*, pages 31–36. ACM Press, August 2001.

10. G. Debunne and S. Guy. Layered Shells for Fast Collision Detection. To be published, 2002.
11. D. H. Eberly. *3D Game Engine Design: A Practical Approach to Real-Time Computer Graphics*. Morgan Kaufmann, 2000.
12. L. France, A. Angelidis, P. Meseure, M. P. Cani, J. Lenoir, F. Faure, and C. Chaillou. Implicit representations of the human intestines for surgery simulation. In *Modelling and Simulation for Computer-aided Medicine and Surgery*, Rocquencourt, France, November 2002.
13. L. France, J. Lenoir, P. Meseure, and C. Chaillou. Simulation of a minimally invasive surgery of intestines. In *Virtual Reality International Conferenc*, Laval, France, May 2002.
14. S. Gottschalk, M. C. Lin, and D. Manocha. OBBTree: a hierarchical structure for rapid interference detection. In *Proc. SIGGRAPH '96*, pages 171–80. ACM Press, 1996.
15. E. Grinspun, P. Krysl, and P. Schröder. CHARMS: A Simple Framework for Adaptive Simulation. In *Proc. SIGGRAPH '02*, pages 281–290. ACM Press, July 2002.
16. D. L. James and D. K. Pai. Artdefo - accurate real time deformable objects. In *Proc. SIGGRAPH '99*, pages 65–72. ACM Press, August 1999.
17. D. L. James and D. K. Pai. DyRT: Dynamic Response Textures for Real Time Deformation Simulation with Graphics Hardware. In *Proc. SIGGRAPH '02*, pages 582–585. ACM Press, July 2002.
18. M. C. Lin and J. F. Canny. Efficient Collision Detection for Animation. In *Proc. of the 3rd Eurographics Workshop on Animation and Simulation*, 1992.
19. J. C. Lombardo, M. P. Cani, and F. Neyret. Real-time Collision Detection for Virtual Surgery. In *Proc. Computer Animation '99*, May 1999.
20. E. Lyard and F. Faure. Impulsion springs: a fast and stable integration scheme dedicated to the mass-spring model. To be published, 2002.
21. D. W. Marhefka and D. E. Orin. Simulation of contact using a nonlinear damping model. In *Proc. IEEE ICRA*, pages 1662–68, 1996.
22. P. Meseure and C. Chaillou. A deformable body model for surgical simulation. *Journal of Visualization and Computer Animation*, 11(4):197–208, September 2000.
23. M. Müller, J. Dorsey, L. McMillan, R. Jagnow, and B. Cutler. Stable real-time deformations. In *ACM SIGGRAPH Symposium on Computer Animation*, pages 49–54. ACM Press, July 2002.
24. I. J. Palmer and R. L. Grimsdale. Collision detection for animation using sphere trees. *Computer Graphics Forum*, 14(4):105–16, May 1995.
25. G. Picinbono, J. C. Lombardo, H. Delingette, and N. Ayache. Improving realism of a surgery simulator: linear anisotropic elasticity, complex interactions and force extrapolation. *Journal of Visualization and Computer Animation*, 13(3):147–167, 2002.
26. J. C. Platt and A. H. Barr. Constraints methods for flexible models. In *Proc. SIGGRAPH '88*, pages 279–88. ACM Press, 1988.
27. D. Terzopulous, J.C. Platt, K. Fleischer, and A. H. Barr. Elastically deformable models. In *Proc. SIGGRAPH '87*, pages 205–14. ACM Press, 1987.
28. G. van den Bergen. Efficient Collision Detection of Complex Deformable Models using AABB Trees. *Journal of Graphics Tools*, 2(4):1–14, 1997.

Modelling of Facial Soft Tissue Growth for Maxillofacial Surgery Planning Environments

Patrick Vandewalle, Filip Schutyser,
Johan Van Cleynenbreugel, and Paul Suetens

Medical Image Computing (Radiology - ESAT/PSI),
Faculties of Medicine and Engineering, University Hospital Gasthuisberg,
Herestraat 49, B-3000 Leuven, Belgium
`Filip.Schutyser@uz.kuleuven.ac.be`

Abstract. When maxillofacial surgery is proposed as a treatment for a patient, the type of osteotomy and its influence on the facial contour is of major interest. To design the optimal surgical plan, 3D image-based planning can be used. However, prediction of soft tissue deformation due to skeletal changes, is rather complex. The soft tissue model needs to incorporate the characteristics of living tissues.

Since surgeon and patient are interested in the expected facial contour some months after surgery when swelling has disappeared, features specific to living tissues need to be modelled. This paper focusses on modelling of tissue growth using finite element methods. This growth is induced by stress resulting from the surgical procedure. We explain why modelling growth is needed and propose a model. We apply this model to 4 patients treated with unilateral mandibular distraction and compare these soft tissue predictions with the postoperative CT image data.

Keywords: soft tissue modelling, maxillofacial surgery simulation

1 Introduction

The soft tissue model explained in this paper fits into our framework of 3D image-based planning systems. This planning environment adheres to a scene-based approach in which image derived visualizations and additional 3D structures (external to the medical image volume) are co-presented and manipulated. This environment includes tools for osteotomy simulation and distraction simulation [1].

Because of the high impact of distraction therapy on the patient's face, prediction of the soft tissue deformation is highly desirable. Therefore, our planning system also includes a soft tissue model of the skin (i.e. the dermis and the underlying structures like fat and muscles).

Fung [2] reports on the biomechanical properties of living tissues. Skin tissues are called quasi-linear viscoelastic materials, meaning that these tissues show creep, relaxation and hysteresis when applying large oscillations around equilibrium, but the characteristics can be well approximated with linear viscoelasticity

N. Ayache and H. Delingette (Eds.): IS4TM 2003, LNCS 2673, pp. 27–37, 2003.

applying small oscillations. However, this modelling implies demanding computations which are in this application area unrealistic. Moreover, this model describes the biomechanical behavior of soft tissues during a short time interval and not the deformations due to e.g. persistent stress over a longer time.

Different approaches have been investigated to model soft tissues. Teschner et al. describe a multi-layer spring model [3], resulting in short simulation times. However no extended validation study is published. The meshing step, based on the approach of Waters [4], is rather tedious and error-prone. Koch et al. [5], Chabanas et al. [6] and Gladilin et al. [7] use finite element methods to model skin tissue. Koch describes skin tissue as an incompressible elastic tissue. Chabanas adds a muscle activation model to animate the face. Gladilin applies a nonlinear elastic model.

In this paper, we develop a finite element model incorporating growth. In section 2, we explain why it is important to incorporate growth in the model by validating a basic linear elastic model. In subsection 2.3 we explain how we have extended this basic model with a growth component. Simulation results for 4 patients are shown in section 3. After a discussion of these results (section 4), concluding remarks finish this paper (section 5).

2 Methods

2.1 Basic Model

Our research for an accurate model for maxillofacial surgery planning starts from a linear elastic soft tissue model which is based on the mechanical equilibrium equations

$$\frac{\partial \sigma_{xx}}{\partial x} + \frac{\partial \tau_{xy}}{\partial y} + \frac{\partial \tau_{xz}}{\partial z} + F_x = 0$$

$$\frac{\partial \tau_{xy}}{\partial x} + \frac{\partial \sigma_{yy}}{\partial y} + \frac{\partial \tau_{yz}}{\partial z} + F_y = 0 \tag{1}$$

$$\frac{\partial \tau_{xz}}{\partial x} + \frac{\partial \tau_{yz}}{\partial y} + \frac{\partial \sigma_{zz}}{\partial z} + F_z = 0$$

with $\sigma_{xx}, \sigma_{yy}, \sigma_{zz}, \tau_{xy}, \tau_{xz}, \tau_{yz}$ the stress components and $\mathbf{F}(F_x, F_y, F_z)$ the volume forces.

The material properties are introduced into these equations through the constitutive equations relating stresses and strains. The soft tissue is modeled as a homogeneous, linear and elastic material, such that we can use Hooke's law:

$$
\begin{bmatrix} \sigma_{xx} \\ \sigma_{yy} \\ \sigma_{zz} \\ \tau_{xy} \\ \tau_{yz} \\ \tau_{zx} \end{bmatrix} = \frac{E}{(1+\nu)(1-2\nu)}
\begin{bmatrix}
1-\nu & \nu & \nu & 0 & 0 & 0 \\
\nu & 1-\nu & \nu & 0 & 0 & 0 \\
\nu & \nu & 1-\nu & 0 & 0 & 0 \\
0 & 0 & 0 & \frac{1-2\nu}{2} & 0 & 0 \\
0 & 0 & 0 & 0 & \frac{1-2\nu}{2} & 0 \\
0 & 0 & 0 & 0 & 0 & \frac{1-2\nu}{2}
\end{bmatrix}
\begin{bmatrix} \epsilon_{xx} \\ \epsilon_{yy} \\ \epsilon_{zz} \\ \gamma_{xy} \\ \gamma_{yz} \\ \gamma_{zx} \end{bmatrix} \Leftrightarrow \boldsymbol{\sigma} = \mathbf{D}\boldsymbol{\epsilon}
$$

$$\tag{2}$$

with strain components $\epsilon_{xx}, \epsilon_{yy}, \epsilon_{zz}, \gamma_{xy}, \gamma_{xz}, \gamma_{yz}$, Young's modulus E and Poisson coefficient ν.

If we define $\{\mathbf{X}\}$ as the initial configuration at time t_0 and $\mathbf{x} = \mathbf{x}(\mathbf{X}, t)$ as the description of the point $\mathbf{X}$ at time t, the displacement vector $\mathbf{u}$ can be defined as $\mathbf{x} = \mathbf{X} + \mathbf{u}$. The Green-Lagrange strain tensor relates the strains ϵ to the displacements $\mathbf{u}$: $\epsilon = \frac{1}{2}(\nabla\mathbf{u} + \nabla\mathbf{u}^T + \nabla\mathbf{u}^T\nabla\mathbf{u})$. We linearize this equation to $\epsilon = \frac{1}{2}(\nabla\mathbf{u} + \nabla\mathbf{u}^T)$.

These equations are discretized using a 3D finite element method. The continuum is modeled as a tetrahedron mesh. For the interpolation between the nodes we use a basic linear, C_0 continuous shape function using 4 nodes for each tetrahedron [8]. The partial differential equations are reduced to a set of linear equations for the vertices of the tetrahedron mesh: $\mathbf{KU} = \mathbf{R}$ with $\mathbf{K} = \int_V \mathbf{B}^T\mathbf{DB}dV$, $\epsilon = \mathbf{Bu} = [\mathbf{B}_q\,\mathbf{B}_r\,\mathbf{B}_s\,\mathbf{B}_t]\mathbf{u}$ and

$$\mathbf{B}_m = \begin{bmatrix} \frac{\partial N_m}{\partial x} & 0 & 0 \\ 0 & \frac{\partial N_m}{\partial y} & 0 \\ 0 & 0 & \frac{\partial N_m}{\partial z} \\ \frac{\partial N_m}{\partial y} & \frac{\partial N_m}{\partial x} & 0 \\ 0 & \frac{\partial N_m}{\partial z} & \frac{\partial N_m}{\partial y} \\ \frac{\partial N_m}{\partial z} & 0 & \frac{\partial N_m}{\partial x} \end{bmatrix} \tag{3}$$

for the 4 tetrahedron vertices $(m = q, r, s, t)$ with shape functions N_m.

The stiffness submatrix $\mathbf{K}_{mn}^i$ is computed for each pair of 2 nodes m and n of an element i. It can be computed from the shape function and the material properties as $\mathbf{K}_{mn}^i = \mathbf{B}_m^T\mathbf{D}^i\mathbf{B}_n V^i$ with V^i the volume of tetrahedron i. The global stiffness matrix $\mathbf{K}$ for the entire model can then easily be assembled from all these element stiffness matrices $\mathbf{K}_{mn}^i$.

The derived set of equilibrium equations is constrained by a set of Dirichlet boundary conditions. They force certain displacements to a fixed value. In our facial model there are 2 types of such boundary conditions. First there are the displacements at the border between bone and soft tissue, prescribed by the planned bone displacements. The other boundary conditions are obtained from the assumption that the soft tissue above the eyes and behind the ears will not be affected by the surgery and can thus be supposed to have a fixed position. These Dirichlet boundary conditions can be easily introduced in the equations, as they can all be formulated as $\mathbf{tu} = \mathbf{r}$.

In this first model there are no volume or body forces defined on the tetrahedron mesh. This means that $\mathbf{R} = 0$, except for the values introduced by the boundary conditions.

2.2   Validation

In the development of a soft tissue model for surgery planning it is very important to be able to see how closely an approximation matches the real postoperative situation. This is the only way to make a good comparison between two implementations and to see if the approximation goals are reached.

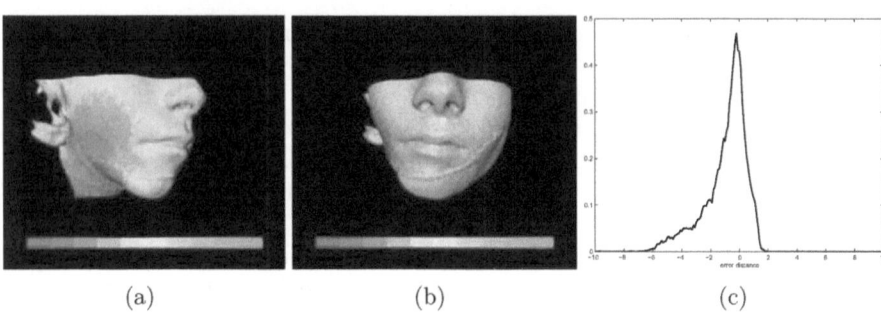

Fig. 1. (a, b): Color-encoded differences between the real post-operative image and the planned image with the basic model. The color scale covers the interval [-10mm ... 10mm]. (c): Histogram of the distances between the predicted and the real post-operative image.

A procedure also has to be validated with data from different patients. This is needed to make sure that the procedure works not only for one specific patient record, but gives good results for any patient.

Procedure.

1. 4 months after surgery, the patient gets a CT scan. This post-operative CT scan is rigidly registered to the pre-operative CT data, using maximization of mutual information [9] on an unaltered subvolume as registration method. From these co-registered post-operative data, a surface representation of the skin is generated using the Marching Cubes algorithm [10].
2. Next, a skin surface is created from the pre-operative CT image using the same algorithm.
3. This pre-operative surface is deformed according to the results of our FEM computations. We then have a planned post-operative surface of the skin. This can be compared with the real post-operative surface as they are both registered to the same pre-operative data.
4. For the vertices of this planned post-operative surface the closest distance to the real post-operative surface is computed and visualized using color-coding. For this coding the normals to the surface are used. They are set to the exterior of the skin surface. When the normal to the real post-operative surface intersects the planned surface in the positive direction, green colors are used for positive values. When it intersects the planned surface in the negative direction, the errors get red colors and negative values.
   This results in an easily interpretable image of the errors over the entire surface (figure 1:a,b).
5. In a last step we discard the positional information and an error histogram is made from all the differences (errors) between planned and real post-operative data (figure 1:c). Because we would like all errors to be zero, the ideal histogram is a Dirac pulse δ. Therefore the average error μ and the standard deviation σ need to be as small as possible.

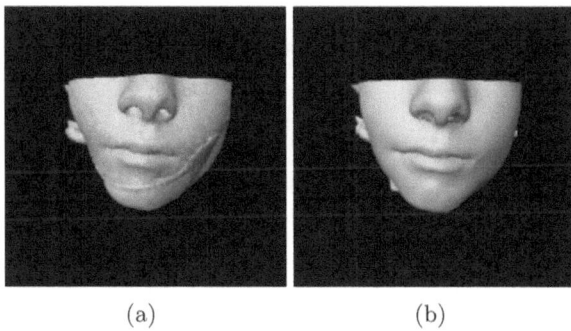

(a)   (b)

Fig. 2. Comparison between predicted and real post-operative facial skin envelopes. (a) Predicted post-operative image (The "wrinkle" you notice, is the result of a fixation bandage applied during the acquisition. This should be avoided during the CT examination); (b) Real (registered) post-operative image.

Validation of the Basic Model. We use this procedure to validate the results with the basic model. Figure 2 shows a direct comparison between the predicted (planned) post-operative image and the registered real post-operative image. It is very difficult, even nearly impossible, to make a good comparison between these two images as they are represented here. Therefore, it is more effective to display the differences between the predicted and the real data according to steps 4 and 5 of the validation procedure (see figure 1).

From this histogram we can see that the average error is negative, showing a 'lack of material'. If we analyze the error image (figure 1) and derive positional information (which has disappeared in the histogram), it is clear that mainly in the parts with high stress due to the displacements (close to the bone displacements), large negative errors occur. In these areas the basic model cannot predict reality sufficiently.

2.3   Soft Tissue Growth

Model Improvement. When using the basic model, a relatively large residual error is found in the area with large tissue stress (mainly the boundary between displaced and fixed tissue). From Fung [2] and Rodriguez [11] we know that soft tissue will grow under stress conditions. As we try to model the soft tissue envelope about 4 months after surgery, it is very important to include this tissue growth into the model.

As a hypothesis, we state that tissue growth can be modeled as an internal volume force in the tetrahedra, similar to the force caused by a thermal expansion. We replace the previous $\mathbf{R} = 0$ by a term

$$\mathbf{R} = \int_V \mathbf{B}^T \mathbf{D} \alpha \boldsymbol{\sigma}_{growth} \mathrm{d}V. \tag{4}$$

with $\boldsymbol{\sigma}_{growth}$ the growth stress and α a stability factor (see 2.3). $\boldsymbol{\sigma}_{growth}$ is a new degree of freedom, which is introduced to fully control the growth. It is a stress induced by the growth process.

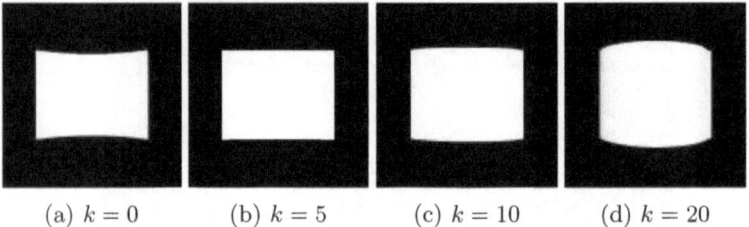

(a) $k = 0$ (b) $k = 5$ (c) $k = 10$ (d) $k = 20$

Fig. 3. A cube is stretched to the left and to the right, and the resulting stress in the cube causes growth with different equilibrium states for $k = 0, 5, 10$ and 20.

The tissue growth is caused by the stress introduced by the bone displacements. Therefore we can state that $\boldsymbol{\sigma}_{growth}$ is the sum of the stress due to the displacements $(\boldsymbol{\sigma}_d)$ and some extra stress $\boldsymbol{\sigma}_g$ generated by the growth process itself (which is induced by the initial displacement stress $\boldsymbol{\sigma}_{d0}$). Therefore we can say $\boldsymbol{\sigma}_{growth} = \boldsymbol{\sigma}_d + \boldsymbol{\sigma}_g(\boldsymbol{\sigma}_{d0})$. We assume that $\boldsymbol{\sigma}_g = k\boldsymbol{\sigma}_{d0}$ if $\boldsymbol{\sigma}_{d0} > 0$ and $\boldsymbol{\sigma}_g = 0$ otherwise. k is the parameter which determines how large the soft tissue growth has to be (figure 3). We then have the following formulation:

$$\mathbf{R} = \int_V \mathbf{B}^T \mathbf{D} \alpha \boldsymbol{\sigma}_{growth} \mathrm{d}V \tag{5}$$

$$= \int_V \mathbf{B}^T \mathbf{D} \alpha (\boldsymbol{\sigma}_d + \boldsymbol{\sigma}_g) \mathrm{d}V \tag{6}$$

$$= \int_V \mathbf{B}^T \mathbf{D} \alpha (\boldsymbol{\sigma}_d + k\boldsymbol{\sigma}_{d0}) \mathrm{d}V \tag{7}$$

These equations cause the tissue model to grow in an iterative process, which is described in the next paragraph. Iterations are made until an equilibrium state is reached, which happens when $\boldsymbol{\sigma}_d = -k\boldsymbol{\sigma}_{d0}$ if $\boldsymbol{\sigma}_{d0} > 0$ and $\boldsymbol{\sigma}_d = 0$ otherwise.

Iterative Procedure. To initialize the iterative process, the stresses introduced by the bone displacements are computed. During the first iteration, these stresses are used to compute the 'growth' forces. When the resulting displacements from this first iteration are computed, the remaining stresses are computed again. In the subsequent iterations we always use the stress remaining from the previous iteration to compute $\boldsymbol{\sigma}_d$. In this way the residual stress in the soft tissue model is reduced in every iteration until the stresses (and thus also the growth per iteration) is smaller than a certain threshold value.

Stability Factor α. In order to keep the iterative growth process controlled and stable, a stability factor α was introduced in the volume force term $\mathbf{R}$. This factor guarantees a safe and stable evolution of the element stresses towards their equilibrium values. $\alpha = 0.0005$ is a good value for α, as can be derived from figure 4, where the number of iterations needed for convergence is plotted for different values of α. For values of $\alpha > 0.0006$ the growth process diverges and no stable equilibrium state can be reached any more.

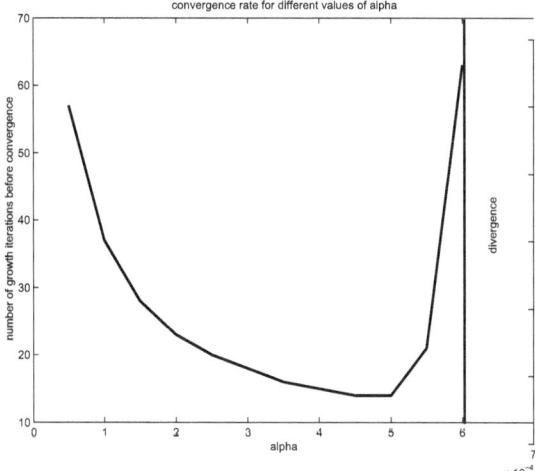

Fig. 4. The number of iterations needed for growth convergence is plotted for different values of α. The optimal value (14 iterations in this case) is reached for $\alpha = 0.00045$ to 0.0005.

Parameter Values. The material parameters E and ν for living soft tissues are hard to measure in practice and they are not described in literature. Young's modulus E is not important for our application, as it appears once in all terms of the equations and can thus be removed from the equations. Contrary to this, the Poisson coefficient ν has a larger implication on the soft tissue deformation.

Good values for this coefficient ν and for the growth parameter k can be obtained through an analysis of the error values obtained from the validation on different patient data. The average error and the standard deviation for different values of ν is shown in figure 5:(a,b), and the same is done for different values of the growth parameter k in figure 5:(c,d).

We analyzed the parameter values on the data of 4 different patients and we noted that the optimal values for the different patients are very close to each other. From this we state that these values for the Poisson parameter ν and the growth parameter k can be used for any arbitrary patient. The parameters (k, ν) are optimal for ($k = 20$, $\nu = 0.25$).

3   Results

We illustrate this method for 4 patients suffering from unilateral microsomia. All these patients have been treated with intraoral unilateral mandibular distraction. The distraction device had two degrees of freedom: unidirectional translation and angular rotation.

Using the validation procedure we described earlier, we compare the results obtained with the improved model to the results obtained with the basic model. Predictions resulting from both methods are compared with the real post-operative image. The error distances are shown in figure 6.

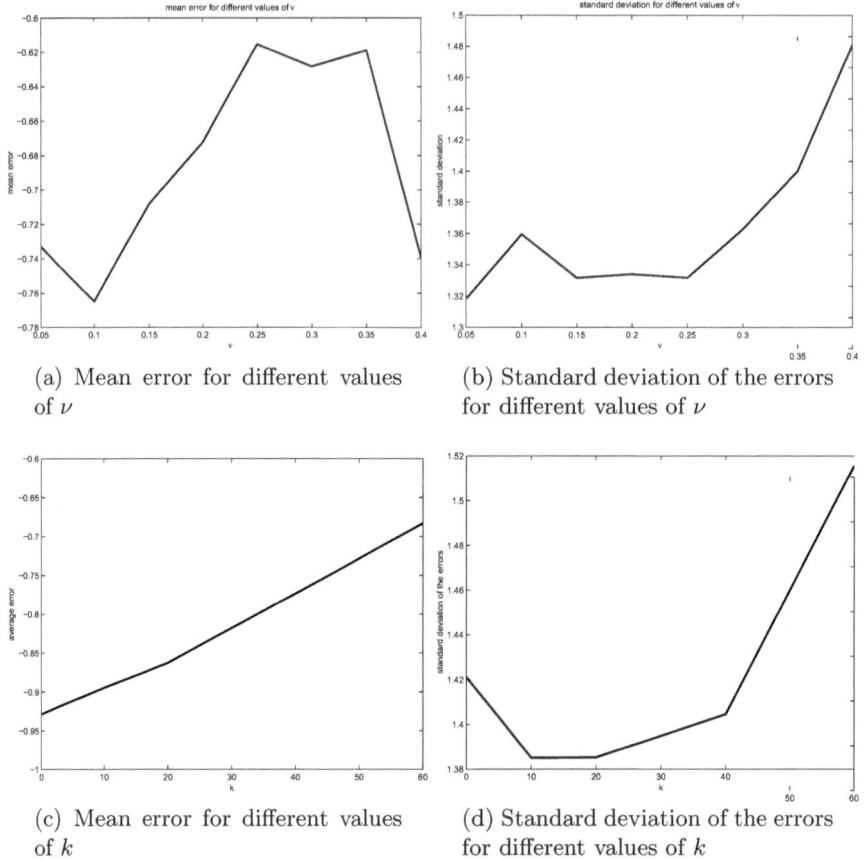

(a) Mean error for different values of ν

(b) Standard deviation of the errors for different values of ν

(c) Mean error for different values of k

(d) Standard deviation of the errors for different values of k

Fig. 5. Mean and standard deviation of the error distribution over a patient's complete face of the predictions for different values of the Poisson parameter ν and growth parameter k.

Although the approximation is still not exact, it can immediately be seen that the prediction for the right cheek is much better in the extended model. This can also be seen in the histogram plot for the two error images (figure 7:a). The mean error $\mu = -0.57$ and the standard deviation $\sigma = 1.43$ both have smaller values than for the basic model ($\mu = -1.05$, $\sigma = 2.17$) and thus make a better approximation of the Dirac pulse δ.

As was indicated in the description of the validation procedure, the new method then needs to be validated on data of different patients, to make sure the results are also better in an arbitrary case. For this goal, the simulations were run with both the basic and the improved model on the data of 4 different patients. The error histograms compared to the real post-operative images were put together and averaged (figure 7:b).

The results are less remarkable (because of the averaging operation), but it is still clear that the results with the improved model have errors (when compared

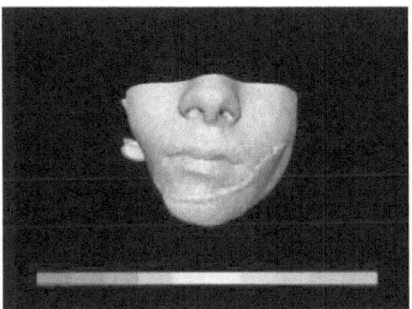 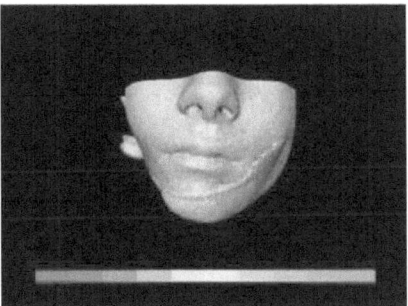

(a) Errors using the basic model   (b) Errors using the model includ-
                                     ing tissue growth (k=20)

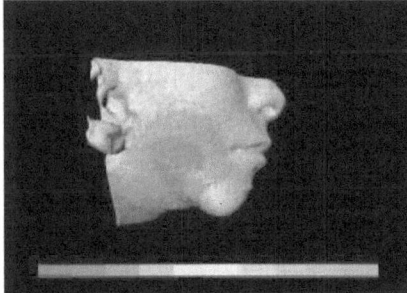 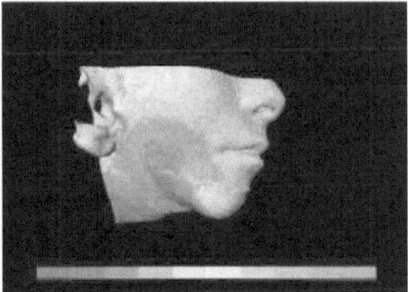

(c) Errors using the basic model   (d) Errors using the model includ-
                                     ing tissue growth (k=20)

Fig. 6. Color-encoded differences between the real post-operative image and the planned image with the basic model (left) and with the extended model including tissue growth (right). The color scale covers the interval [-10mm ... 10mm].

to the real post-operative images) which are more centered around 0 ($\mu = -0.06$ compared to $\mu = -0.55$) and have a smaller standard deviation ($\sigma = 2.15$ compared to $\sigma = 2.66$) than when the basic model is used.

4 Discussion

An important goal of maxillofacial surgery planning is to give the surgeon and the patient an accurate idea about what the face will look like as a result of the surgery. People are interested in a prediction of the facial outlook when swelling etc. has gone. They want to know what the patient will look like a few months after surgery, on the long term.

Therefore we need to incorporate long-term tissue behavior - like tissue growth - into the model used. Because of the large stresses induced by surgery, soft tissue grows considerably and has a large effect on the facial skin surface.

The validation results show the improvements made to the soft tissue model by including tissue growth. This definitely increases the accuracy by which maxillofacial surgery can be modelled. On the other hand, the validation also revealed

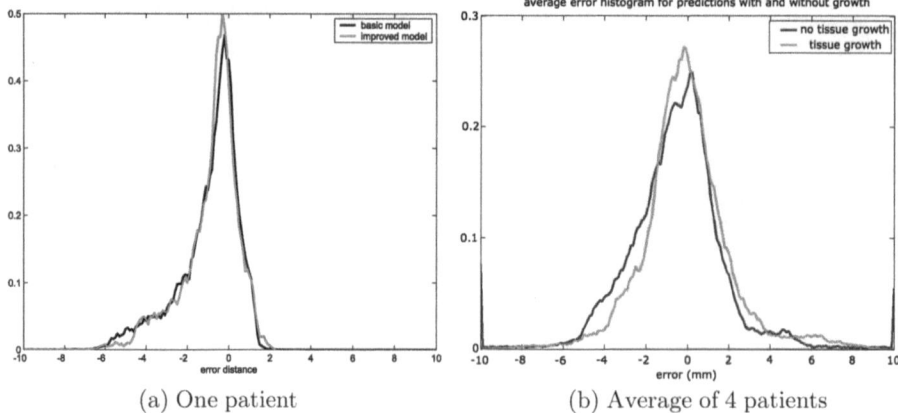

(a) One patient   (b) Average of 4 patients

Fig. 7. Histogram data, comparison between the basic model and the extended model (growth parameter $k = 20$).

that there are still relatively large areas in which the skin surface is modelled incorrectly.

The type of improvements to be made in an next step is not a simple question. Most people can easily say whether two faces are different (even with small differences) or not, but have difficulties to point the differences. In a similar way we see that our models are still not precise enough but it is difficult to find the refinements needed to make this last step.

Intuitively a first refinement could be made by discriminating muscles and fat as different soft tissue types with different characteristics. A next challenge will then certainly be to determine the characteristics of those different soft tissue types, as they are still largely unknown.

The boundary conditions form another field where important improvements could be made. It would be very useful to know precisely if the soft tissue next to the displaced bone elements stays fixed to it, or what happens exactly. And which parts of the soft tissue do not deform given a certain type of osteotomy? These are all questions which are still unanswered, but which can dramatically influence the quest for facial soft tissue models.

5   Conclusion

We developed a new soft tissue deformation prediction model for maxillofacial surgery planning. Starting from a simple, linear elastic 3-dimensional finite element model, a new model was constructed which includes the soft tissue growth during the months after surgery. An extensive validation procedure was also developed, which enabled us to objectively compare results with different methods.

The results obtained with this new model including tissue growth show a significant improvement over the results without tissue growth. The planned post-operative images match the real post-operative images much better now, making the surgery planning more useful to surgeons and patients. But before they can be really applied in practice, some further refinements still need to be made, as proposed in the discussion.

Acknowledgments

The work discussed here belongs to a grant for research specialization from the Flemish Institute for stimulation of the scientific-technological research in the industry (IWT) to Filip Schutyser. This work also partly belongs to the Flemish government IWT GBOU 020195 project on Realistic Image-based Facial Modeling for Forensic Reconstruction and Surgery Simulation.

References

1. F. Schutyser, J. Van Cleynenbreugel, M. Ferrant, J. Schoenaers, P. Suetens: Image-based 3D planning of maxillofacial distraction procedures including soft tissue implications. Proceedings 3rd international conference on medical image computing and computer-assisted intervention - MICCAI2000, lecture notes in computer science, vol. 1935, pp. 999–1007, October 11-14, 2000
2. Y.C. Fung: Biomechanics: Mechanical properties of Living Tissues. 2nd edition. Ch. 7, Springer-Verlag, 1993, p. 242–320
3. M. Teschner, S. Girod, B. Girod: Optimization Approaches for Soft-Tissue Prediction in Craniofacial Surgery Simulation. Proceedings 2nd international conference on medical image computing and computer-assisted intervention - MICCAI'99, lecture notes in computer science, vol. 1679, p. 1183–1190, September 19-22, 1999
4. K. Waters: A physical model of facial tissue and muscle articulation derived from computer tomography data. SPIE Vol. 1808, Visualization in Biomedical Computing, volume 21, p. 574–583, 1992.
5. R.M. Koch, M.H. Gross, F.R. Carls, D.F. von Büren, G. Fankhauser, Y. Parish: Simulating Facial Surgery Using Finite Element Methods. SIGGRAPH 96 Conference Proceedings, 1996, p. 421–428
6. M. Chabanas, Y. Payan: A 3D Finite Element Model of the Face for Simulation in Plastic and Maxillo-Facial Surgery. Proceedings of MICCAI 2000, pp. 1068–1075
7. E. Gladilin, S. Zachow, P. Deuflhard, H.-C. Hege: Adaptive Nonlinear Elastic FEM for Realistic Soft Tissue Prediction in Craniofacial Surgery Simulations, SPIE Medical Imaging, San Diego, 2002.
8. O.C. Zienkiewicz, R.L. Taylor: The finite element method. Part 1: The basis, Butterworth Heinemann Oxford, 2000
9. F. Maes, A. Collignon, D. Vandermeulen, G. Marchal, P. Suetens: Multimodality image registration by maximization of mutual information, IEEE Transactions on Biomedical Imaging, April 1997, Vol. 16:2, pp 187–198
10. W. E. Lorensen, H.E. Cline: Marching Cubes: a High Resolution 3D Surface Construction Algorithm, ACM Computer Graphics & Applications, November 1991, Vol. 11:6, pp 53–62
11. E. K. Rodriguez, A. Hoger, A. D. McCulloch: Stress-dependent finite growth in soft elastic tissues, Journal of Biomechanics, 1994, Vol. 27:4, pp 455–467
12. S. Balay, K. Buschelman, W. D. Gropp, D. Kaushik, L. C. McInnes, B. F. Smith: PETSc 2.0 Portable Extensible Toolkit for Scientific Computations, http://www.mcs.anl.gov/petsc, 2001

A Physically-Based Virtual Environment Dedicated to Surgical Simulation

Philippe Meseure[1], Jérôme Davanne[1], Laurent Hilde[1], Julien Lenoir[1],
Laure France[2], Frédéric Triquet[1], and Christophe Chaillou[1]

[1] ALCOVE, INRIA Futurs
LIFL (Computer Science Laboratory), CNRS UMR 8022, Bât M3
Université des Sciences et Technologies de Lille
59655 Villeneuve d'Ascq CEDEX, France
[2] SYSCOM, Université de Savoie, Chambéry
meseure@lifl.fr
http://www.lifl.fr/~meseure/SPORE

Abstract. In this paper, we present a system dedicated to the simulation of various physically-based and mainly deformable objects. Its main purpose is surgical simulation where many models are necessary to simulate the organs and the user's tools. In our system, we found convenient to decompose each simulated model in three units: The mechanical, the visual and the collision units. In practice, only the third unit is actually constrained, since we want to process collisions in a unified way. We choose to rely on a fast penalty-based method which uses approximation of the objects depth map by spheres. The simulation is sufficiently fast to control force feedback devices.

1 Introduction

Physics can greatly enhance realism in virtual environments and is necessary in surgical simulations. Whereas systems for manipulating rigid bodies exist (and some are marketed[1]), no system, to our knowledge can handle together different deformable models such as tissue, finite elements, mass/spring nets and particles. Surgical environments require the use of very different models to represent both the biologic tissue (organs, membranes, fluids such as blood or water) and the tools (bag, thread).

We intend to design a flexible environment able to simulate the bodies required in a simulation. Two main problems appear. First, we have to unify all the models in a common interface for the simulation to process them transparently. Second, we have to deal with Collision Detection (CD) between models in a unified way, in order to avoid all specific couples of different colliding models.

Besides, since practionners use both visual and haptics feedbacks during an operation, the surgical tools in a simulator are generally dotted with actuators to provide haptic sensation. Consequently, the simulation must be adapted to the high refresh rate of haptic devices (typically 300-1000Hz).

---

[1] http://www.havoc.com, http://www.mathengine.com, http://www.cm-labs.com

N. Ayache and H. Delingette (Eds.): IS4TM 2003, LNCS 2673, pp. 38–47, 2003.

This paper is organized as follows: After a state of the art of physically-based models and collision detection in section 2, our three-unit model is presented in section 3 and our collision detection in section 4. We show in section 5 the implementation and the results. In section 6, we discuss about the alternative of using special effects instead of physical simulation, and we conclude.

2   Previous Work

2.1   Previous Work on Physically-Based Models

In the field of surgical simulation, many models are based on the Finite-Element Method (FEM) [6][27][23][2][31], sometimes focused on the surface element [3][16]. The challenge here is to reduce the inner complexity of the FEM in order to make the resolution in real-time. Continuous models also include dynamic splines [26][13]. A second family, called "discrete" models or particle systems can also handle a large variety of bodies. Not only elastic bodies [8][21], but also fluids, clay or smoke can be reproduced [20] whereas continuous methods provide non real-time computations. Some models including Finite Difference Methods [10] and Mass/tensor systems [7] [24] try to conciliate the relative efficiency of mass/spring systems and the fidelity of the continuous models to the reality.

Discrete and continuous models only differ by the way that the equation are computed, but result in the same system:

$$\overline{\overline{M}} \frac{d^2\mathbf{x}}{dt^2} + \overline{\overline{D}} \frac{d\mathbf{x}}{dt} + \overline{\overline{K}}(\mathbf{x})\mathbf{x} = \mathbf{R} \qquad (1)$$

where $\mathbf{x}$ is the position or displacement vector of the node coordinates, $\overline{\overline{M}}$ the generalized mass, $\overline{\overline{D}}$ the damping and $\overline{\overline{K}}$ the rigidity matrices. To avoid the integration of this time-dependent differential equation, only the static component of the equations can be considered:

$$\overline{\overline{K}}(\mathbf{x})\mathbf{x} = \mathbf{R} \qquad (2)$$

The result does not reproduce dynamic phenomena such as movement or friction, but the system is less complex to solve and the result is stable. This stability is desirable, but the lack of dynamic phenomena can sometimes penalizes the realism of the simulation. We thus decided to rely on both static and dynamic models.

2.2   Previous Work on Collision Detection

Whereas CD between rigid bodies has been well studied over the past years, computing the collision reponse between general deformable bodies remains an issue. As stated by Fisher et al. [12], no physical theory can today decide how two general deformable objects will deform under contact. Instead, penalty-based methods [22] which apply forces to prevent interpenetration are well adapted. We should only ensure that the detection phase occurs at a high rate and objects do not move too fast to prevent objects from going through one another.

Nevertheless, penalty-based methods require a measure of the intersection zone. Many algorithms focus on the overlap computation of two polyhedra but are restricted to convex cases [30]. Kim et al. [18] used the graphics hardware to compute a global penetration depth between non-convex polyhedra, but this method is still time consuming and remains incompatible with deformable bodies. Another approach relies on the depth map of the bodies [12], but its current computation time is still prohibitive.

To speed up CD, Bounding Volume (BV) and in particular spheres are often used. It is indeed very fast to know if two spheres intersect [15] and their overlap depth is also computed easily. They have been intensively used for bounding distance and penetration between non-convex objects, in a hierarchical way [17] [25]. Another interest of spheres is that they do not require any orientation and only their position has to be updated during movement.

3 The Generic Physically-Based Model

Since our system is intended to handle various physically-based models, a generic description for simulated bodies is needed. The geometric and mechanical requirements are often opposite. On one hand, the geometric part needs a lot of vertices for a highly detailed representation. On the other hand, the mechanical part should be simple enough to ensure a real-time computation. A usual solution to the opposite requirements of visual and mechanical representations is to skin a simple mechanical model with a complex geometric representation. For instance, Cani used implicit surfaces [4] whereas Chadwick relied on Free-Form Deformations [5]. The main idea is to build a rich visual representation over a simple mechanical model. This gives the appearance of complexity even if the underlying model is coarse.

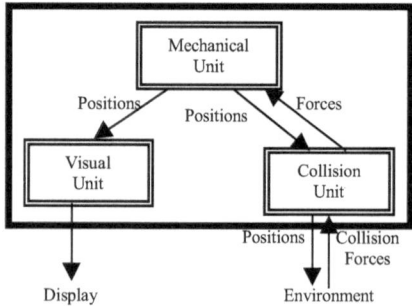

Fig. 1. The 3-unit generic description and its working principle

However, if geometric complexity is desirable for the visual representation, it must be avoided for CD purpose, since this phase is known to be time-consuming. That is why we have to rely on a decomposition of the bodies into three units: the mechanical unit, the visual unit, and the collision unit. The decomposition of a body into three units are presented on *Fig. 1*. The mechanical part provides the geometric part with positions[2] to allow it to compute the object shape, and the same information is provided to the collision unit to detect overlap with the other objects. Then, the collision unit feeds the mechanical unit with the necessary collision forces.

The master part of this generic model is the mechanical unit. It is responsible for the computation of equations of motion, which can be static or dynamic. Dynamic equations require an integration phase where the positions and velocities are computed from the applied forces. We rewrite equation (1) in the Cauchy form as:

$$
\begin{cases}
\dfrac{d\mathbf{v}}{dt} = \overline{\overline{M}}^{-1}(\mathbf{R} - \overline{\overline{D}}\mathbf{v} - \overline{\overline{K}}(\mathbf{x})\mathbf{x}) \\
\dfrac{d\mathbf{x}}{dt} = \mathbf{v}
\end{cases}
\tag{3}
$$

All the degrees of freedom (positions and velocities) of the simulated bodies are grouped together in a state vector **y**. All the implemented models *i* are asked to provide a function $\mathbf{f}_i$ (acceleration and velocity computations) such that:

$$
\frac{d\mathbf{y}_i}{dt} = \mathbf{f}(\mathbf{y}_i, t)
\tag{4}
$$

These ODE can be solved by various explicit integration schemes. Nevertheless, to guarantee the convergence and stability of the ODE solving, implicit integration scheme is often recommended [1]. Unfortunately, the implicit Euler requires to solve a non linear system. Since the usual Newton method is not fast enough, we rely on the Broyden method which is fast to solve non linear systems [14].

Care must be taken for the third unit which is responsible of all the interactions between the bodies of the simulation. For our generic model, a unified CD must be found. It is described in the next section.

4   Collision Detection and Depth Evaluation

We chose to base our penetration depth computation neither on polyhedra nor polygons. Instead, we found convenient to use spheres to represent the volume of the bodies. Similarly to [12] and [11], we compute a scalar function for each object which characterizes the depth of a point inside the object. This field is then approximated by summing elementary depth maps generated by spheres (see *Fig. 2*). Since this construction is too slow to be performed at run time, it is only applied on unmovable or rigid bodies in the preparation phase. For each deformable model, an appropriate construction method has to be found. If a volumetric body is decomposed into tetraedra for its mechanics, the collision model is obtained by surrounding each tet-

---

[2] And velocities, in specific cases.

raedron with a sphere. For 2D body, we only sample the surface with a sufficient number of spheres (the sampling can be adaptive) (See *Fig. 3.*).

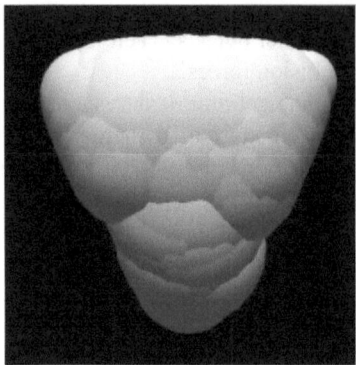 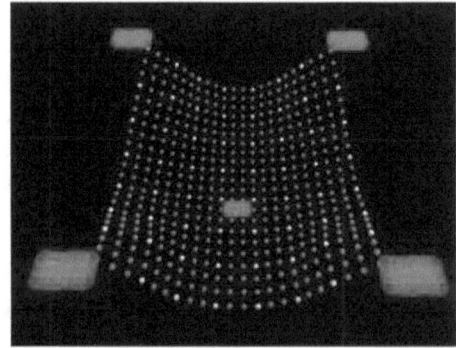

Fig. 2. A uterus modeled with 110 spheres   **Fig. 3.** Sampling of a tissue with about 500 spheres

During each simulation step, all the object spheres are put in a regular grid [29]: Only spheres in the same cell are checked for interpenetration. By checking spheres of the same body, this kind of CD can deal with self-collisions very easily. Several optimizations have been used such as time-stamping techniques to avoid emptying the grid or to prevent from generating the same colliding sphere couple twice. Moreover, the spheres of the unmovable bodies are put in the grid once for all, at the beginning of the simulation. Though not implemented yet, it would be convenient to use a hierarchical grid, to take into account various sphere radii.

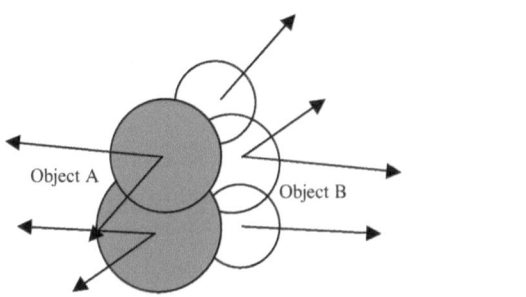

Fig. 4. The colliding spheres and the generated forces between two objects

The previously seen method generates a list of colliding spheres. A collision between two bodies will generally induce several sphere overlaps. A local spring is then inserted between colliding spheres, which generates on their centers a force proportional to the interpenetration. Each body is submitted to the sum of all the forces acting on its spheres (see *Fig. 4*). This collision processing scheme has been adapted to design an intermediate model for the control of haptic devices. More details can be found in [9].

5 Implementation

5.1 Architecture of the System

The previous sections presented the fundamental ideas of our simulation testbed, called SPORE (Simulation of Physically-based Objects for Real-time Environments). It is implemented in C++ and operational on different platforms (Windows/Unix).

In order to be as flexible as possible, SPORE relies on a minimal kernel in charge of all common processes, which include the integration of ODE, collisions and linking constraints (we use springs for such constraints). Apart from the kernel, a collection of physically-based models is proposed. *Fig. 5* shows the software architecture. The bodies are grouped in three main categories:

1. **Unmovable bodies:** These objects cannot be moved, but can prevent objects from moving as they become obstacles. We can take benefit of this property in CD.
2. **Active bodies:** The position of these bodies are controlled by the user in a straightforward way.
3. **Passive bodies:** These objects are animated by the physical laws. They are called passive because they only undergo and react to external constraints and cannot move on their own.

The process of the kernel lies in the following simple loop:

```
while (not_done) {
        forall active bodies get their position
        detect collisions
        forall colliding spheres
           compute collision response
        forall linking constraints compute their forces
        forall active bodies
           compute their force feedback
        forall passive objects
           sum their forces
           compute accelerations and velocities
        Integrate the state vector ODE
        Compute the new state vector of the system
}
```

5.2 Results

We have tested a number of mechanical bodies in our system, in the framework of a lapararoscopic surgical simulator for gynaecology. We currently simulate rigid bodies (see *Plate 1*), fluids as particles with Lennard-Jones interaction (see *Plate 3*), 2D mass/spring nets (see *Plate 2*), volumetric bodies with a deformable surface [21] (see *Plate 4*), dynamic splines [19] (see *Plate 5*). The computation time of a simulation loop is shown in *Table 1*: These times have been measured on a 1GHz Pentium III (see http://www.lifl.fr/~meseure/SPORE for videos).

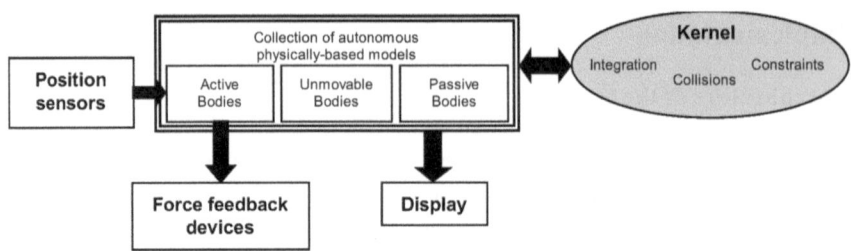

Fig. 5. Architecture of the SPORE environment

Table 1. Simulation times of various objects including CD with the tools and the environment.

Rigid Body	Fluid	Mass/ spring	Deform- able body	1D dyn. Spline	2D dyn. Spline
1ms	2ms	4ms	2ms	8ms	16ms

These times include the mechanical simulation and all the collisions of the scene (including tools and camera). The implicit integration scheme is always chosen except when it is useless (for fluids for instance). On *Plate 6.*, we show the simulation of an intestine in the abdominal cavity. We also control laparoscopic forceps with force feedback.

6   Physical Simulation vs Special Effects

All the algorithms used in our library have been optimized to provide the most efficient simulation. However, the computation time tends to be prohibitive as the simulation becomes more complex. We have found convenient for some complex phenomena to use only visual effects instead of heavy physical simulations. For instance, we use textures to represent the coagulation steam, since the simulation and the visual representation of smoke is generally very time-consuming. For sucking up the blood and other liquids, we do not actually simulate the fluid flow, but instead we only discard some particles of the model. Other tools such as clamps are not physically simulated as well.

We are searching for some criteria allowing us to decide between a special effect and a physical simulation. For instance, if a phenomenon does not imply behavioral alterations of the surrounding bodies, it can be only represented visually. However, if a realistic behavior is crucial, a physically-based approach is necessary.

7   Conclusion

In this paper we have presented an environment which enables the simulation of various deformable bodies. It handles the mechanical behavior, the visualization and the interaction between bodies through a decomposition of models in three units. Our library provides a powerful tool to design surgical simulators. We are currently working on adding volumetric objects such as finite difference and/or finite element models. We are specially interested in multiresolution to trade-off between accuracy and computation time.

Acknowledgements

This project has received a grant from ANVAR, the RNTL and the ACI 2001. The design of the intestine has been studied within a collaboration with the EVASION project of INRIA Rhône-Alpes which is supported by an ARC INRIA.

References

1. Baraff, D., and Witkin, A., "Large Steps in Cloth Simulation" *Siggraph'98, Computer Graphics annual conference series*, Orlando, 19-24 July 1998, 43-54.
2. Berkley, J., Weghorst, S., Gladstone, H., Raugi, G., Berg, D., and Ganter, M., "Banded Matrix Approach to Finite Element Modelling for Soft Tissue Simulation" *Virtual Reality: Research, Development and Applications*, 4, 203-212.
3. Bro-Nielsen, M., and Cotin, S., "Real-time Volumetric Deformable Models for Surgery Simulation using Finite Elements and Condensation" *Eurographics'96, Computer Graphics Forum*, 15 (3), Poitiers, 28-30 August 1996, 57-66.
4. Cani, M.P., "Layered Deformable Models with Implicit Surfaces" *Graphics Interface'98*, Vancouver, 18-20 June 1998, 201-208.
5. Chadwick, J.E., Hauman, D.R., and Parent R.E., "Layered Construction for Deformable Anima-ted Characters" *SIGGRAPH'89, Computer Graphics*, 23(3), Boston, 31 July- 4 August 1989, 243-252.
6. Cotin, S., Delingette, H., and Ayache, N., Real-Time Elastic Deformations of Soft Tissues for Surgery Simulation" *IEEE Trans. on Visualization and Computer Graphics*, 5 (1), January-March 1999, 62-73.
7. Cotin, S., Delingette, H., and Ayache, N., "A hybrid elastic model allowing real-time cutting, deformation and force-feedback for surgery training and simulation" *The Visual Computer*, 16 (8), 2000, 437-452.
8. Cover, A.S., Ezquerra, N.F., O'Brien, J., Rowe, R., Gadacz, T., and Palm, E., "Interactively Deformable Models for Surgery Simulation" *IEEE Computer Graphics & Applications*, 13 (6), November 1993, 68-75.
9. Davanne, J., Meseure, P., and Chaillou, C., "Stable Haptic Interaction in a Dynamic Virtual Environment" *IEEE/RSJ IROS 2002*, Lausanne, 1-3 October 2002.
10. Debunne, G., Desbrun, M., Cani, M.P., and Barr, A.H., "Dynamic Real-Time Deformations using Space & Time Adaptive Sampling" *Siggraph'01, Computer Graphics annual conference series*, Los Angeles, August 2001.
11. Eberhardt, B., Hahn, J., Klein, R., Strasser, W., and Weber, A., "Dynamic Implicit Surfaces for Fast Proximity Queries in Physically Based Modeling", *Graphisch-Interaktive Systeme (WSI/GRIS)*, Universität Tübingen, 2000.
12. Fisher, S., and Lin, M.C., "Fast Penetration Depth Estimation for Elastic Bodies using Deformed Distance Fields" *IEEE/RSJ IROS, 2001*.

13. France, L., Lenoir, J., Meseure, P., and Chaillou, C., "Simulation of a Minimally Invasive Surgery of Intestines", *VRIC 2002*, Laval, 17-23 June 2002.
14. Hilde, L., Meseure, P., and Chaillou; C., "A fast implicit integration method for solving dynamic equations of movement" *VRST'200*, Banff, 15-17 November 2001, 71-76.
15. Hubbard, P.M., "Approximating Polyhedra with Spheres for Time-Critical Collision Detection" In *ACM Trans. on Graphics*, 15 (3), July 1996, p 179-209.
16. James, D.L., and Pai, D.K., "ARTDEFO: Accurate Real Time Deformable Objects" *Siggraph'99, Computer Graphics annual conf. series*, Los Angeles, 8-13 August 1999, 65-72.
17. Johnson, D.E., and Cohen, E., "Bound Coherence for Minimum Distance Computations" *IEEE ICRA*, Detroit, 10-15 Mai 1999, 1843-1848.
18. Kim, Y., Otaduy, M., Lin, M., and Manocha, D., "Fast Penetration Depth Computation for Physically-based Animation", *ACM Symp. on Computer Animation*, 21-22 July 2002.
19. Lenoir, J., Meseure, P., Grisoni, L., and Chaillou, C., "Surgical Thread Simulation" *MS4CMS*, Rocquencourt, 12-15 November 2002.
20. Luciani, A., Habibi, A., Vapillon, A., et Duroc, Y., "A Physical Model of Turbulent Fluids" *EUROGRAPHICS workshop on Animation and Simulation*, Maastricht, 2-3 september 1995, 16-29.
21. Meseure, P., and Chaillou, C., "A Deformable Body Model for Surgical Simulation" *Journal of Visualization and Computer Animation*, 11 (4), September 2000, 197-208.
22. Moore, M., and Wilhelms, J., "Collision Detection and Response for Computer Animation" *Siggraph'88, Computer Graphics*, 22, 4, Atlanta, 1-5 August 1988, 289-298.
23. Nienhuys, H.W., and van der Stappen, A.F., "Combining Finite Element Deformation with Cutting for Surgery Simulations", *Eurographics'00 (short presentations)*, Interlaken, 20-25 August 2000, 43-51.
24. Picinbono, G., Delingette, H., and Ayache, N., "Non-Linear and Anisotropic Elastic Soft Tissue Models for Medical Simulation" *IEEE ICRA*, Seoul, Korea, 2001.
25. Quinlan S., "Efficient Distance Computation between Non-Convex Objects" *IEEE ICRA*, 1994.
26. Rémion, Y., Nourrit, J.M., and Nocent, O., "Dynamic Animation of n-Dimensional Déformable Objects" *WSCG'2000*, Plzen, 7-11 February 2000.
27. Sagar M. A., Bullivant D., Mallinson G. D., Hunter P. J, and Hunter J. W., "A Virtual Environment and Model of the Eye for Surgical Simulation" *Siggraph'94, Computer Graphics annual conf. series*, Orlando, 24-29 July 1994, 205-212.
28. Triquet, F., Meseure, P., and Chaillou, C., "Fast Polygonization of Implicit Surfaces" *WSCG'01 Conference*, Plzen, 5-8 February 2001, 283-290.
29. Turk, G., *Interactive Collision Detection for Molecular Graphics*, Master Thesis, Technical Report TR90-014, University of North Carolina, 1989.
30. Van den Bergen G., "Proximity Queries and Penetration Depth Computation on 3D Game Objects" *Game Developper Conference*, 2001.
31. Wu, X., Downes, M., Goketin, T., and Tendick, F., "Adaptive Nonlinear Finite Elements for Deformable Body Simulation using Dynamic Progressive Meshes" *Eurographics'01, Computer Graphics Forum*, 20 (3), Manchester, 4-7 September 2001.

Plates

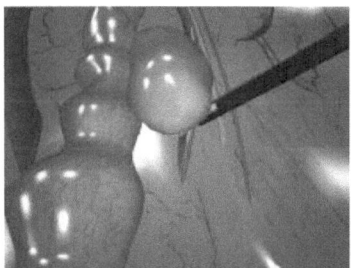

Plate 1. Simulation of a rigid ovary.

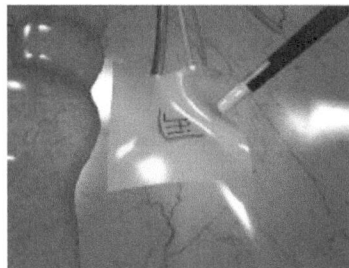

Plate 2. Simulation of a tissue.

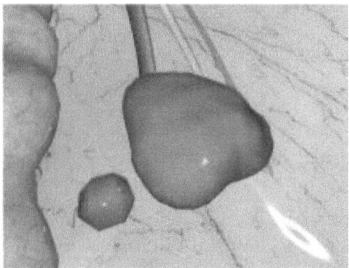

Plate 3. Simulation of blood as a particle system.

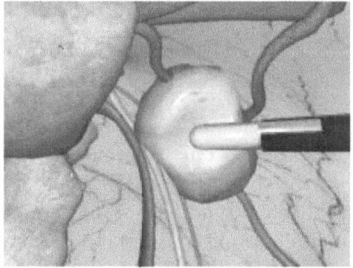

Plate 4. Deformation of a surface body (an ovary).

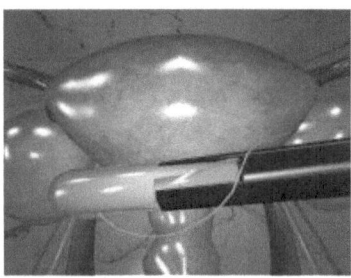

Plate 5. Simulation of a piece of string modeled by a 1D dynamic spline.

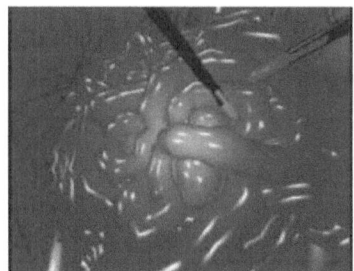

Plate 6. Simulation of an intestine.

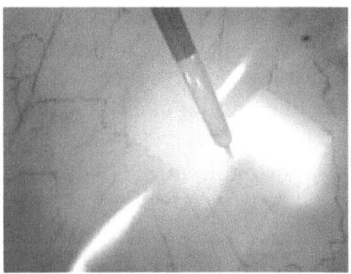

Plate 7. Coagulation steam.

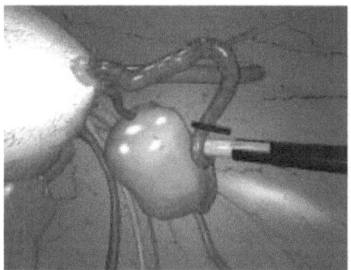

Plate 8. Clamping of a fallopian tube.

Soft-Tissue Simulation
Using the Radial Elements Method

Remis Balaniuk[1,2] and Kenneth Salisbury[1]

· Department of Surgery, School of Medicine
Stanford University
Stanford, CA 94305-5655, USA
{remis,jks}@robotics.stanford.edu
http://www.stanford.edu
· Universidade Católica de Brasília
Brasilia, Brazil
http://www.ucb.br

Abstract. This paper introduces the Radial Elements Method – REM for the simulation of deformable objects. The REM was conceived for the real time, dynamic simulation of deformable objects. The method uses a combination of static and dynamic approaches to simulate deformations and dynamics of highly deformable objects. The real time performance of the method and its intrinsic properties of volume conservation, modeling based in material properties and simple meshing make it particularly attractive for soft tissue modeling and surgery simulation.

1 Introduction

Physically based simulation of deformable objects is a key challenge in Virtual Reality (VR). A major application area for this research is the simulation of bio-materials. Physical modeling of biomaterials has a broad range of applications ranging from understanding how soft and hard tissue respond under loading, to patient-specific planning of reconstructive procedures, to the training of surgical skills, and much more. Real-time interaction with these models is necessary in order to impose conditions of interest, to examine results and alternative solutions, to learn surgical procedures by performing them in simulation. Modeling the biomechanics of muscles, tissues, and organs is intrinsically a computationally difficult undertaking - doing so at haptically real-time rates requires significant computational resources and algorithmic finesse. The development of sufficiently real-time modeling of biomaterials is a topic of intense interest to a worldwide research community, and methods to simulate deformable materials are at the center of this research.

1.1 State of the Art

A number of methods have been proposed to simulate deformable objects ranging from non-physical methods, where individual or groups of control points or shape

N. Ayache and H. Delingette (Eds.): IS4TM 2003, LNCS 2673, pp. 48–58, 2003.

parameters are manually adjusted to shape editing and design, to methods based on continuum mechanics, which account for material properties and internal and external forces on object deformation.

Simulation methods can be classified as Dynamic or Static. Dynamic methods simulate the time evolution of a physical system state. The bodies are assumed to have mass and energy distributed throughout. Differential equations and a finite state vector define each model. Numerical integration techniques approximate the system state (position and velocity) at discrete time steps. In static methods equilibrium equations or closed form expressions describe the system and these equations are solved to find a static solution at each time step. In static and quasi-static methods the time and the system state are usually not considered (e g spring-only models respond immediately to load change). For a more comprehensive and accurate simulation of deformable objects a state-based approach should be used, given the significant internal dynamics of the deformable media, as well as the dynamics of the movements (translations and rotations) of the object in space. Nevertheless, static and quasi-static methods can be useful in applications where the objects deform but do not move, or move slowly, and the deformable media is highly damped. If the problem permits, a static solution seems more attractive than a dynamic one. Dynamic solutions usually pose various problems, which depend on the type of method used.

For a survey of deformable modeling in computer graphics the reader is referred to [4]. Other methods are the "Finite Elements and Condensation" [12], "Geometric Nonlinear FEM" [6], "Boundary Element Method" [5] and "Space and Time Adaptive Sampling" [3]. Some examples of medical simulators are [9] and [10].

The REM is an evolution of some concepts we introduced on a previous method, the LEM - Long Elements Method. The LEM approach [1] [2] introduced two new concepts to the simulation of deformable objects: the meshing based on long elements and the combination of static and dynamic approaches to simulate the same object. The long element (Fig. 1) is an elastic spring. It has the geometry of a deformable slender rod, defined by its length and the area of its cross-section. The meshing of an object using long elements reduces the number of elements on the model in one order of magnitude for the same granularity if compared to a standard tetrahedral or cubic meshing. Deformations are simulated using the static solution while movements in space, energy and gravity are simulated separately on a dynamic engine. Large deformations that rapidly change the entire shape of the object can be stably simulated in just one time step because of this separation.

2   Method Presentation

Both the LEM and the REM are based on a static solution for elastic deformations of objects filled with uncompressible fluid. The volumes are discretised on a set of long elements (Fig. 1), and an equilibrium equation is defined for each element using bulk variables. The set of static equations plus the Pascal principle

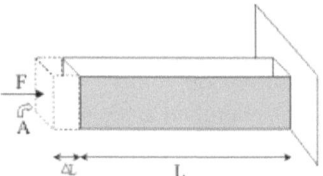

Fig. 1. Long element.

and the volume conservation are used to define a system that is solved to find the object deformations and forces. The forces obtained are then integrated to simulate movement.

2.1   Static Solution

The static deformation engine simulates the long element as a mass-less linear spring, defined by its length, area and elasticity. These values are defined for each element based on the material properties of the simulated media [11]. An equilibrium equation is defined for each long element relating its stress (internal and external pressures) to its strain, or deformation (change in length). The static equilibrium condition states that the forces, or pressures, inside the element should be equal to the external forces, or pressures, applied externally: $P_{int} = P_{ext}$.

The external pressure P_{ext} on the surface is affected by the atmospheric pressure and by the stress when an elongation exists, so $P_{ext} = P_{atm} + s$. For small applied forces, the stress s in a material is usually linearly related to its deformation (its change in length in our long elastic object). Defining elasticity E as the variable relating stress and the fractional change in length: $\Delta L/L$, it is possible to write: $s = E\Delta L/L$

The internal pressure (P_{int}) is formed by the pressure of the fluid (without gravity) and the effect of the gravity acceleration (g), so: $P_{int} = P_{fluid} + dgh$ where h is the distance between the upper part of the fluid and the point where the pressure is calculated and d is the density of the fluid. From the last three equations, a continuous equation can be obtained as:

$$E\Delta L/L - \Delta P = dgh \tag{1}$$

where $\Delta P = P_{fluid} - P_{atm}$.

Another external pressure to be considered comes from contacts between the object and its environment. To obey the action-reaction law, at the points on the object surface in contact with the environment the pressure applied to the external contact and to the object must have the same magnitude. It means that the external pressure P_{env} applied by the contact must be equal to $P_{env} = P_{ext} - P_{int}$. For the elements corresponding to these points a term is added to the right side of equation 1:

$$E\Delta l/l - \Delta P = dgh + P_{env} \tag{2}$$

External contacts can also be modeled as displacements imposed to some surface vertices. The deformation ΔL is defined by the penetration of the contact in order to make the surface follow the contact position (y). In this case the equation 1 can be rewritten as:

$$\Delta L = y. \tag{3}$$

To simulate an object its volume is filled (meshed) using long elements and an equilibrium equation is defined for each element based on the stated principles. To make the connection between the elements two border conditions are applied:

1. Pascal's principle says that *an external pressure applied to a fluid confined within a closed container is transmitted undiminished throughout the entire fluid.* Mathematically:

$$\Delta P_i = \Delta P_j \text{ for any } i \text{ and } j. \tag{4}$$

2. The fluid is considered incompressible and consequently the volume conservation must be guaranteed when the object deforms. The volume dislocated by a contact will affect the entire object instantaneously. To implement the volume conservation one equation is added to the system, stating that the sum of the deformations of all long elements (the changes in lengths) must be 0: $\sum_{i=1}^{N} A_i \Delta l_i = 0$ where N is the total number of elements. This equation ensures a global conservation of volume defined by the whole set of elements. Note that a controled increase or decrease of the volume of the object can also be simulated by defining the desired volume change at the right side of this last equation.

The surface tension also affects the external pressure on the elements. Deformations cause a change on the object surface area, even if its volume is kept constant. This change creates forces on the surface generating a surface tension. One of the effects of these forces on a deformable object is to make the contours of the surface smoother. To reproduce these forces due to the change in the surface area we include linear elastic connections between neighbor elements coupling their changes in length. A number of terms will be added to the right side of the equation of the external pressure corresponding to the neighborhood considered around the element. These terms are of the form $P = F/A = kx/A$, where x is the difference in deformation between an element and its neighbor and k is a local spring constant. For a given element i the term relating its deformation to the deformation of its neighbor j is: $k_{ij}(\Delta l_i - \Delta l_j)/A_i$.

Equations 2, 4 and the superficial tension equation define the final equation for an element i. Considering a typical configuration with 6 neighbors j_1 to j_6:

$$(E_i/l_i + 6kA_i)\Delta l_i - kA_i(\Delta l_{j_1} + \ldots + \Delta l_{j_6}) - \Delta P = d_i g h_i + P_{env_i} \tag{5}$$

where the superficial spring constant k was done constant for all elements to make easier the notation.

Fig. 2. REM meshing.

The equilibrium equations (5 and 3) plus the volume conservation equation define a typical numerical problem of type $A.x = B$. A is a sparse matrix and the system is solved using standard numerical methods to estimate the inverse matrix A^{-1}. The system always has a solution. If we limit the use of equation 3 to a fixed set of points (usually points attached to some sort of internal or external frame) and unless some topological change occurs during the simulation, the system needs to be solved just once. The vector of deformations x can be estimated multiplying the vector of external pressures and gravity B by A^{-1}. B is always defined and it is obtained based on collision detection between the object and its environment and the orientation of the object in space.

2.2   Meshing Geometry

The REM meshing is based on a spherical geometry (see section 2.3). Fig. 2 shows the meshing of a 3D object. The mesh is based on radial elements, starting at the object's center of mass and ending at the surface of the object. The extremities of the elements define the surface vertices, that are connected to form triangles defining the rendering surface. Meshing based on radial elements was used previously on other simulation methods as in [7] and [8].

The radial elements are an extension of the concept introduced by the long elements. Instead of deforming just by changing its length, the radial element can also deform by rotating as a torsional spring. The radial element can be defined as a 3D vector, having magnitude and orientation. Four deformation values are simulated for each radial element, corresponding to a change in length and 3 scalar angular displacements (Euler angles). The changes in length are estimated by the static method presented in section 2.1. The rotations are simulated using the method presented at section 2.4. The initial length and orientation of each radial element is defined by a spherical meshing algorithm . Using the radial elements we are able to simulate complex deformations, caused by normal and tangential forces applied to the object.

2.3   Meshing Algorithm

The meshing algorithm is very simple and straightforward. A surface model of the object is supposed to be known. A central point inside the object must be

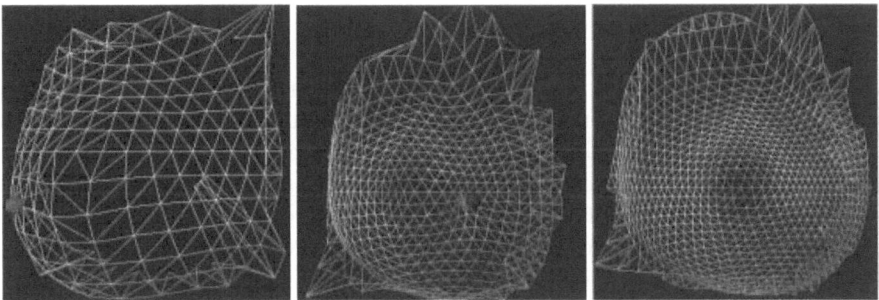

Fig. 3. Breast modeled using the REM at different resolutions.

chosen. To be physically correct it should be the object's center of mass, but the choice of any point close to the geometric center of the volume will not cause significant differences on the simulation. The granularity of the model needs to be defined as well. The volume will be "sliced" to define the radial elements and the number of slices determines the number of radial elements and consequently the resolution of the model. The right image on Fig. 2 shows the slices used to mesh the liver. The total number of radial elements created will be $n^2 - n$ where n is the number of slices. To distribute the radial elements around the center homogeneously we use the following algorithm:

```
// 'center' and 'direction' are vectors
// 'center' and 'slices' were chosen beforehand
steps = (slices/2)-1
for( i=0; i<= steps; i++) {
    angle1=i*0.5*Pi/steps;
    direction.x = sin(angle1);
    for(j=0;j<= (steps-i)*4 ;j++) {
        angle2=j*0.5*Pi/(steps-i) ;
        direction.y = sin(angle2)*cos(angle1);
        direction.z = cos(angle2)*cos(angle1);
        // shootLine draws a line from 'center' on
        // the direction of 'direction' and returns
        // the distance from 'center' to the object's surface
        length = shootLine (center , direction);
        new radialElement (center, direction, length);
        if(i>0) {
            length = shootLine (center , -direction);
            new radialElement (center, -direction, length);
        }
    }
}
```

Fig. 3 shows the meshing of a model of the breast in 3 different resolutions.

2.4   Torsional Deformations

The radial elements are now treated as torsional springs. The rotations of the
elements are caused by the tangential forces been applied to the object. Typically
a force vector F applied to a point at the object will be decomposed in force
vectors applied to the closest radial element. A first decomposition of the force
vector defines the normal force that will stretch or compress the element: $F \cdot p$
where p is the unitary orientation vector of the radial element. The torque vector
is obtained by $T = F \times r$ where r is the moment arm defined by the radial element
(3D vector having magnitude defined by the length of the element).

Torsional springs generally follow the constitutive law: $T = K\Theta$, where T is
the applied torque, K is the stiffness of the spring and Θ is the angular displacement (in radians) of the spring. To enforce continuity, neighbor radial elements
are linked by linear elastic connections coupling their rotations. A torque is created when the rotations of two neighbors are not the same. To consider this
coupling torque a number of terms are added to the right side of the torsional
spring equation corresponding to the neighborhood considered around the element. For a given element i the term relating its rotation to the rotation of its
neighbor j is: $k_{i,j}(\Theta_i - \Theta_j)$.

Considering the typical configuration with 6 neighbors j_1 to j_6 the final
equation describing the rotational spring i is:

$$K_i * \Theta_i + k_{i,j_1}(\Theta_i - \Theta_{j_1}) + \ldots k_{i,j_6}(\Theta_i - \Theta_{j_6}) = T_i \qquad (6)$$

External contacts can also be modeled as rotations imposed to some radial
elements. The rotation Θ_i is defined by the position of the contact in order to
make the surface follow the contact position (y). In this case the equation 6 can
be rewritten as:

$$\Theta i = N(y - C) - p \qquad (7)$$

where C is the center of the mesh, N() returns the normalized unitary vector for
a given input vector and p is the unitary orientation vector of the radial element.

Torques and angular displacements are vectors. For each element the equations 6 or 7 define 3 scalar equations, one for each Euler angle. The system defines
a typical numerical problem of type $A.x = B$. A is a sparse matrix and the system is solved once using standard numerical methods to estimate the inverse
matrix A^{-1}. If we limit the use of equation 7 to a fixed set of points and unless
some topological change occurs during the simulation, the vector of rotations
x can be estimated multiplying the vector of scalar torques and scalar imposed
rotations B by A^{-1}. Each radial element obtains 3 scalar angular displacements,
from 3 different equations, corresponding to the rotations about each of the 3
axis of its reference frame.

In summary, the deformation engine consists typically of 2 matrix X vector
multiplications that in one time step of the simulation estimate the changes in
length and the 3D rotations of all radial elements having as input vectors of
pressure and torques being applied to the object.

2.5   Dynamic Solution

This method chooses a simplified approach to handle the dynamics of the deformable objects. One single point is considered, corresponding to the center of the radial mesh. This choice strongly simplifies the simulation and ensures stability, even at low update rates. Translations, rotations, velocity, gravity, energy and mass can be simulated, although some other dynamic phenomena, as viscosity, cannot. To simulate the dynamics of the mesh the same mass-less elements used by the deformation engine are simulated as energy-storing elements. Elastic potential energy due to compression, stretching or rotation is simulated to create forces and torques and to derive a state for the simulated object. The radial elements are now modeled as dynamic springs attached to the object center of mass. These springs are relaxed when the solid is not touched (not deformed). The deformation of a radial element creates a force and a torque at the center of mass. The forces are created by the changes in length of the element and the torques by its rotations.

The force due to each element has magnitude given by $F_i = K_i \Delta L_i$ on the direction of the element. The spring constant K_i depends on its length: $K_i = A_i E_i / L_i$. E_i is the elasticity of the element. The total force being applied to the object at one time step is the sum of all forces applied by the elements to the center of mass. This force is integrated on time using an explicit method to estimate the translations of the object in space. The mass of the object is supposed to be known.

The tangential forces applied to the object create a torque at the center of the mesh, causing the whole object to rotate. The sum of the torques applied by the radial elements at the center of mass defines the object's torque that is integrated to define angular acceleration and velocity for the object. The moment of inertia of the object is estimated considering that each radial element define a slender rod pivoting at its end. So its moment of inertia is defined as $I = 1/3ML^2$, where M is the mass of the element and L is its length.

The total mechanical energy of the object is defined by $E = E_e + E_g + E_k$ where E_e is the elastic potential energy stored as a result of the deformations of the radial elements, E_g is the gravitational potential energy defined by the height of the object, and E_k is the kinematic energy defined by the translational and rotational motion energy of the object. The total energy of the object is controled to ensure the stability of the simulation.

2.6   Results and Future Directions

A non-optimized prototype of the REM was implemented. On a dual Pentium III 2.0 GHz desktop PC one iteration of the simulation loop takes about 0.05 seconds for a 800 elements mesh. Fig. 4 shows some examples of deformable objects, simulated with REM, being touched through a haptic interface. The breast was implemented using a set of radial element attached to a fixed frame (the back of the model) while the other elements (the front) are free. The fixed elements have their length and orientation imposed by the position and orientation of

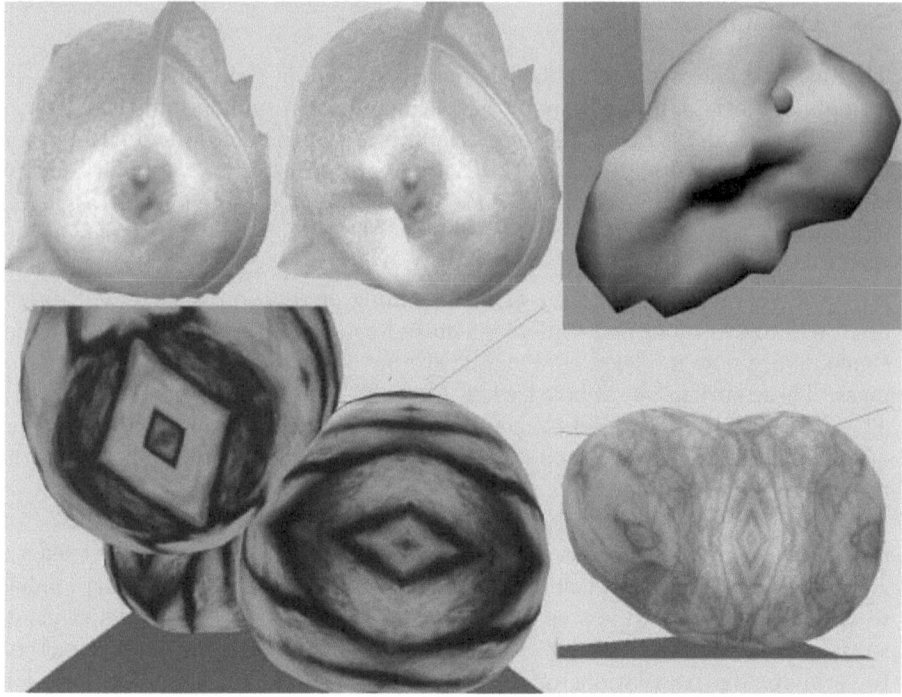

Fig. 4. Deformable objects simulated with REM.

the frame, while the free elements are deformed by external pression applied through the haptic interface. The 3 balls on a box show how the method can stably simulate persistent contact between the object and its environment.

Many aspects of our approach remain to be explored. The study of the errors added by the first order approximations and linear elements is necessary in order to define to what extent is it possible to mimic a given material. Comparisons between the REM and other existing methods in terms of speed in accuracy are needed. We are currently studying strategies for meshing complex objects by combining multiple spherical meshes inside the same model. We are also working on algorithms for real time remeshing of objects to enable interactive topological changes (cutting, removing parts, suturing, etc) (Fig. 5).

2.7   Conclusion

The Radial Elements Method has some interesting properties for the real time simulation of deformable objects. Large deformations are estimated in just one time step. Unless topological changes occur the deformations can be estimated using only multiplications of matrices by vectors, without matrix inversions. The dynamics of the object are simple and stable, enabling any kind of interaction between the object and its environment, including persistent contact with other

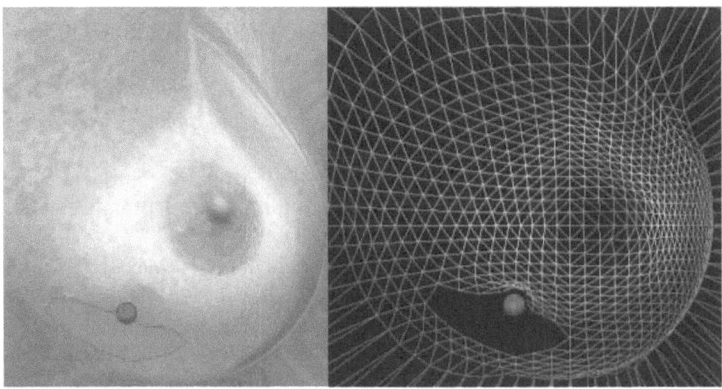

Fig. 5. Cutting on a REM model.

objects in constrained spaces. The simplicity of the method and the reduced number of elements makes it fast and scalable, while enabling a full simulation of deformations and dynamics.

On the other hand, the method has limitations. Internal dynamics and viscosity cannot be simulated given the one point dynamic model. The shapes that can be modeled are restricted to star-shaped objects, given that all radial elements need to start at the center and get to the surface of the object.

Acknowledgments

The authors wish to gratefully acknowledge the support that has made this work possible including contributions from the NIH/NLM Grant LM07295-01 and the Departments of Computer Science and Surgery at Stanford University (Stanford, CA). We would also like to thank Prof. Ivan Costa and Prof. Federico Barbagli for their contributions to this work.

References

1. Costa, Ivan Ferreira and Balaniuk, Remis "LEM - An approach for real time physically based soft tissue simulation " *Proceedings of the IEEE International Conference on Robotics and Automation ICRA2001*, May 2001, Seoul, Korea.
2. Balaniuk, Remis and Costa, Ivan Ferreira. "LEM - An approach for physically based soft tissue simulation suitable for haptic interaction". *Conference Paper, Fifth PHANTOM Users Group Workshop - PUG00*, Aspen, USA, October 2000.
3. Debunneet al., "Dynamic Real-Time Deformations using Space and Time Adaptive Sampling" *Proceedings of the SIGGRAPH'01* .
4. Sarah F. F. Gibson and Brian Mirtich. "A survey of deformable models in computer graphics." *Technical Report TR-97-19*, Mitsubishi Electric research Laboratories, Cambridge, MA, November 1997. (http://www.merl.com/reports/TR97-19/index.html).

5. D. James and D. Pai, "Artdefo Accurate Real Time Deformable Objects", *Computer Graphics*, vol. 33, pp: 65–72, 1999.
6. Yan Zhuang and John Canny. "Haptic Interaction with Global Deformations". *Proceedings of the IEEE International Conference on Robotics and Automation - ICRA 2000*, 2428-2433, 2000.
7. P. Meseure and C. Chaillou "A Deformable Body Model for Surgical Simulation", *Journal of visualization and Computer Animation, 11, 4*, Septembre 2000, pp 197-208.
8. Gascuel, Verroust, Puech, "A modelling system for Complex deformable bodies suited to animation and collision processing", *Journal of Visualisation and Computer animation vol 2, number 3*, 1991.
9. Murat Cenk Cavusoglu, Frank Tendick, Micahel Cohn and S. Shankar Sastry. "A Lapraroscopic Telesurgical Workstation". *IEEE Transactions on Robotics and Automation*, 15(4), 728-739, 1999.
10. Stephane Cotin, Herve Delingette and Nicholas Ayache. "Real-time Elastic Deformations of Soft Tissues for Surgery Simulation". *IEEE Transaction on Visualization an Computer Graphics*, 5(1), 62-73, 1999.
11. Y.C. Fung. *Biomechanics - Mechanical Properties of Living Tissues*. 2ed. Springer, 1993.
12. M. Bro-Nielsen and S. Cotin, "Real-time Volumetric Deformable Models for Surgery Simulation using Finite Elements and Condensation". In *Proceedings of Eurographics'96 - Computer Graphics Forum*, pages 57-66, 1996.

GeRTiSS: A Generic Multi-model Surgery Simulator

Carlos Monserrat, Oscar López, Ullrich Meier, Mariano Alcañiz,
Carmen Juan, and Vicente Grau

Medical Image Computing Laboratory (MedICLab),
Universidad Politécnica de Valencia, Spain
cmonserr@dsic.upv.es

Abstract. The construction of surgery simulators will be a key tool in the development and diffusion of minimally invasive surgery. Nowadays, most simulators are oriented to training surgeons in only one surgery technique. Most of them only permit the modelling of tissues with only one kind of deformable model. In this paper, we present our generic surgery simulator for minimally invasive surgery. In this surgery simulator, surgeons can construct any surgery scenario that they want to practice. Our surgery simulator permits the surgeons to select the deformable model that best adjusts to the biomechanical properties of each organ. Once the surgeon has finished the training, our surgery simulator can generate a report that contains an assessment evaluation of that training.

1 Introduction

One of main challenges in new technologies applied to medicine is the construction of virtual environments for surgical training. Environments of this kind are called surgery simulators. These simulators require [6][3]:

- the internal organs of a patient, to be visualised as realistically as possible;
- the organs to react realistically in real time to user interactions and to the established environmental restrictions;
- these virtual organs to react, to typical surgeon gestures (like cauterisation, cutting or clipping) through realistic geometric and topologic modifications,.

While 3D visualisation techniques are sufficiently developed to accomplish the first requirement, the last two are very difficult to complish. This limitation is due to the fact that current computer power is too slow with respect to the computational requirements of surgery simulators to simulate the biomechanical behaviour of internal organs of virtual patients.

Surgery simulators can be grouped into three technologically sequential generations [1][21]:

- first generation: these surgery simulators only consider the geometric nature of human anatomy [8][9];
- second generation: these include all surgery simulators that also permit physical interactions with anatomical structures [4][14][17] [23];
- third generation: apart from the characteristics of second generation surgery simulators, these consider the functional nature of the organs.

N. Ayache and H. Delingette (Eds.): IS4TM 2003, LNCS 2673, pp. 59–66, 2003.

Current research and development is being carried out mainly in second generation surgery simulators. The simulators included in this second generation can be sorted into four levels according to their complexity [22]:

- Needle-type simulators with simple visual objects and haptics with a minimum degree of freedom (only one edge) [10][13].
- Exploration-type or catheter installation-type [2] [11][12].
- Simulators that permit training in only one surgical task [5][23].
- Full simulators that permit surgeons to practice complete surgical operations (i.e. laparoscopy [7][18], endoscopy [14], arthroscopy [15] or intraocular surgery [20]).

In this paper, we present a second generation generic surgery simulator (for minimally invasive surgery) which can be classified as a full simulator. Our simulator permits the users to construct their own surgical scenario from real and/or synthetic organs with desired pathologies. All of these work on a low-cost cluster of PCs.

2   The Generic Real Time Surgery Simulator (GeRTiSS)

GeRTiSS is a surgery simulator for minimally invasive surgery, a second-generation full simulator. Its generality is due to its separation in two well differentiated modules: the Scene Generator and the Surgery Simulator (see Figure 1).

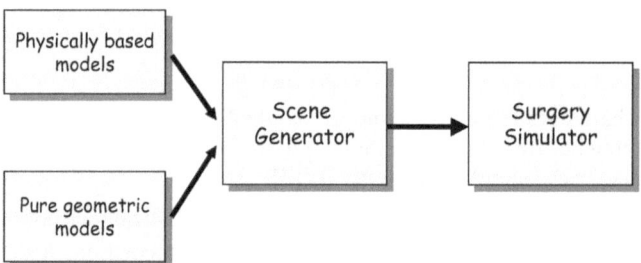

Fig. 1. Modules of the surgery simulator.

2.1   The Scene Generator

The main objective of the Scene Generator is to allow the user to select the tools and the organs needed for the simulation of the operation in an as simple and comfortable way as possible. Organs can be real and/or synthetic organs with or without pathologies. A user can also associate each organ with any of the deformable models implemented using deformation parameters which are adjusted to the biomechanical characteristics of the organ. In Figure 2, we show a laparoscopy and arthroscopy scenario that are being constructed using the Scene Generator.

To construct scenarios, the Scene Generator has implemented the following tools:

- Tools for loading synthetic and/or real organs into the scenario
- Tools for establishing the input points of surgical instruments.
- Tools for associating different physical properties to different organs.

- Tools for establishing boundary conditions.
- Tools for linking tissues.
- Tools for adding especial tissues (i.e. peritoneum).
- Tools for associating textures to organs that appear in the scenario.

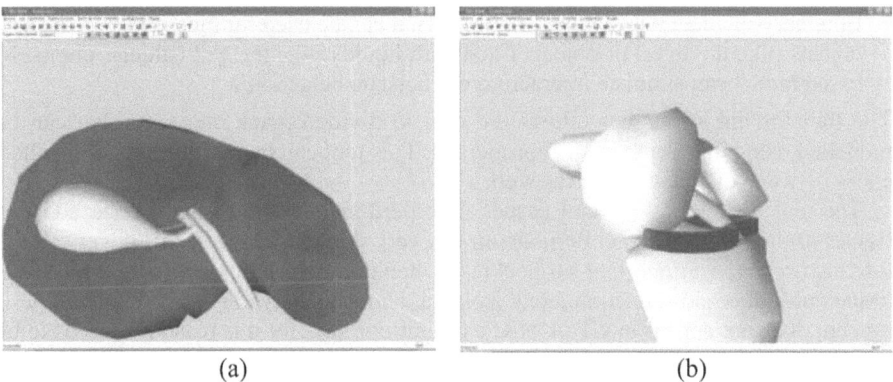

Fig. 2. Laparoscopy (a) and arthroscopy scenes (b).

To load organs, we have implemented tools that can read a set of standard and "ad-hoc" standard 3D-file formats. Once the organs have been loaded, various tools allow users to apply geometric transformations to the organs in order to adjust them to the spatial characteristics of the surgical scene that is being constructed.

The tools for establishing the input points of the surgical instruments allow users to set the points through which these instruments come into the surgical scenario (see Figure 3). They also allow users to set the initial input angle of each instrument and to assign each input point to an instrument or to a camera. Finally, they allow the users to select the tools to be used during the operation.

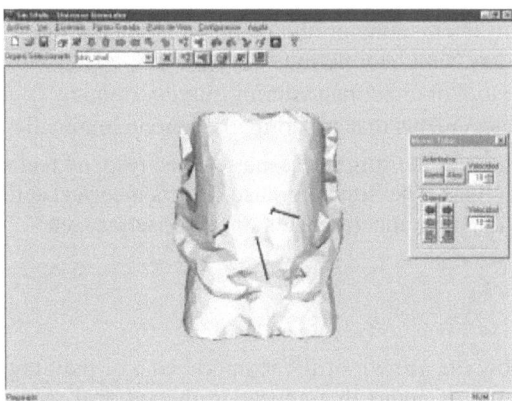

Fig. 3. Establishing surgical instrument input points.

The association of different physical properties to different organs permits, on the one hand, to associate to each organ one of the deformable models that our Simulator has implemented and, on the other, to set the deformable parameters that will govern the tissue deformation. The deformable models that our Simulator has currently implemented are [19][16]:

- Mass-spring model: That can be used for modelling elastic surfaces (like peritoneum) and quasi-one-dimensional elements (like arteries and veins).
- Boundary-element based model (BEM): That can be used for modelling volumetric objects (like the liver) or objects filled with liquids (like the gall balder). The BEM based model can simulate interactive viscoelastic behaviour.

The tool for linking tissues allows the user to divide organs into parts that can be modelled using different physical properties. This tool can be used for creating adherences between different organs as well.

The tool for adding especial tissues, like peritoneum, allows the user to add this tissues to the surgical scene. Peritoneum is a very thin elastic tissue that covers all the internal organs. An important surgical task when doing a laparoscopy is to dissect this tissue until all organs are completely isolated. Unfortunately, due to its thinness, peritoneum does not appear in CT or MRI explorations and, for this reason, it needs to be added later.

The result of the Scene Generator is a "project". This project contains all the files that our Surgery Simulator needs. The Scene Generator can obtain as many projects as needed for training surgeons (with different levels of difficulty). In addition, the Scene Generator can be used to generate a database of surgical scenes and surgeons need only select the project which contains the virtual surgical scene required.

2.2 Surgery Simulator

Our Surgery Simulator takes a scene created with Scene Generator as input and allows the user to train on this scene. Once training has finished, the Surgery Simulator can generate a report with an assessment of the training.

The Surgery Simulator allows the users (see figure 4):

- To have different interactions on the organs that appear in the surgical scene. At present, the Surgery Simulator permits to cut, to cauterise, to drag and to clip.
- To exchange the tool of either hand during surgery training.
- To train all the movements that are normally done in minimally invasive surgery.

The deformable organs of the surgical scene selected respond realistically and interactively to user interactions. The Simulator also detects incorrect actions during surgery, such as inadequate clipping, incorrect cuts or cauterisations, etc.

3 Results

In Figure 5, we show the growth of the temporal cost of the Surgery Simulator with respect to the number of nodes that appear in the surgical scene. For these experiments, we used a 450 Mhz PC Pentium III with 256 MB of main memory.

As can be observed, the temporal cost grows linearly with the number of nodes. For complex scenes, it is necessary to use the parallel version of the deformable model libraries, which is the bottleneck for all surgery simulators. The example shown in Figure 4 has less than 2000 nodes.

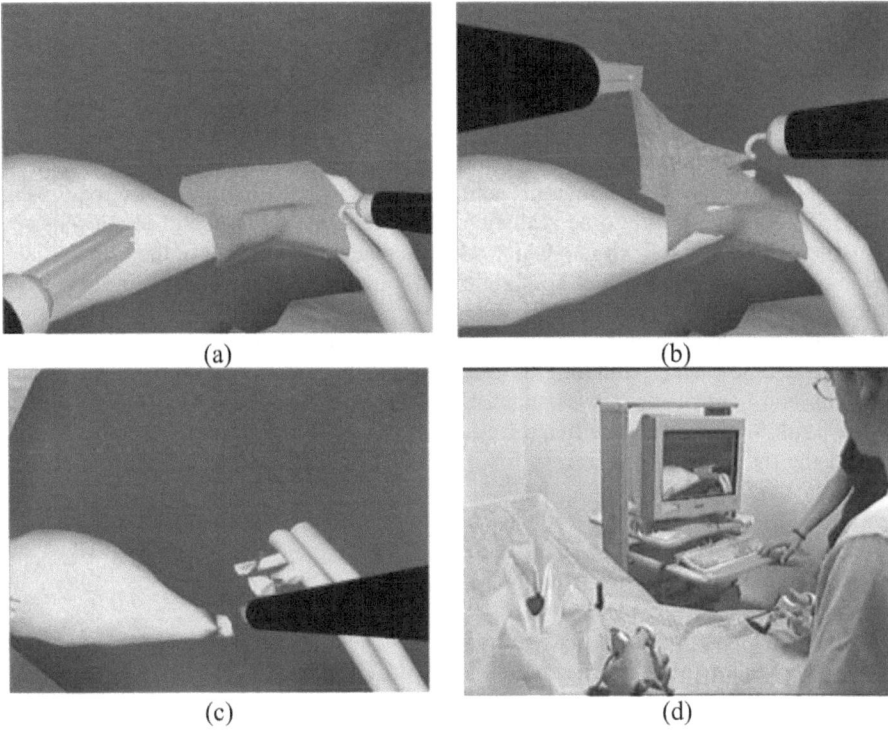

Fig. 4. A sequence of laparoscopy (a,b,c) and the user interface (d) of our Surgery Simulator.

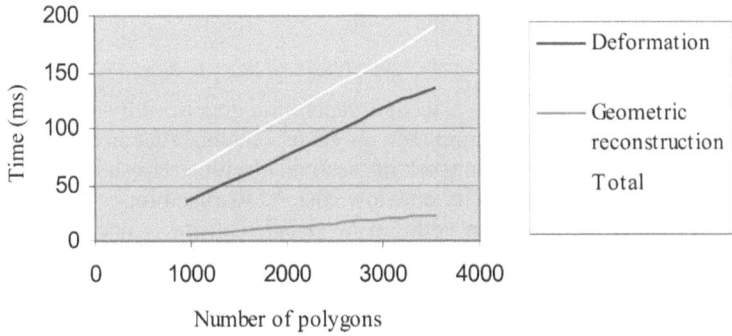

Fig. 5. Computational costs of our Surgery Simulator.

4   Discussion

4.1   The Need for Sense of Touch

In all surgical procedures, touch is the second most important sense following sight. In open surgery, sense of touch allows surgeons to identify the different tissues and to detect possible pathologies that appears in organs (i.e. tumours). In minimally invasive surgery, the sense of touch allows surgeons to infer the 3D space from the two-dimensional images of surgical scenarios that the camera shows in real surgery. For this reason, the sense of touch in surgery simulators can increase the educational and training opportunities.

However, including sense of touch in a surgery simulator needs much more computational power than that shown in section 3. The sense of sight normally requires at least 15Hz of refresh rate to be realistic and immersive. The sense of touch requires more then 500 Hz of refresh rate which must be taken into account in the Simulator.

Obviously, this implies that both sight and touch refresh rates needs an exclusive processor dedicated to each task. To solve this problem in our surgery simulator, we use a cluster of PCs which is connected through a Fast-Ethernet using a Master-Slave connection. The master contains the Surgery Simulator and must refresh the visual interface and control the Slaves. Each Slave can simultaneously control up to two haptics with force feedback with a refresh rate above 500 Hz. The refresh of haptics is done using the messages of refresh forces sent by the Master and historical information that each Slave collects.

4.2 Problems with Haptics with Force Feedback

One of the biggest problems that we have seen when we have tried to introduce our Surgery Simulator to medical scenarios is the cost and fragility of haptics with force feedback. The high cost of force feedback devices make the Surgery Simulator four times more expensive than without force feedback. Also, the fragility of the haptics makes frequent repairs necessary during normal use.

5 Conclusions

In this paper we have presented a second-generation generic full simulator for minimally invasive surgery. In this simulator, surgeons can construct any surgical scene in which to train. Our simulator can work interactively (with a refresh rate higher than15 Hz) in scenarios of low complexity on a low cost PC workstation.

If realistic sense of touch, in addition to visual realism, is needed, haptics with force feedback can be connected to our simulator. These haptics require a processor with exclusive dedication for computing feedback forces to users due to the high refresh rate requirements. The problem with these devices is their cost and fragility.

Our experience has shown that the successful introduction of a surgery simulator in medical environment requires the reduction of the cost of the haptics with force feedback and an increase in their strength.

References

1. N.J. Avis, N.M. Briggs, F. Kleinermann, D.R. Hose, B.H. Brown, M.H. Edwards, *Anatomical and Physiological Models for Surgical Simulation*, Proc. Medicine Meets Virtual Reality (MMVR: 7), J.D. Westwood et al. (eds.), Studies in Health Technology and Informatics, 62, pp. 23-29, IOS Press, Amsterdam, 1999

2. C. Baur, D. Guzzoni, O. Georg, *Virgy: A Virtual Reality and Force Feedback Based Endoscopic Surgery Simulator*, Proc. Medicine Meets Virtual Reality (MMVR: 6), J.D. Westwood et al. (eds.), Studies in Health Technology and Informatics, 50, pp. 111-116, IOS Press, Amsterdam, 1998
3. O.S. Bholat et al., *Defining the Role of Haptic Feedback in Minimally Invasive Surgery*, Proc. Medicine Meets Virtual Reality (MMVR: 7), J.D. Westwood et al. (eds.), Studies in Health Technology and Informatics 62, pp. 62-66, IOS Press, Amsterdam, 1999
4. M. Bro-Nielsen, D. Helfrick et al., *VR Simulation of Abdominal Trauma Surgery*, Proc. Medicine Meets Virtual Reality (MMVR: 6), J.D. Westwood et al. (eds.), Studies in Health Technology and Informatics, 50, pp. 117-123, IOS Press, Amsterdam, 1998
5. J. Brown, K. Montgomery, J.-C. Latombe, M. Stephanides, *A Microsurgery Simulation System*, Proc. Medical Image Computing and Computer-Assisted Intervention (MICCAI: 4), W.J. Niessen, M. Viergever (eds.), Lecture Notes in Computer Science, 2208, pp. 137-144, Springer, Berlin, 2001
6. H. Delingette, *Toward Realistic Soft-Tissue Modelling in Medical Simulation*, Proc. IEEE, 86 (3), pp. 512-523, 1998
7. M. Downes et al., *Virtual Environments for Training Critical Skills in Laparoscopic Surgery*, Proc. Medicine Meets Virtual Reality (MMVR: 6), J.D. Westwood et al. (eds.), Studies in Health Technology and Informatics, 50, pp. 316-322, IOS Press, Amsterdam, 1998
8. B. Geiger, R. Kikinis, *Simulation of Endoscopy*, Proc. Computer Vision and Robotics in Medicine (CVRMed), N. Ayache et al. (eds.), pp. 277-281, Berlin, 1995
9. L. Hong et al., *3D Virtual Colonoscopy*, Proc. Biomedical Visualization, M. Loew et al. (eds.), pp. 26-32, IEEE Computer Society Press, 1995
10. Immersion Corp., *CathSim® Vascular Access Trainer*, http://www.immersion.com/, San Jose, 2002
11. Immersion Corp., *AccuTouch® Endovascular Trainer*, http://www.immersion.com/, San Jose, 2002
12. Immersion Corp., *AccuTouch® Endoscopy Trainer*, http://www.immersion.com/, San Jose, 2002
13. H. Kataoka, T. Washio, M. Audette, K. Mizuhara, *A Model for Relations between Needle Deflection, Force, and Thickness on Needle Penetration*, Proc. Medical Image Computing and Computer-Assisted Intervention (MICCAI: 4), Lecture Notes in Computer Science, 2208, pp. 966-974, Springer, Berlin, 2001
14. U. Kühnapfel, H.K. Çakmak, H. Maaß, *Endoscopic Surgery Training Using Virtual Reality and Deformable Tissue Simulation*, Computers & Graphics, 24, pp. 671-682, 2000
15. A.D. McCarthy, R.J. Hollands, *A Commercially Viable Virtual Reality Knee Arthroscopy Training System*, Proc. Medicine Meets Virtual Reality (MMVR: 6), J.D. Westwood et al. (eds.), Studies in Health Technology and Informatics, 50, pp. 302-308, IOS Press, Amsterdam, 1998
16. Monserrat, C. et al., "A fast Real Time Tissue Deformation Algorithm for Surgery Simulation", CAR'97, Elsevier Publishers.
17. C. Monserrat, M. Alcañiz, U. Meier, J.L. Poza, M.C. Juan, V. Grau, *Simulador para el Entrenamiento en Cirugías Avanzadas*, Proc. Congreso Internacional de Ingeniería Gráfica (INGEGRAF: 12), CD-ROM, Secretaría del XII Congreso Internacional de Ingeniería Gráfica, Valladolid, 2000
18. C. Monserrat, J.L. Poza, U. Meier, M.C. Juan, M. Alcañiz, V. Grau, *Sistema de Laparoscopia Virtual para el Entrenamiento de Cirujanos*, Proc. Congreso Español de Ingeniería Gráfica (CEIG: 10), R.J. Arinyo et al. (eds.), pp. 327-339, Castellón, 2000
19. Monserrat, C.; Meier, U.; Alcañiz, M.; Chinesta, F.; Juan, M.C.; "A new approach for the real-time simulation of tissue deformation in surgery simulation", Computer Methods and Programs in Biomedicine, v. 64, pp. 77-85, 2001

20. M.A. Sagar, D. Bullivant et al., *A Virtual Environment and Model of the Eye for Surgical Simulation*, Proc. SIGGRAPH, pp. 205-212, 199621.   R.M. Satava, *Health Care in the Information Age*, Medical Virtual Reality (Chapter 12), IOS Press and Ohmsa, pp. 100-106, 1996.
22. R.M. Satava, S.B. Jones, *Current and Future Applications of Virtual Reality for Medicine*, Proc. IEEE, 86 (3), pp. 484-489, 1998
23. M.A. Schill, C. Wagner, M. Hennen, H.-J. Bender, R. Männer, *Eye-Si – A Simulator for Intra-ocular Surgery*, Proc. Medical Image Computing and Computer-Assisted Intervention (MICCAI: 2), C. TaylorL, A Colchester (eds.), Lecture Notes in Computer Science, 1679, pp. 1166-1174, Springer, Berlin, 1999.

Simulation for Preoperative Planning and Intraoperative Application of Titanium Implants

Oliver Schorr, Jörg Raczkowsky, and Heinz Wörn

Institute for Process Control and Robotics, Universität Karlsruhe (TH),
Engler-Bunte-Ring 8,
76133 Karlsruhe, Germany
{schorr,rkwosky,woern}@ira.uka.de
http://wwwipr.ira.uka.de

Abstract. Towards semi-automated transfer of preoperatively planned bone re-positionings in the operation theatre, this paper presents a solution by performing preoperative bending of common titanium plate implants. The approach is based on our simulation system KasOp which defines preoperatively bone cut trajectories and bone repositionings. According to physiological matters the segments are repositioned and the computer models of the implants are bended in respect to physical constraints following the physiological shape of a bone of same age and sex. We are in process of developing a bending device which can bend standard titanium implants corresponding to the simulation. Therefore, precast implants can be used during surgery speeding up the fixation procedure of the cut bone segments.

1 Introduction

Mostly, systems in Computer Assisted Surgery can be divided into two different groups: Those assisting the preoperative, preparative part of a surgical intervention and those supporting the intraoperative surgical process. Preoperative devices range from simple radiological tomography viewers [1] up to advanced simulation and planning systems with force feedback [2], 3D or 4D visualization and complex planning tools. Navigation systems [3], surgical robots for minimal invasive [4] or open surgery [5] and augmented or virtual reality visualization describe the intraoperative environment.

However, these two groups are separated by the question of how to get the preoperatively planned information into the operation room for use with intraoperative devices. Often, approaches lack of this possibility and surgeons only can simulate an intervention rather than execute the preoperative plan. On the other hand, there is an increasing number of systems addressing this problem by combining their systems with navigation devices or surgical robots [6][7].

The Medical Robotics Research Group of IPR developed in the framework of the collaborative research center SFB414 an entire solution for cranio maxillo facial surgery [8]. Its process chain starts with a patient related tomography data acquisition

N. Ayache and H. Delingette (Eds.): IS4TM 2003, LNCS 2673, pp. 67–73, 2003.

and 3D-modelling of the patients situs and is continued by our planning workstation KasOp. KasOp has a defined interface which allows us to transfer trajectories for osteotomies to the surgical robot system ROBACKA in the OR at the end of our process chain [9]. Therefore, we are able to perform very precise cuts corresponding to preoperatively defined osteotomy lines. Thus, intraoperative bone cutting could be automated and we are able to plan it preoperatively.

On the other hand, the fixation of the osteotomized bone segments is still a manual process. The segments are repositioned in respect to the functional aspects and the aesthetic sensation of the surgeon in order to reach a physiological shape of the patients head after surgery. This goal is partially reached by moving bone segments from their original anatomical positions to the target positions of the surgical plan. Overall, it is a difficult, time consuming, iterative task of repeated segment repositioning and bending of titanium implants for fixation (see figure 1) in a trial and error like manner whose complexity is increased by the number of different bone segments. This methodology is imprecise, not quantifiable and its outcome depends on the abilities and the aesthetic sensation of the surgeon.

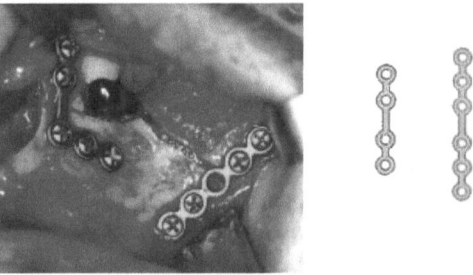

Fig. 1. Titanium osteothesis implants. During surgery the implants are used for fixation of bone segments (left). On the right two implant types within the focus of this approach.

Our novel approach is to assist the surgeon during the intraoperative fixation of the bone segments making it possible to plan the new position of each segment in advance of an intervention. We assist the fixation by introduction of a bending simulation and bending machine for standard titanium implants. The surgeon can use the bending machine which produces the precast implants according to the preoperative simulation result. Afterwards, the implants define the correlation of the bone segments and the remaining skull. Therefore intraoperative bending of the titanium plates is shortened and the surgeon needs to do fine tuning only.

2   Methods

Our approach is based on our planning framework KasOp which allows us to cut out and reposition bone segments. In order to reach a physiological shape of the patients head we include a physiological patient model of same age and sex into our planning environment. According to this shape our simulation bends the plates.

2.1 Framework KasOp

KasOp has a generic architecture concept which allows to use different visualization techniques, input devices or execution systems (i.e. robots). The base system can be extended by different modules either by linking them directly into the program or connecting them via a CORBA interface. Figure 2 shows the graphical user interface of the KasOp-system. Essential part of the KasOp framework is the so called Input-Server which manages different input devices. Currently, we are using 2D and 6D mouse as well as the force feedback device PHANToM.

The patient data we use for simulation is 3D triangle surfaces generated from segmentation and triangulation of tomography data. The simulation is totally based on tissue (bone, skin, etc.) surfaces. No volume information is currently used thus.

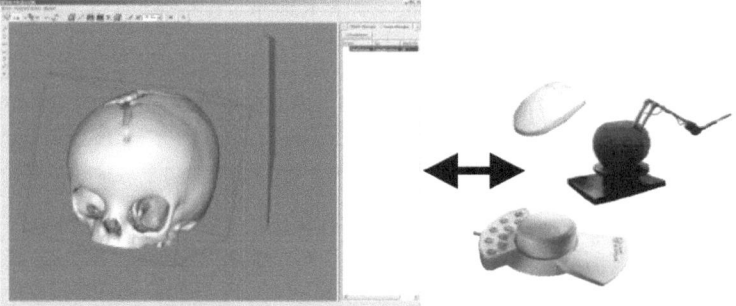

Fig. 2. Graphical user interface (left) of KasOp showing a patient model. For user interaction KasOp currently uses 2D and 6D mouse and a force feedback device.

KasOp enables the surgeon to define landmarks, lines (i.e. in order to measure distances), polygons and trajectories. Trajectories are used on the one hand to preoperatively simulate the cutting and repositioning of bone segments. On the other hand, they are used for the execution of the cutting through our surgical robot system. Surgeons define trajectories by base points located on the surface of the patient model. Intermediate points between the base points are calculated by the system in respect to anatomical features of the surface. The result is two sets of points, one for each surface describing the volume (figure 3).

Figure 4 shows an example for a bone cutting simulation. The bone is cut along the defined trajectory and the remaining objects are identified. Both objects can be treated and visualized independently. In order to achieve the desired physiological shape of the patients skull, a physiological dataset of the same age and sex as a reference model is included in the planning environment. The bone segment is repositioned by the surgeon so that it approximates this shape best. This process is supported by semitransparent visualization of the physiological dataset.

As the coordinate systems of physiological and patient dataset are different, we need a correlation method for the inclusion of the physiological surface model. This correlation can be achieved by a landmark based matching algorithm. It is essential for the result of this matching process that the landmarks defined for the matching are not located in pathological regions of the patient.

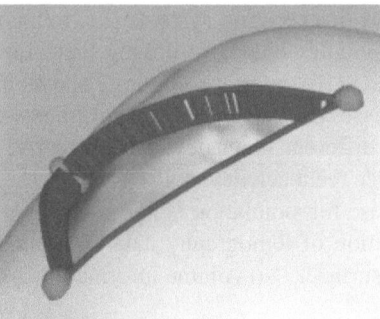

Fig 3. A defined trajectory in the planning system KasOp is represented by two point sets, one for the inner and one for the outer surface of the bone.

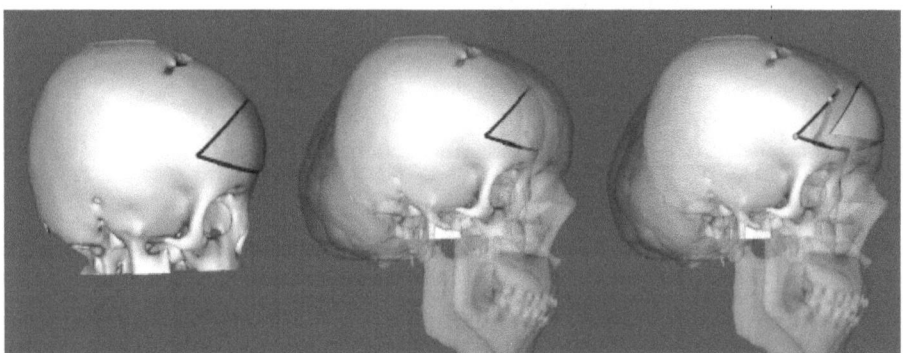

Fig 4. From left to right: Definition of trajectory, inclusion of a physiological bone structure, repositioning of bone segments according to the shape of the physiological data set.

The accuracy of the matching is increased by the number of used landmarks. We found that increasing accuracy is also possible by combination of a landmark based method with a iterative closest point (ICP) surface matching method. Therefore, we use the landmark based method for initial transformation and the ICP algorithm for an accurate fine tuning of the physiological position.

2.2 Bending Simulation

Our bending simulation works on surface models and requires the following steps to be performed in advance of the simulation:

- the patient surface model is cut along the trajectory/trajectories producing different bone segments
- a physiological surface model is included and matched to best fit with the patient
- the bone segments are repositioned according to surgical matters (physiological or anatomical restraints) along the physiological surface model

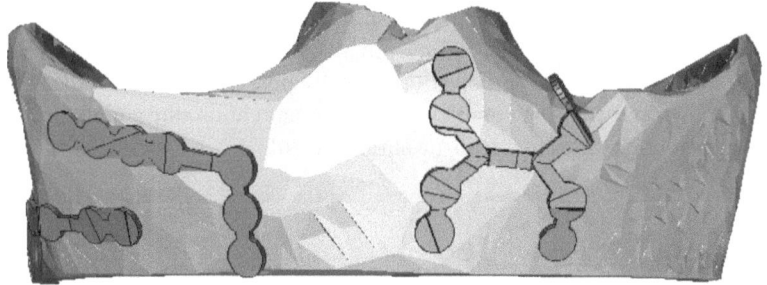

Fig. 5. The implants are bended in order to best fit to the underlying triangle surface.

The bending method is based on a previously presented method [10] including the modeling of the titanium implants by three dimensional convex polygons. The implants are bended according to a reference position and a corresponding triangulated surface. Figure 5 shows some examples for different implant types.

For computation, each type of implant is modeled as surface model with height zero as we do not have to take forces or stresses into account during the simulation process. The basic idea of the bending algorithm is to compute a set of possible bending edges first and then determine the edges reaching the best fit. This means minimizing the volume between the deformed implant and the corresponding surface.

This method is currently expanded for taking into account that the implant needs to be bended in respect to the aligned physiological dataset if there is space between the bone segments.

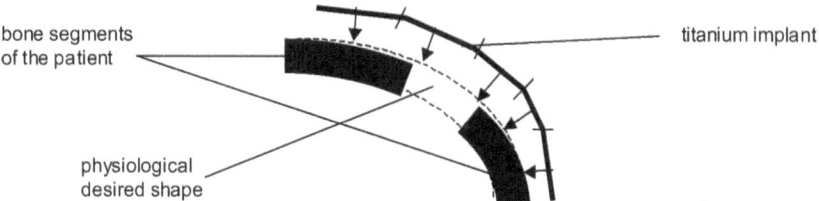

Fig. 6: Bending principle. The bone segments of the patient are repositioned according to a desired physiological shape. The implant is bended at specified positions in order to approximate the physiological shape best.

2.3   Bending Machine

A device for bending of the implants using the results of our simulation is under development. We identified two approaches for realizing such a device: totally manual or motor driven. We are focussing the manual solution at the moment.

Our experience with the simulation showed the necessity to restrict the simulation to certain bending angles as well as implant types in order to decrease the degrees of freedom necessary for the bending device. Figure 1 shows two linear implant types focused by our machine development.

The planning system includes a model of the bending machine in order to describe the restrictions of the physical bending process. During planning, the user will be guided to an individualized, optimized, physically possible implant shape. Integrated in the planning process of the complete surgical intervention, the simulation of physical implant bending relates to a realistic surgical result.

3 Discussion

The bending algorithm is currently under redesign to provide a solution for the gaps between repositioned bone segments. It is essentially important that the shape of the physiological bone is taken into account during this process.

We also detected the necessity of developing methods to simulate deformation of the bone segments. Intraoperative deformation of bone segments is already part of cranio-maxillo-facial surgery on children. It is necessary for better fitting the segments to the physiological shape, as cutting itself does not change the curvature of the repositioned segment.

Another issue required to be addressed in future will be an automated collision detection during the reposition process. At the moment, the surgeon is able to reposition the segments at arbitrary locations in the simulation environment. However, in order to avoid incorrect bended implants it is necessary to evaluate that all segments do not intersect each other in advance of the bending simulation.

4 Conclusion

Our planning environment KasOp enables surgeons to preoperatively define complex trajectories, to simulate bone cuts hereafter and perform bone repositionings. This process is supported by inclusion of physiological patient data of same age and sex in order to get information about a physiological shape. Together with our surgical robot system were are able to perform exactly these preoperatively defined trajectories intraoperatively. This means, we can not only simulate but also execute the planned data, resulting in intraoperatively provided bone segments previously defined in the simulation environment. In contrast to that, manual fixation of the bone segments is still a time consuming process and there exists no approach to assist the surgeon during this procedure.

Our research presented in this paper aims to support the surgeon during fixation. We introduce a preoperative bending simulation of standard titanium implants which is limited to physical possible bending edges. The bending simulation connects the bone segments using computer models of the titanium implants. Moreover, the simulation is forced to follow the physiological shape of the bone if there is space between the segments.

In order to provide the bending information intraoperatively and enable the surgeon to follow exactly his preoperatively defined plant we are developing a bending device. This device will be able to use the simulation data and will bend the implants according to this information. Intraoperatively, the surgeon simply has to use the precast implants and fix the bone segments at the predefined positions making intra-operative bending unnecessary. Moreover, the postoperative result will get preoperatively plannable. Overall goal of our research is to shorten the time which is intraoperatively necessary to reposition and fix bone segments shorting also the time the patient will have to be narcotized.

References

1. eHealthEngines Inc.: OpenMed Viewer, http://www.worldcare.com.my /zope/WorldCare/ menu_frame/technology/technology/omv_desc (Nov 2002)
2. M. Komori, R. Yoshida, T. Matruda, T. Takahashi: User Haptic Measurement for Design of Medical VR Applications, in proceedings of Computer Assisted Radiology and Surgery CARS, p.17-22 (2000)
3. BrainLab Inc: VectorVision, http://www.slackinc.com/techupdate/brainlab.htm (Nov. 2002)
4. J. Marescaux, J. Leroy, M. Gagner, F. Rubino, D. Mutter, M. Vix, S. E. Butner, M. K. Smith : Transatlantic robot-assisted telesurgery, Nature Vol 413, Macmillan Magazines Ltd (2001)
5. A.. Lahmer, M. Börner, A. Bauer: Experiences with an image-guided planning system (ORTHODOC) for cementless hip replacement, in proceedings of the First Joint Conference on Computer Vision, Virtual Reality and Robotics in Medicine and Medical Robotics and Computer Assisted Surgery (CVRMed-MRCAS), Grenoble (1997)
6. O. Schermeier, T. Lueth, C. Cho, D. Hildebrand, M. Klein, J. Bier: The precision of the RoboDent system - an in vitro study, in proceedings of Computer Assisted Radiology and Surgery CARS (2001)
7. L. Zamorano, A. Pandya, Q. H. Li, R. Pérez-de la Torre, P. Pittet, F. Badano, V. Robert: The Clinical Use and Accuracy of the NeuroMate Robot for Open Neurosurgery, in proceedings of Computer Assisted Radiology and Surgery (CARS), p. 185-190 (2000)
8. D. Engel, A. Pernozzoli, O. Schorr, J. Brief, T. Heurich, J. Raczkowsky, S. Hassfeld, H. Woern, J. Muehling: Evaluation of a Computer Aided Planning and Surgical Robot System for Craniofacial Surgery, in proceedings of Computer Assisted Radiology and Surgery (CARS), p. 1080 (2002)
9. D. Engel, J. Raczkowsky, H. Wörn: Sensor-aided Milling with a Surgical Robot System, in proceedings of Computer Assisted Radiology and Surgery (CARS), p. 212-217 (2002)
10. C. Burghart, K. Neukirch, S. Hassfeld, U. Rembold, H. Woern: Computer Aided Planning Device for Preoperative Bending of Osteosynthesis Plates, in proceedings of Medicine Meets Virtual Reality (MMVR), Newport Beach (2000)

Deformable Tissue Parameterized by Properties of Real Biological Tissue

Anderson Maciel, Ronan Boulic, and Daniel Thalmann

Virtual Reality Lab, École Polytechnique Fédérale de Lausanne,
CH 1015 Lausanne, Switzerland
{anderson.maciel,ronan.boulic,daniel.thalmann}@epfl.ch
http://vrlab.epfl.ch

Abstract. Realistic mechanical models of biological soft tissues are a key issue to allow the implementation of reliable systems to aid on orthopedic diagnosis and surgery planning. We are working to develop a computerized soft tissues model for bio-tissues based on a mass-spring-like approach. In this work we present several experiments towards the parameterization of our model from the elastic properties of real materials.

1 Introduction

Biomechanics literature provides a number of mathematical models that approximate the behavior of bio-tissues and bio-structures. Most of them are typically conceptual. They are used in Biomechanics to understand the complex behavior of materials that are heterogeneous, anisotropic, viscoelastic, and which properties may drastically change due to environmental and use conditions. However, to be applied on graphical computer simulation of biomechanical systems like the musculoskeletal system, existent models have to be converted and simplified.

We are currently working on a generalized approach towards functional modeling of human articulations. We plan to use such articulation model in medical applications to aid on diagnosis of joint disease and planning of surgical interventions. Our approach is based on a biomechanical model of the tissues present in the joint. This model relies on the mechanical and physical properties of biomaterials to provide correct motion, contact and deformation. The present work, that is a continuation of [1], presents part of our biomechanical model. It deals with the important issue of representing the mechanical properties of real biological soft tissues in their virtual models.

1.1 Objectives

A number of computational methods to simulate soft tissues have been proposed. In this work we use a generalized mass-spring model, called molecular model [2], to

N. Ayache and H. Delingette (Eds.): IS4TM 2003, LNCS 2673, pp. 74–87, 2003.

simulate the behavior of cartilage and ligament. However, though mass-spring is a classical deformation approach, to set up the parameters of the model is not trivial. This problem is still more intricate if the behavior of a complex real material is intended to be represented.

The main goal of this work, thus, is to find a configuration of the elasticity coefficients of all springs in our molecular system, such that the elasticity of the whole piece of virtual tissue corresponds to the elasticity of the material it is supposed to be made of. It is important to say that we are mostly interested now in the elastic part of the material behavior. Non-linear parts of the stress-strain curve are out of the scope of the present paper.

1.2 Contents

Section 2 brings an overview of bio-tissues properties, particularly tissues of joints. Next, in section 3, we present the computational models most widely applied to soft tissues modeling. Following in the same chapter, the molecular system we are developing is introduced and the issues related to the goal of this work are discussed. Section 4 brings our implementation of the model and a collection of comparative tests performed on the top of it. Finally, our conclusions and future directions are rendered in the section 5.

2 Bio-tissues Properties

The composition and behavior of bones, cartilages and ligaments has been studied for many years. However, though we know much about these tissues, newer and better measurement techniques continuously update the available data. In this section, a brief presentation of the material properties of the main tissues involved in a joint is done. Muscles and tissues of superior layers will not be discussed here because of their active role (muscles produce force) and their less significant influence on the joint range of motion. So, let us see the properties of bone, cartilage and ligament.

2.1 Bone

Bone is identified as either cancellous (also referred to as trabecular or spongy) or cortical (also referred to as compact) [3, 4], see Fig. 1. The basic material comprising cancellous and compact bone appear identical, thus the distinction between the two is the degree of porosity and the organization. The porosity of cortical bone ranges from 5 to 30% while cancellous bone porosity ranges from 30 to 90%. Bone porosity is not fixed and can change in response to altered loading, disease, and aging.

Cancellous bone is actually extremely anisotropic and inhomogeneous. Cortical bone, on the other hand, is approximately linear elastic, transversely isotropic and relatively homogenous. The material properties of bone are generally determined using mechanical testing procedures; however, ultrasonic techniques have also been employed. Force-deformation (structural properties) or stress-strain (material properties) curves can be determined by means of such tests. Bone shows a linear range in which the stress increases in proportion to the strain. The slope of this region is defined as Young's Modulus or the Elastic Modulus. However, the properties of bone and most biological tissues depend on the freshness of the tissue. These properties can change within a matter of minutes if allowed to dry out. Cortical bone, for example, has an ultimate strain of around 1.2% when wet and about 0.4% if the water content is not maintained.

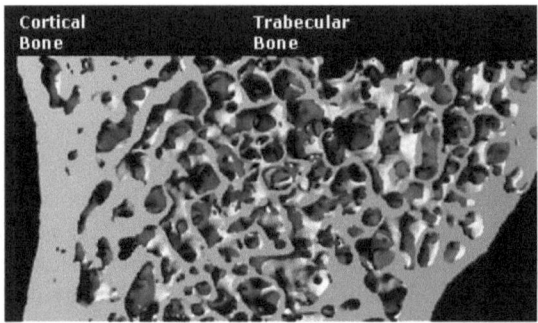

Fig. 1. Bone structure with its different layers. (Photograph by Dr. Yebin Jiang [5])

2.2 Cartilage

Articular cartilage, also called hyaline cartilage, is made of a multiphasic material with two major phases: a fluid phase composed by water (68-85%) and electrolytes, and a solid phase composed by collagen fibrils (primarily type II collagen) (10-20%), proteoglycans and other glycoproteins (5-10%), and the chondrocytes (cartilaginous cells) [6]. The cartilaginous tissue is extremely well adapted to glide. Its coefficient of friction is several times smaller than the one between the ice and an ice skate. There are electrostatic attractions between the positive charges along the collagen molecules and the negative charges that exist along the proteoglycan molecules. Hydrostatic forces also exist as forces are applied to cartilage and the fluid tries to move throughout the tissue. It is the combined effect of all these interactions that give rise to the mechanical properties of the material.

Like bone, the properties of cartilage are anisotropic. The anisotropy results in part from the structural variations. Because of its structure, cartilage is rather porous, allowing fluid to move in and out of the tissue. When the tissue is subjected to a compressive stress, fluid flows out of the tissue. Fluid returns when the stress is removed. The mechanical properties of cartilage change with its fluid content, thus making it important to know the stress-strain history of the tissue to predict its load carrying capacity. The material properties also change with pathology. The compressive aggregate modulus for human articular cartilage correlates in an inverse manner with the water content and in a direct manner with proteoglycan content per wet weight. There is no correlation with the collagen content thus suggesting that proteoglycans are responsible for the tissue's compressive stiffness. See Fig. 2 for elastic behavior of cartilage.

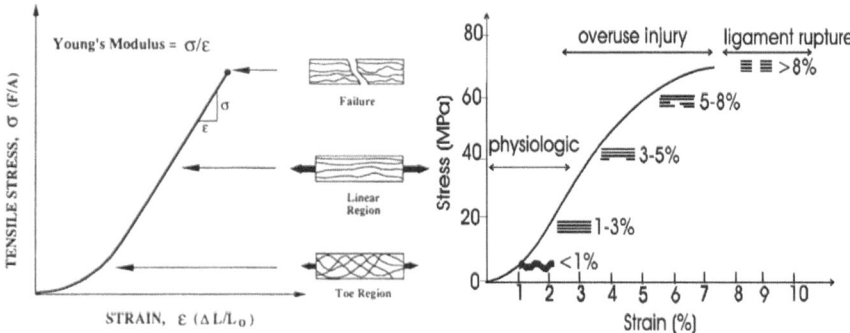

Fig. 2. Cartilage stress-strain curve (adapted from [7]), and Ligament stress-strain relationship. (modified from [8]). Young's modulus is defined by the slope of the linear region.

2.3 Ligament

Although significant advances have been made in the biology, biochemistry, and mechanics of soft tissue biomechanics, there is limited information available on in-vivo tissue mechanical characteristics and behavior. Experimental works, like the one of Stewart et al. [9] on ligaments of the hip joint capsule, bring important information that reveals the structural irregularity of that tissue. However, without accurate values of in-vivo information, such extrapolations from human – and also animal – insitu bone-ligament-bone testing to the function of intact human ligaments can not be made confidently. Currently, we know that ligaments are composite, anisotropic structures exhibiting non-linear time and history-dependent viscoelastic properties. Described in this section are the mechanical behavior of ligamentous tissue, the physiological origin of this behavior, and the implications of such properties to ligament function during normal joint motion.

The force-elongation curve represents rather structural properties of the tissues. Material properties, in turn, are more generally expressed in terms of a stress-strain relationship. See Fig. 2 for stress-strain relationship of ligament.

Ligaments have characteristics of strain rate sensitivity, stress relaxation, creep, and hysteresis. They exhibit significant time- and history-dependent viscoelastic properties. Time-dependent behavior means that, during daily activities, ligaments are subjected to a variety of load conditions that affect their mechanical properties. For example, they become softer and less resistant after some minutes of running, returning to normal hardness when the exercise is interrupted. History-dependency, in turn, means that frequent intense activities will change the tissue properties in a medium term basis. For example, the ligaments of an athlete, after 6 months of daily training, will become softer and thus more adapted to the intense exercise, even when he is not training. In the same way, if the activities are interrupted for some months, the ligament properties will go back to normal levels. Ligaments are also temperature and age sensitive.

2.4 Conclusion

Biological tissues are materials of a very complex behavior. So the harder (bone) as the softer ones (cartilage and ligament) present non-linear mechanical properties that vary from sample to sample, are dependent on structure and composition, and are time and history dependent. As a consequence of such wide set of variants, existent measured properties are not reliable and just barely describe the general behavior of these materials. Despite of that, specific situations can be delimited in which we are able to predict behavior from a reduced set of input parameter values. One of these situations is the linear elastic deformation of soft bio-tissues. Here is where this work puts its focus.

3 Deformation Model

Approaches for modeling object deformation range from non-physical methods to methods based on continuum mechanics [10]. In this section, we focus on physically based approaches specifically used for modeling soft tissues. In this category, we present the two widest used, the classical mass-spring systems and finite element methods, and also our approach. Other physically based methods used to model deformable objects, like implicit surfaces and particles systems are not considered here.

3.1 Related Work

Mass-Spring Systems. Mass-spring is a physically based technique that has been widely and effectively used for modeling deformable objects. An object is modeled as a collection of point masses connected by springs in a lattice structure. Springs connecting point masses exert forces on neighboring points when a mass is displaced from its rest position.

Mass-spring systems have been widely used in facial animation. Terzopoulos and Waters used a three-layers mesh of mass points associated to three anatomically distinct layers of facial tissue (dermis, subcutaneous fatty tissue, and muscle) [11]. To improve realism, Lee et al. added further constraints to prevent penetrations between soft tissues and bone [12]. In biomechanical modeling, mass-spring systems were used by Nedel [13] to simulate muscle deformation. Muscles were represented at two levels: action lines and muscle shape. This shape was deformed using a mass-spring mesh. Aubel [14] used a similar approach with a multi-layered model based on physiological and anatomical considerations. Bourguignon and Cani [15] proposed a model offering control of the isotropy or anisotropy of elastic material. The basic idea of their approach is to let the user define, everywhere in the object, mechanical characteristics of the material along a given number of axes corresponding to orientation of interest. All internal forces will be acting along these axes instead of acting along the mesh edges. Mass spring systems have also been used for cloth motion [16] and surgical simulation [17].

Mass-spring models are easy to construct, and both interactive and real-time simulations of mass-spring systems are possible even with desktop systems. Another well-known advantage is their ability to handle both large displacements and large deformations. However, mass-spring systems have some drawbacks. Since the model is tuned through its spring constants, good values for these constants are not always easy to derive from measured material properties. Furthermore, it is difficult to express certain constraints (like incompressibility and anisotropy) in a natural way. Another problem occurs when spring constants are large. Such large constants are used to model nearly rigid objects, or model non-penetration between deformable objects. This problem is referred as "stiffness". Stiff systems are problematic because of their poor stability, which requires small time steps for numerical integration resulting in slow simulation [10].

Finite Elements Method. Whereas mass-spring models start with a discrete object model, more accurate physical models consider deformable objects as a continuum: solid bodies with mass and energies distributed throughout. Though models can be discrete or continuous, the method used for solving it is discrete. Finite element method is used to find an approximation for a continuous function that satisfies some equilibrium expression. In FEM, the continuum (object) is divided into elements joined at discrete node points. A function that solves the equilibrium equation is found for each element.

The basic steps in using FEM to compute object deformations are [10]:

1. Derive an equilibrium equation from the potential energy equation of the deformable system in terms of material displacement over the continuum.
2. Select the appropriate finite elements (generally, tetrahedra) and corresponding interpolation functions for the problem. Subdivide the object into elements.
3. For each element, re-express the components of the equilibrium equation in terms of the interpolation function and the element's node displacements.
4. Combine the set of equilibrium equations for all the elements in the object into a single system. Solve the system for the node displacement over the whole object.
5. Use the node displacements and the interpolation functions of a particular element to calculate displacements or other quantities of interest (such as internal stress or strain) for points within the element.

Debunne et al. used a space and time adaptive level of detail, in combination with a large displacement strain tensor formulation [18]. To solve the system, explicit FEM was used where each element is solved independently through a local approximation, which reduces computational time. Hirota et al. [19] used FEM in simulation of mechanical contact between nonlinearly elastic objects. The mechanical system used as a case study was the Visible Human right knee joint and some of its surrounding bones, muscles, tendons and skin. The approach relied on a novel penalty finite element formulation based on the concept of material depth to compute skin, tendons and muscles deformation. To achieve real-time deformation, reducing computing time is necessary. Bro-Nielsen and Cotin studied this problem using a condensation technique [20]. With this method, the computation time required for the deformation of a volumetric model can be reduced to the computation time of a model only involving the surface nodes of the mesh. The Boundary Element Method (BEM), used by James and Pai [21] to attain interactive speeds and accuracy at the same time for linear deformation, is also based on the FEM. The difference is that the BEM considers only the elements on the surface and use precomputed Green's functions that are combined later in real-time. Drawbacks of this method are that material properties must be homogeneous through all modeled object, and it is only suitable for surface-based applications.

FEM provide a more realistic simulation than mass-spring methods but are computationally less efficient. In addition, the linear elastic theory used to derive the potential energy equation assumes small deformation of the object, which is true for materials such as metal. However, for soft biological material, objects dimensions can deform in large proportions so that the small deformation assumption no longer holds. Because of this change, the amount of computation required at each time is greatly increased.

3.2 Molecules-Based System

Our approach to model soft tissues is presented in this section. It is based on a work of Jansson et al. [2] that has been used in computer-aided design. Their work exploits a generalized mass-spring model – which they call molecular model – where mass points are, in fact, spherical mass regions called molecules. Elastic forces are then established between molecules by a spring-like connection. In the present work we aim at integrating properties of materials to define the stiffness of such spring-like connections.

The Force Model. The model is described by two sets of elements: E, a set of spherical elements (molecules), and C, a set of connections between the elements in E (Eq. 1).

$$E = \{e_1, e_2, \cdots e_n\}; C = \{C_{e_1}, C_{e_2}, \cdots C_{e_n}\}; C_{e_i} = \{c_1, c_2, \cdots c_m\} \tag{1}$$

The model's behavior is determined by the forces produced on each element of E by each connection of C and some external forces.

$$\vec{F}_e = \vec{F}_G + \vec{F}_L + \vec{F}_C + \vec{F}_{collisions} \text{ , where:} \tag{2}$$

F_G: gravity;
F_L: ambient viscous friction;
F_C: connection elastic forces, see [2] for more details.

3.3 Setting up Springs Stiffness

The rheological standard to define the elasticity of a material is Young's modulus. Young's modulus is a property of a material, not of an object. So it is independent of the object's shape. However, when you discretize an object by a set of springs, the stiffness of every spring must be proportional to the fraction of the volume of the object it represents.

Towards a generic method to calculate all spring constants we tested a number of hypotheses. These tests have been done on an implementation of our model where objects were deformed due to a homogenously distributed force on one of their faces. To verify the correspondence between the deformation obtained with our model and the deformation that the object should undergo with respect to its Young's modulus we used the following relation:

$$E = \frac{F \cdot l_0}{\Delta l \cdot A} \; , \tag{3}$$

where E is the Young's modulus, F is the applied force, Δl is the object's elongation, l_0 is the length of the object in rest conditions, and A is the cross-sectional area of the object. So, we started by the simplest linear topology case and went through many variations until the generic random topology. Section 4 and conclusions compare the effectiveness of every method presented below.

Linear Case. In the case where all springs produce force in only one direction and one wants to measure elasticity in the same direction, one can easily calculate the constant k of the springs directly from the geometrical information of the object and the Young's modulus of its material. This is done using Eq. 4. The problem of this very specific case is that its rectangular topology does not allow the construction of stable objects in two or three dimensions (Fig. 3). To overcome this limitation, we propose two strategies. The first using tetrahedral discretization and the second including diagonal springs to the rectangular topology. Both methods allow us to obtain stable objects. However, diagonal springs can make the number of springs grows up to $(n^2 - n) / 2$, where n is the number of molecules. This increases computation time.

$$k = \frac{EA}{l_0} \tag{4}$$

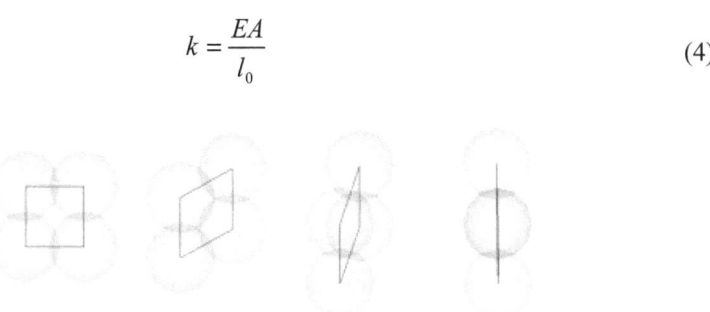

Fig. 3. These four situations are stable states with a rectangular topology. These and many other different relative positions between spheres are allowed for the same elongation of springs.

Tetrahedral Regions. A straightforward path in order to reach object stability is creating a tetrahedral mesh of springs. Gelder [22] presented a method to calculate stiffness for elastic edges of triangular meshes. His approach is based on the area of the triangles formed by the edge. He also proposed the extension of this approach to 3D, where volumes of neighboring tetrahedra are used to calculate the k for an edge. The main drawback of this method is that it only works if the mesh of springs has a tetrahedral topology. k is calculated as shown in Eq. 5, where E is the Young's Modulus, T_e is a tetrahedron incident upon c, and c is the edge to which we are calculating k.

$$k_c = \frac{E \sum_e vol(T_e)}{|c|^2} \qquad (5)$$

Diagonal Springs. In order to avoid tetrahedral meshes we have chosen to create diagonal spring connections in every face of a rectangular mesh of springs, as shown in Fig. 4, or yet, create also diagonals of parallelepipeds (Fig. 5). It offered the desired result but gave rise to a difficulty in terms of computing spring constants.

If we use the standard linear method to compute them, the diagonal springs will produce an extra force to linear movement that will causes elongation of the whole object to be shorter than expected. The object will become abnormally stiffer, and to avoid this some method should be developed to reduce springs k's. The straightforward solution is to divide all linearly calculated k's by a constant C. We tested arbitrary values to C and concluded that even for the same topology, when other parameters change, such constant must change to preserve the same elasticity.

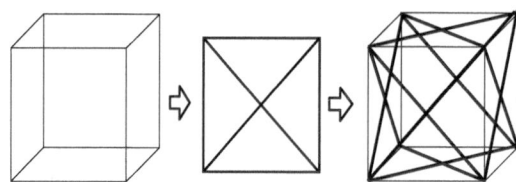

Fig. 4. Diagonals of faces.

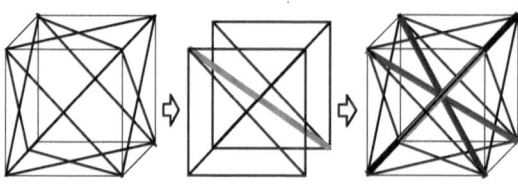

Fig. 5. Diagonals of parallelepipeds.

Angles Solution. As we do not want to work manually to find which constant C best fits each new object or topology, we implemented a method that automatically calculate new k's based on diagonal's angles.

Bringing the problem to 2D to make explanation easier, we see that force produced in vertical direction is increased by diagonals springs according to the angles between them. As we have right triangles in this topology, the relationship between a vertical connection and a diagonal one is proportional to cosine and the final k is given by the relation:

$$k_f = k_0 \cos_0 + k_1 \cos_1 + ... + k_n \cos_n \text{,} \tag{6}$$

where index 0 represents the connection to which we are calculating k, and indexes 1 to n the connections that share an extremity with 0. $\cos_i$ is the cosine of the angle between connections 0 and i. Considering a homogeneous object, all k/l_i (where l_i is the length of connection i) are equal. So, from this and Eq. 6 we deduce:

$$k = \frac{k_f}{(l_0 + l_1 \cos_1 + ... + l_n \cos_n)} . \tag{7}$$

This approximates the actual value of k. However, angles change along simulation and should be calculated for every iteration, which causes cost increase. Even worse, the right triangles deform along simulation becoming arbitrary triangles, which invalidates this method.

General Case. As we have just seen, during deformation the molecules may change their positions so that diagonals' angles are no longer right angles. Besides, arbitrary initial topologies may have some angles that are not right angles from the beginning. To handle these cases we have developed a novel method to calculate spring constants. It is a statistical method inspired on general concepts of Quantum Mechanics that heuristically estimates new values for k from the number of connections around a molecule. As the number of connections of a molecule increases, smaller is the portion of the object's volume that each connection represents, and its spring constant must be smaller too. So, though we do not calculate exactly the volume represented by a connection, we can probabilistically guess which topological case we have, just counting how many connections share a molecule. Fig. 6 illustrates some typical 2D situations.

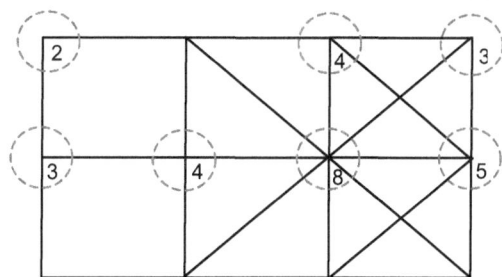

Fig. 6. Different situations and their numbers of connections.

We considered the 3D case and the diagonal on faces topology, and stated the following algorithm to calculate Hooke's constant for every connection of an object. This algorithm should present better results when larger and larger numbers of molecules are involved. It would happen because statistically, the finer the sampling the smaller the error.

Algorithm that calculates Hooke's constant based in our statistical method.

```
N1 = number of connections of molecule 1
N2 = number of connections of molecule 2
D1; //estimated num. of directions among conn. of m. 1
D2; //estimated num. of directions among conn. of m. 2
if N1 < 8
   D1 = N1
else if ( N1 >= 9 ) AND ( N1 < 12 )
   D1 = N1 - 1
else if ( N1 >= 13 ) AND ( N1 < 20 )
   D1 = N1 - 4
else if N1 >= 20
   D1 = N1 / 2
end
if N2 < 8
   D2 = N2
else if ( N2 >= 9 ) AND ( N2 < 12 )
   D2 = N2 - 1
else if ( N2 >= 13 ) AND ( N2 < 20 )
   D2 = N2 - 4
else if N2 >= 20
   D2 = N2 / 2
end
area1 = ( 2 * cross-sectional area ) / D1
area2 = ( 2 * cross-sectional area ) / D2
hooke = ( Youngs Modulus of molecule 1 * area1
        + Youngs Modulus of molecule 2 * area2 )
        / ( 2 * nominal distance )
```

Iterative Solution. The division by a constant approach mentioned earlier can be extended by automatically calculating the value of C according to the obtained deformation. The elongation in a specified dimension of an object can be applied in Eq. 3 to verify, for a given force, what is the effective Young's modulus (E) of the object at each time step. Doing that iteratively in the simulation loop we can minimize the difference between the obtained and the specified E changing the value of C according to it.

4 Experiments

We implemented the model described in Sect. 3.3 in the form of a framework that will be used in a future work to develop medical applications. The code has been written in C++ on PC platform. In this implementation the forces are integrated along time to produce new molecules positions in a static simulation.

4.1 Test Scenarios

A number of experiences have been performed to test the validity of the hypotheses of Sect. 3.3. Four deformable objects have been created which represent the same volume in the space (Fig. 7). Though they do not have exactly the same volume, the regions considered here have exactly the same vertical length and nearly the same cross sectional area. The vertical length of the considered volume, in the initial state is equal to 15 mm and its cross-sectional area is around 9 cm^2. The superior extremity of the objects is fixed and a tension force of 1 N is applied on the inferior extremity. The elasticity of the material is specified as 5000 N/m^2 (= 5 kPa).

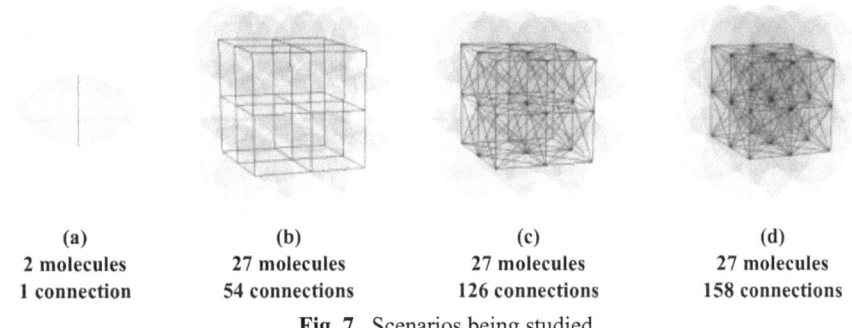

(a)	(b)	(c)	(d)
2 molecules	27 molecules	27 molecules	27 molecules
1 connection	54 connections	126 connections	158 connections

Fig. 7. Scenarios being studied.

Table 1. Resultant elasticity obtained with the different methods for the 4 cases of Fig. 7. Shaded cells indicate correct values or with a small acceptable error.

Method	Young's Modulus (N/m^2) / error (%)							
	(a)		(b)		(c)		(d)	
Linear case	4'994	0.12	4'996	0.08	12'465	149.30	20'678	313.56
Statistical	10'004	100.08	6'098	21.96	3'637	27.26	4'904	1.92
Angles	4'997	0.06	5'000	0.00	4'916	1.68	5'903	18.06
Angles (update)	4'999	0.02	4'996	0.08	11'550	131.00	18'621	272.42
Iterative	5'000	0.00	5'000	0.00	5'000	0.00	5'000	0.00

For each object we computed all methods to calculate k presented in Sect. 3.3. Table 1 compares the results obtained for Young's modulus when each method is used for calculation of k. The method based on tetrahedral regions has not been implemented here; we rely on the work of Gelder [22]. Later, to inspect the behavior of the methods when the number of molecules grows drastically, we created the situation of Fig. 8 where 1000 molecules and 7560 connections represent a volume of approximately 1 m^3. The vertical length of the considered volume, in the initial state is equal to 86 cm and its cross-sectional area is 1 m^2. One extremity of the object is fixed and a tension force of 1000 N is applied on the other one. The elasticity of the material is specified as 10^5 N/m2 (= 100 kPa). The obtained results are in Table 2.

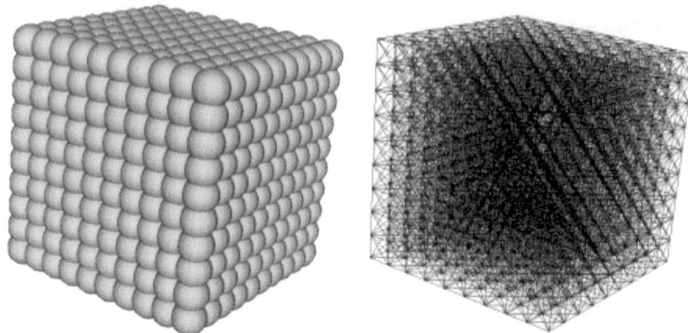

Fig. 8. Test with an increased number of molecules and connections. 1000 molecules and 7560 connections are used to model this box.

Table 2. Tension tests results for a cube of elasticity expected to be 10^5 N/m^2.

Method	Elongation (mm)	Young's Modulus (N/m^2)	Error (%)
Linear case	2.300	373'913	274
Statistical method	11.384	75'544	24
Angles	5.890	146'010	46
k / 3.75	8.603	100'000	0
Iterative	8.600	100'000	0

5   Discussion and Conclusions

None of the methods tested here works unconditionally for any case. Though the iterative solution provided our best results so far, it requires a strategy to specify the direction in which E will be measured and some cases may present numerical instabilities, which may disturb convergence. Only after new tests we can use it as a generic solution.

Standard method always works if mesh topology is rectangular, angles method if the diagonals form right triangles, and statistical method approximates the correct value in any case but usually gave objects softer than we expected. Both angles and statistical method can still be improved and we will keep working in this direction. We will also consider implementing tetrahedral method to apply it onto an intermediate tetrahedral mesh that can be constructed from arbitrary meshes, and then use the calculated values into the same arbitrary mesh. All our tests have been done applying tensile forces, but others, like shear test, must be done in order to validate a method for our purposes. Even if a spring configuration gives good results in tensile tests, we cannot affirm shear tests will also be correct. Finally, many important properties of tissues have not been taken into consideration in this work, like anisotropy and viscosity of tissues. Once these features become part of our deformation model, properties of materials will change over time, and any method used to configure our spring-like connections must take this factor into account to properly update all parameters during simulation.

References

1. Maciel, A., Boulic, R., Thalmann, D. *Towards a Parameterization Method for Virtual Soft Tissues based on Properties of Biological Tissue* In. 5th IFAC 2003 Symposium on Modelling and Control in Biomedical Systems, Melbourne (Australia). Elsevier (to appear).
2. Jansson, J. and Vergeest, J. S. M. A discrete mechanics model for deformable bodies. Computer-Aided Design. Amsterdam (2002).
3. Fung, Y-C. *Stress-Strain History Relations of Soft Tissues in Simple Elongation.* In Biomechanics: Its foundations and Objectives, Prentice Hall, Englewood-Cliffs (1972).
4. Fung, Y-C. *Biomechanics: Mechanical Properties of Living Tissues.* Second Edition, Springer-Verlag, New York (1993).
5. Washington University Departments web site. *Bone Structure.* Available at: http://depts.washington.edu/bonebio/bonAbout/structure.html.
6. Mow, V. C. and Hayes, W. C. *Basic Orthopaedic Biomechanics.* Second Edition. Lippincott-Raven Publishers (1997).
7. Mow V. C. et al. *Structure and Function of Articular Cartilage and Meniscus.* Basic Orthopaedic Biomechanics, Raven Press, New York (1991).
8. Butler, D.L. et al. *Biomechanics of ligaments and tendons.* Exercise and Sports Science Reviews, 6, 125-181 (1984).
9. Stewart, K. J. et al. *Spatial distribution of hip capsule structural and material properties.* In: Journal of Biomechanics. n. 35, pp. 1491-1498. Elsevier (2002).
10. Gibson, S. F. F. and Mirtich, B. A *Survey of Deformable Modeling in Computer Graphics.* Technical Report No. TR-97-19, Mitsubishi Electric Research Lab., Cambridge, (1997).
11. Terzopoulos, D. and Waters, K. *Physically-Based Facial Modeling, Analysis, and Animation.* The Journal of Visualization and Computer Animation, Vol.1, pp. 73-80, (Dec. 1990).
12. Lee, Y. and Terzopoulos D. *Realistic Modeling for Facial Animation.* Proceedings of SIGGRAPH 95, Computer Graphics Proceedings, Annual Conference Series, pp. 55-62, Los Angeles (1995).
13. Nedel, L. P. and Thalmann, D. *Real Time Muscle Deformations Using Mass-Spring Systems.* Computer Graphics International 1998, pp. 156-165, Hannover (1998).
14. Aubel, A. and Thalmann, D. *Interactive Modeling of Human Musculature.* Computer Animation 2001, Seoul (2001).
15. Bourguignon, D. and Cani, M-P. *Controlling Anisotropy in Mass-Spring Systems.* Computer Animation and Simulation 2000, pp. 113-123, Interlaken (2000).
16. Baraff, D. and Witkin, A. *Large Steps in Cloth Simulation.* Proceedings of ACM SIGGRAPH'98, ACM Press (1998), pp. 43-54.
17. Brown, J. et al. *Real-Time Simulation of Deformable Objects: Tools and Applications.* In Computer Animation 2001, Seoul (2001).
18. Debunne, G. et al. *Dynamic Real-Time Deformations Using Space & Time Adaptive Sampling.* Proceedings of SIGGRAPH 2001, pp. 31-36, Los Angeles (USA), 2001.
19. Hirota, G. et al. *An Implicit Finite Element Method for Elastic Solids in Contact.* In Computer Animation 2001, Seoul (2001).
20. Bro-Nielsen, M. and Cotin, S. *Real-time Volumetric Deformable Models for Surgery Simulation using Finite Elements and Condensation.* Computer Graphics Forum (Eurographics'96), 15(3), pp. 57-66 (1996).
21. James, D. L. and Pai, D. K. *ArtDefo – Accurate Real Time Deformable Objects.* Proceedings of SIGGRAPH 99, Computer Graphics Proceedings, Annual Conference Series, pp. 65-72, Los Angeles (1999).
22. Gelder, A. *Aproximate Simulation of Elastic Membranes by Triangulated Spring Meshes.* In: Journal of Graphics Tools. v. 3, n. 2, pp. 21-42. A. K. Peters, Ltd. Natick (1998).

Analysis of Myocardial Motion and Strain Patterns Using a Cylindrical B-Spline Transformation Model

Raghavendra Chandrashekara[1], Raad H. Mohiaddin[2], and Daniel Rueckert[1]

. Visual Information Processing Group, Department of Computing, Imperial College of Science, Technology and Medicine, 180 Queen's Gate, London, SW7 2BZ, UK
. Cardiovascular Magnetic Resonance Unit, Royal Brompton and Harefield NHS Trust, Sydney Street, London, SW3 6NP, UK

Abstract. We present a novel method for tracking the motion of the myocardium in tagged magnetic resonance (MR) images of the heart using a nonrigid registration algorithm based on a cylindrical free-form deformation (FFD) model and the optimization of a cost function based on normalized mutual information (NMI). The new aspect of our work is that we use a FFD defined in a cylindrical rather than a Cartesian coordinate system. This models more closely the geometry and motion of the left ventricle (LV). Validation results using a cardiac motion simulator and tagged MR data from 6 normal volunteers are also presented.

1 Introduction

Magnetic resonance (MR) imaging is unparalleled in its ability to obtain high-resolution cine volume-images of the heart, and with the use of tissue tagging [5] detailed information about the motion and strain fields within the myocardium can be obtained. This is clinically useful since cardiovascular diseases such as ischemic heart disease affect the motion and strain properties of the heart muscle in localized regions of the myocardium. The absence of automated tools to assist clinicians with the analysis of tagged MR images has made it difficult for MR tagging to become a valuable diagnostic tool for routine clinical use. The main difficulties encountered are the loss of contrast between tags as the heart contracts and the need to estimate through-plane motion [7]. The technique presented in this paper is a fully automated one that uses an image registration algorithm [23] combined with a cylindrical free-form deformation (FFD) model to extract the motion field within the myocardium from tagged MR images. We take account of through-plane motion of the myocardium by using both short-axis (SA) and long-axis (LA) images of the LV to recover the complete 3D motion of the myocardium over time. The transformations obtained can be used to directly calculate various clinically relevant parameters like strain.

1.1 Background

MR tagging relies on the perturbation of the magnetization of the myocardium in a specified spatial pattern at end-diastole. This pattern appears as dark stripes or

N. Ayache and H. Delingette (Eds.): IS4TM 2003, LNCS 2673, pp. 88–99, 2003.
© Springer-Verlag Berlin Heidelberg 2003

grids when imaged immediately after the application of the tag pattern. Because the myocardium retains knowledge of the perturbation, the dark stripes or grids deform with the heart as it contracts, allowing local deformation parameters to be estimated.

B-Splines and Active Contour Models. The most popular method for the tracking tag stripes in spatial modulation of magnetization [6,5] (SPAMM) MR images is through the use of active contour models or snakes [14]. Amini *et al.* [3,1,2] used B-snakes and coupled B-snake grids to track the motion of the myocardium in radial and SPAMM tagged MR images while Huang *et al.* [13] used a Chamfer distance potential to build an objective function for fitting a 4D B-spline model to extracted tag line features. These B-spline models were defined in a Cartesian coordinate system. Young *et al.* [27] used the tag displacements obtained from active contour models in a finite element model to calculate deformation indices. Displacement data obtained using active contour models were used in a class of deformable models defined by Park *et al.* [20,21] to characterize the local shape variation of the LV such as contraction and axial twist.

Various packages have been developed (Kumar and Goldgof [15] and Guttman *et al.* [12]) for tracking tags in SPAMM images. These have been used by researchers to both validate and fit specific models of the LV that reflect its geometry (O'Dell *et al.* [17], Declerck *et al.* [8], Ozturk and McVeigh [19], Denney and Prince [9]).

Optical Flow Methods. Conventional optical flow (OF) methods are insufficient for tagged MR image analysis since the contrast between the tags changes, due to T1 relaxation, during the cardiac cycle. A number of methods have been proposed to model this variation in contrast. These include the variable brightness optical flow (VBOF) method of Prince and McVeigh [22] and the local linear transformation model of Gupta and Prince [11]. A key difficulty faced here is the need to model the MR imaging parameters such as the longitudinal relaxation parameters of the imaged tissue. Dougherty *et al.* [10] circumvented the problem of the modelling the brightness variation of tagged MR images by preprocessing the images with a series of Laplacian filters to estimate the motion between two consecutive images. Work is still being done to extend OF methods to 3D.

HARP. Harmonic phase (HARP) MRI [18] is another technique which can be used to derive motion patterns from tagged MR images. The method is based on the fact that the Fourier transforms of SPAMM images contain a number of distinct spectral peaks. Using a bandpass filter to extract a peak in the Fourier domain yields a complex harmonic image, consisting of a harmonic magnitude image which describes the change in heart geometry as well as the image intensity changes, and a harmonic phase image which describes the motion of the tag pattern in the myocardium. The advantage of HARP imaging is that it can be used to directly calculate the strain from the images. This work has been applied primarily to 2D tagged MR images and its extension to 3D tagged MR images is still an active area of research.

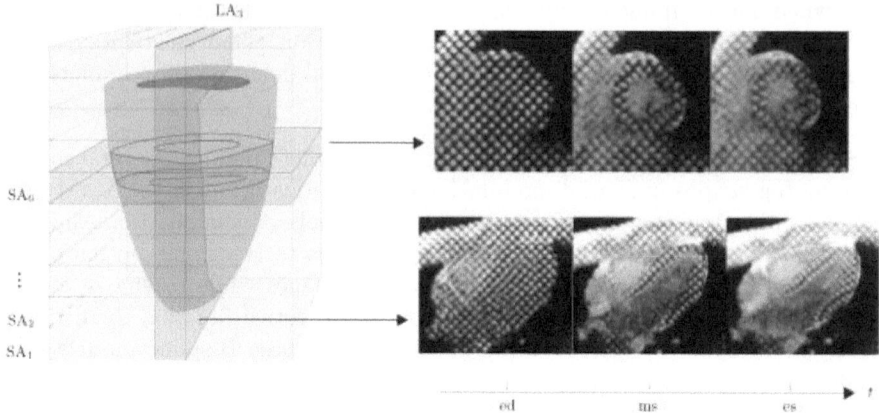

Fig. 1. A typical configuration of imaging planes required to fully reconstruct the deformation field consists of both short-axis (SA) planes as well as long-axis (LA) planes. The images on the right show a set of SPAMM images for the corresponding SA and LA imaging planes at three time points, end-diastole (ed), mid-systole (ms), and end-systole (es).

The work presented in this paper is related to our previous research on myocardial motion tracking using nonrigid image registration [7]. Here have extended our method to include a cylindrical free-form transformation which models more closely the geometry and motion of the LV.

2   Method

The normal human left ventricle undergoes a number of different types of deformation as it pumps blood out to the rest of the body. These include contraction, twisting, and a shortening of the LV. To accurately measure these modes of motion we need to acquire multiple short-axis (SA) and long-axis (LA) images of the LV. A typical configuration which could be used is given in figure 1, which consists of a series of contiguous SA and LA image planes. If we consider a material point P in the myocardium at a position $\mathbf{u} = (x, y, z)$ at time $t = 0$ (corresponding to end-diastole) that moves to another position $\mathbf{u}'$ at time $t = i$, our task is to find the transformation $\mathbf{T}$ such that:

$$\mathbf{T}(\mathbf{u}, t) = \mathbf{u}' \tag{1}$$

We define $\mathbf{T}(\mathbf{u}, t)$ using a series of cylindrical free-form deformations as described in the next section:

$$\mathbf{T}(\mathbf{u}, t) = \sum_{h=1}^{t} \mathbf{T}^h_{\text{local}}(\mathbf{u}) \tag{2}$$

2.1   Cylindrical Free Form Transformations

Free-form deformation (FFD) [24] of an object is achieved by embedding it within a lattice of control points. By moving the control points defining the FFD the object can be deformed. A cylindrical free-form deformation (CFFD) is defined on a domain Ω:

$$\Omega = \{(r, \theta, z) | 0 \leq r < R, 0 \leq \theta < 2\pi, 0 \leq z < Z\} \tag{3}$$

corresponding to the volume of interest, by a mesh of control points $n_r \times n_\theta \times n_z$, as shown in figure 2, where n_r is the number of control points in the radial direction, n_θ is the number of control points in the θ-direction (tangential direction), and n_z is the number of control points in the z-direction (LA direction). The cylindrical control point grid is aligned with the left ventricle by calculating the center of mass of the myocardium in the apical and basal short-axis image slices; the line joining the apex to the base then defines the LA of the left ventricle. A shearing and translation transformation, $\mathbf{S}$, is calculated which aligns this axis with the axis of the cylindrical control point grid. Thus, each $r\theta$-plane in the cylindrical coordinate system is aligned with a short-axis image plane. The shearing and translation transformation is given by the matrix:

$$\mathbf{S} = \begin{pmatrix} 1 & 0 & -\frac{a_x - b_x}{a_z - b_z} & \frac{a_x b_z - a_z b_x}{a_z - b_z} \\ 0 & 1 & -\frac{a_y - b_y}{a_z - b_z} & \frac{a_y b_z - a_z b_y}{a_z - b_z} \\ 0 & 0 & 1 & -a_z \\ 0 & 0 & 0 & 1 \end{pmatrix} \tag{4}$$

where $\mathbf{a} = (a_x, a_y, a_z)^{\mathrm{T}}$, and $\mathbf{b} = (b_x, b_y, b_z)^{\mathrm{T}}$ represent the positions of the apex and the base of the left ventricle respectively. We calculate the coordinates of a point, $\mathbf{u} = (u_x, u_y, u_z)$, in the myocardium in the cylindrical coordinate system by first multiplying $(u_x, u_y, u_z, 1)^{\mathrm{T}}$ by $\mathbf{S}$ to obtain:

$$\begin{pmatrix} s_x \\ s_y \\ s_z \\ 1 \end{pmatrix} = \mathbf{S} \begin{pmatrix} u_x \\ u_y \\ u_z \\ 1 \end{pmatrix} \tag{5}$$

and then converting to cylindrical polar coordinates:

$$r = s_x^2 + s_y^2 \tag{6}$$
$$\theta = \arctan2(s_y, s_x) \tag{7}$$
$$z = s_z \tag{8}$$

where $\arctan2(s_y, s_x)$ is the standard C++ math library function that calculates the arctangent of s_y/s_x, taking into account the quadrant that θ lies in, which depends on the signs of s_x and s_y.

$\mathbf{T}_{\mathrm{local}}^h(r, \theta, z)$ is then defined by:

$$\mathbf{T}_{\mathrm{local}}^h(r, \theta, z) = \sum_{l=0}^{3} \sum_{m=0}^{3} \sum_{n=0}^{3} B_l(u) B_m(v) B_n(w) \phi_{i+l, j+m, k+n} \tag{9}$$

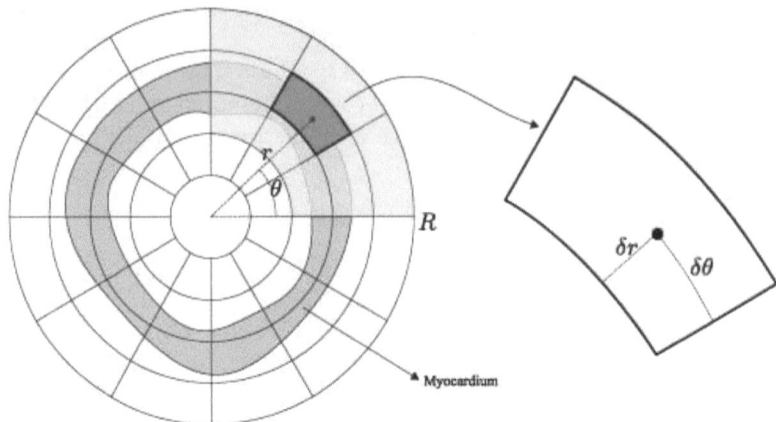

Fig. 2. The control point grid for the cylindrical free-form deformation (CFFD). Here $n_r = 5$ and $n_\theta = 12$. The number of control points in the z-direction is not shown.

where $i = \lfloor (n_r - 1)r/R \rfloor - 1$, $j = \lfloor (n_\theta - 1)\theta/(2\pi) \rfloor - 1$, $k = \lfloor (n_z - 1)z/Z \rfloor - 1$, $u = (n_r - 1)r/R - i - 1$, $v = (n_\theta - 1)\theta/(2\pi) - j - 1$, and $w = (n_z - 1)z/Z - k - 1$.

2.2   Combined Nonrigid Registration of SA and LA Images

The estimation of the deformation field $\mathbf{T}(\mathbf{u}, t)$ proceeds in a sequence of registration steps. Since we are only interested in recovering the motion field within the myocardium we use only voxels inside the myocardium in the first time frame (corresponding to end-diastole) as the images to register to. This not only allows us to produce more accurate results but also to do the registration more quickly. We begin by simultaneously registering the SA and LA volume images taken at time $t = 1$ to the segmented volume images taken at time $t = 0$. This is done by optimizing a cost function based on the weighted sum of the normalized mutual informations (NMIs) of the registered images.

The NMI [25] between two images, A and B, is defined by:

$$\mathcal{C}_{\text{similarity}} = \frac{H(A) + H(B)}{H(A, B)} \tag{10}$$

where $H(A)$ and $H(B)$ are the marginal entropies of images A and B respectively, and $H(A, B)$ is the joint entropy of the two images. Since the SA and LA images we use contain different numbers of voxels we weight the separate components of the normalized mutual information in the similarity measure according to the numbers of voxels in the myocardium in the segmented SA and LA images taken at end-diastole:

$$w_{\text{SA}} = \frac{N(V_{\text{SA},0})}{N(V_{\text{SA},0}) + N(V_{\text{LA},0})} \tag{11}$$

$$w_{\text{LA}} = \frac{N(V_{\text{LA},0})}{N(V_{\text{SA},0}) + N(V_{\text{LA},0})} \tag{12}$$

$N(V_{\text{SA},0})$ and $N(V_{\text{LA},0})$ are the numbers of voxels in the myocardium in the segmented images taken at end-diastole. Thus the similarity measure is given by:

$$\mathcal{C}_{\text{similarity}}(\varPhi) = w_{\text{SA}} \frac{H(V_{\text{SA},0}) + H(\mathbf{T}(V_{\text{SA},t}))}{H(V_{\text{SA},0}, \mathbf{T}(V_{\text{SA},t}))} + w_{\text{LA}} \frac{H(V_{\text{LA},0}) + H(\mathbf{T}(V_{\text{LA},t}))}{H(V_{\text{LA},0}, \mathbf{T}(V_{\text{LA},t}))}$$

(13)

where the $\varPhi$ are the control points defining the local transformation, $\mathbf{T}$, $V_{\text{SA},0}$ and $V_{\text{SA},t}$ are the volume images formed by the SA slices at times 0 and t respectively, $V_{\text{LA},0}$ and $V_{\text{LA},t}$ are the volume images formed by the LA slices at times 0 and t respectively, and $\mathbf{T}(A)$ represents the image A after it has been registered to its corresponding image at time $t = 0$. It is important to note that equation 13 only measures the similarity of the intensities of the SA and LA images between two time points. No extraction of features such as tag lines or intersections is used to calculate the optimal transformation.

Because the similarity measure is coupled to both the SA and LA image sets, we are able to recover the complete 3D motion of the myocardium. This is because a single 3D transformation is optimized which must maximize both the similarity between the SA and LA images. Thus, the through-plane motion that is present in the SA images is described by the transformation because of the presence of the LA images which the transformation must simultaneously register.

After registering the volume V_1 to V_0 we obtain a single FFD representing the motion of the myocardium at time $t = 1$. To register volume V_2 to V_0 a second level is added to the FFD and then optimized to yield the transformation at time $t = 2$. This process continues until all the volumes in the sequence are registered allowing us to relate any point in the myocardium at time $t = 0$ to its corresponding point throughout the sequence.

3   Results

In this section we present validation results using cardiac motion simulator data and data from normal volunteers.

3.1   Cardiac Motion Simulator Data

For the purposes of validation, a cardiac motion simulator as described in Waks *et al.* [26] was implemented. The motion simulator is based on a 13-parameter model of left-ventricular motion developed by Arts *et al.* [4] and is applied to a volume representing the LV that is modelled as a region between two confocal prolate spheres while the imaging process is simulated by a tagged spin-echo imaging equation [22].

To determine how accurately we could track the motion of the myocardium, simulated image sets were generated using the k-parameter values given in figure 4 of [26]. The k-parameter values were derived from a bead experiment on

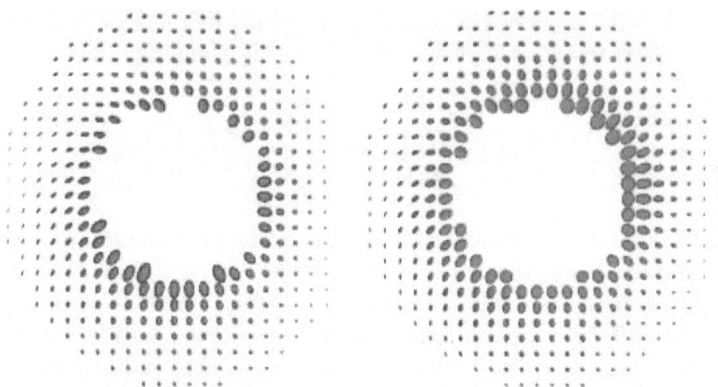

Fig. 3. The actual and estimated principal strains. The axes of the ellipsoids define the principal strain directions.

a dog heart [4]. A series of short- and long-axis images were generated using the imaging parameter values given in [26] for 10 equally spaced time instants between end-diastole and end-systole. The maximum displacement of any point in the myocardium was 21 mm. The root-mean-square (r.m.s.) error between the true and estimated displacements for all points in the myocardium increased from 0.13 mm (just after end-diastole) to 0.64 mm (at end-systole). As strain is very sensitive to errors in the estimated displacement field we also plotted the principal strain directions. Figure 3 shows the true and estimated strain fields for a mid-ventricular short-axis slice where we have used tensor ellipsoids for visualization. The high degree of similarity indicate a very good performance for the method.

3.2   Volunteer Data

Tagged MR data from 6 healthy volunteers was acquired with a Siemens Sonata 1.5 T scanner consisting of a series of SA and LA slices covering the whole of the LV. For two volunteers no LA slices were acquired and for the remaining 4 volunteers 3 LA slices were acquired. A cine breath-hold sequence with a SPAMM tag pattern was used with imaging being done at end expiration. The image voxel sizes were $1.40 \times 1.40 \times 7$ mm, with the distance between slices being 10 mm, and 10–18 images were acquired during the cardiac cycle, depending on the volunteer.

For each of the volunteers the deformation field within the myocardium was calculated using the method presented in section 2 for all times between end-diastole and end-systole. To test the performance of the method, tag-intersection points in four different imaging planes (basal SA slice, mid-ventricular SA slice, apical SA slice, and horizontal LA slice) were tracked manually by a human observer. The r.m.s. error between the estimated and observed displacements of the tag-intersection points are given in table 1. We have also visualized the tag

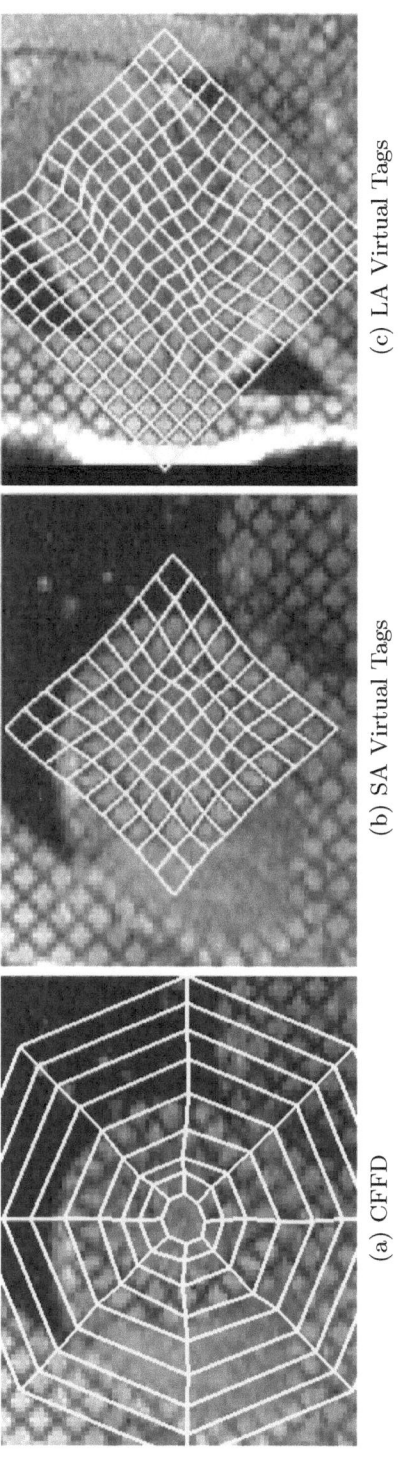

(a) CFFD   (b) SA Virtual Tags   (c) LA Virtual Tags

Fig. 4. The image on the left shows the CFFD control point grid overlaid on a SA view of the heart. The images in the middle and on the right show the deformation of virtual tag grids in SA and LA views of the heart respectively. Animations showing the tracking of tags are available at http://www.doc.ic.ac.uk/~rc3/IS4TM2003

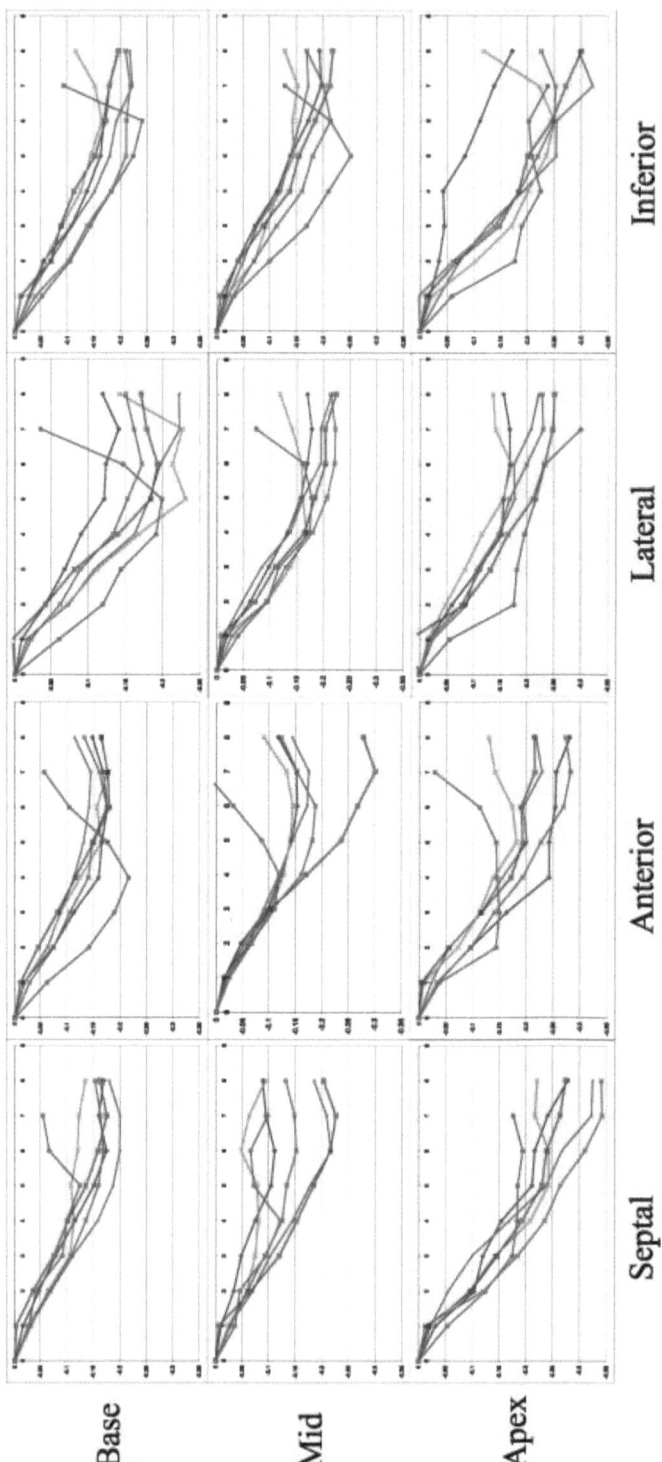

Fig. 5. Variation of the average circumferential strain in the base, middle and apex of the left ventricle against time. The vertical scale is from -0.35 to 0. Larger versions of these graphs can be found at http://www.doc.ic.ac.uk/~rc3/IS4TM2003

Table 1. R.M.S. error between estimated and observed displacements of tag-intersection points for the 6 volunteers.

Volunteer	1	2	3	4	5	6
R.M.S. Error/mm	1.9 ± 1.5	0.96 ± 0.24	1.5 ± 1.0	1.5 ± 0.77	1.3 ± 0.56	1.9 ± 1.4

tracking in the form of virtual tag grids in figure 4. Here a grid has been overlaid on a SA or LA view of the heart at end-diastole and been allowed to deform with the calculated transformations as the heart contracts. As can be seen in the figures the tag tracking has been performed very well since the virtual tag grids follow the actual tag pattern in the images.

From the output transformations we were also able to directly calculate the strain fields within the myocardium. For example in figure 5 we have plotted the variation of the circumferential strain over the whole of the LV. As can be seen in the figure the circumferential strain is uniform around the center of the LV, but increases towards the apex. These results should be compared with the ones reported by Moore et al. [16].

4   Conclusions

We have presented a novel method for the estimation of motion fields within the myocardium using nonrigid image registration and a cylindrical B-spline transformation model. We have validated our results using a cardiac motion simulator and presented measurements of strain fields in normal human volunteers. Future work will focus on building a statistical model of the motion of the myocardium. The principal modes of motion extracted will be incorporated as prior knowledge about the expected types of motion to improve the accuracy and speed of the algorithm.

References

1. A. A. Amini, Y. Chen, R. W. Curwen, V. Mani, and J. Sun. Coupled B-snake grids and constrained thin-plate splines for analysis of 2D tissue deformations from tagged MRI. *IEEE Transactions on Medical Imaging*, 17(3):344–356, June 1998.
2. A. A. Amini, Y. Chen, M. Elayyadi, and P. Radeva. Tag surface reconstruction and tracking of myocardial beads from SPAMM-MRI with parametric B-spline surfaces. IEEE Transactions on Medical Imaging, 20(2):94–103, February 2001.
3. A. A. Amini, R. W. Curwen, and J. C. Gore. Snakes and splines for tracking nonrigid heart motion. In Bernard Buxton and Roberto Cipolla, editors, *Proceedings of the Fourth European Conference on Computer Vision*, volume 1065 of *Lecture Notes in Computer Science*, pages 251–261, Cambridge, UK, April 1996. Springer.
4. T. Arts, W. C. Hunter, A. Douglas, A. M. M. Muijtjens, and R. S. Reneman. Description of the deformation of the left ventricle by a kinematic model. *Biomechanics*, 25(10):1119–1127, 1992.
5. L. Axel and L. Dougherty. Heart wall motion: Improved method of spatial modulation of magnetization for MR imaging. *Radiology*, 172(2):349–360, 1989.

6. L. Axel and L. Dougherty. MR imaging of motion with spatial modulation of magnetization. *Radiology,* 171(3):841–845, 1989.
7. R. Chandrashekara, R. H. Mohiaddin, and D. Rueckert. Analysis of myocardial motion in tagged MR images using nonrigid image registration. In M. Sonka and J. Michael Fitzpatrick, editors, *Proceedings of the SPIE International Symposium on Medical Imaging,* pages 1168–1179, San Diego, California USA, 24–28 February 2002. SPIE.
8. J. Declerck, N. Ayache, and E. R. McVeigh. Use of a 4D planispheric transformation for the tracking and the analysis of LV motion with tagged MR images. In *SPIE Medical Imaging,* vol. 3660, San Diego, CA, USA, February 1999.
9. T. S. Dennney, Jr. and J. L. Prince. Reconstruction of 3D left ventricular motion from planar tagged cardiac MR images: An estimation theoretic approach. *IEEE Transactions on Medical Imaging,* 14(4):1–11, December 1995.
10. L. Dougherty, J. C. Asmuth, A. S. Blom, L. Axel, and R. Kumar. Validation of an optical flow method for tag displacement estimation. *IEEE Transactions on Medical Imaging,* 18(4):359–363, April 1999.
11. S. N. Gupta and J. L. Prince. On variable brightness optical flow for tagged MRI. In *Information Processing in Medical Imaging,* pages 323–334, June 1995.
12. M. A. Guttman, J. L. Prince, and E. R. McVeigh. Tag and contour detection in tagged MR images of the left ventricle. *IEEE Transactions on Medical Imaging,* 13(1), March 1994.
13. J. Huang, D. Abendschein, V. G. Dávila-Román, and A. A. Amini. Spatio-temporal tracking of myocardial deformations with a 4-D B-spline model from tagged MRI. *IEEE Transactions on Medical Imaging,* 18(10):957–972, October 1999.
14. M. Kass, A. Witkin, and D. Terzopoulos. Snakes: Active contour models. *International Journal of Computer Vision,* 1(4):321–331, 1988.
15. S. Kumar and D. Goldgof. Automatic tracking of SPAMM grid and the estimation of deformation parameters from cardiac MR images. *IEEE Transactions on Medical Imaging,* 13(1):122–132, March 1994.
16. C. C. Moore, E. R. McVeigh, and E. A. Zerhouni. Quantitative tagged magnetic resonance imaging of the normal human left ventricle. *Topics in Magnetic Resonance Imaging,* 11(6):359–371, 2000.
17. W. G. O'Dell, C. C. Moore, W. C. Hunter, E. A. Zerhouni, and E. R. McVeigh. Three-dimensional myocardial deformations: Calculation with displacement field fitting to tagged MR images. *Radiology,* 195(3):829–835, June 1995.
18. N. F. Osman, E. R. McVeigh, and J. L. Prince. Imaging heart motion using harmonic phase MRI. *IEEE Transactions on Medical Imaging,* 19(3):186–202, March 2000.
19. C. Ozturk and E. R. McVeigh. Four-dimensional b-spline based motion analysis of tagged MR images: Introduction and in vivo validation. *Physics in Medicine and Biology,* 45:1683–1702, 2000.
20. J. Park, D. Metaxas, and L. Axel. Analysis of left ventricular wall motion based on volumetric deformable models and MRI-SPAMM. *Medical Image Analysis,* 1(1):53–71, 1996.
21. J. Park, D. Metaxas, A. A. Young, and L. Axel. Deformable models with parameter functions for cardiac motion analysis from tagged MRI data. *IEEE Transactions on Medical Imaging,* 15(3):278–289, June 1996.
22. J. L. Prince and E. R. McVeigh. Motion estimation from tagged MR images. *IEEE Transactions on Medical Imaging,* 11(2):238–249, June 1992.

23. D. Rueckert, L. I. Sonoda, C. Hayes, D. L. G. Hill, M. O. Leach, and D. J. Hawkes. Nonrigid registration using free-form deformations: Application to breast MR images. *IEEE Transactions on Medical Imaging*, 18(8):712–721, August 1999.

24. T.W. Sederberg and S. R. Parry. Free-form deformations of solid geometric models. In *Proceedings of SIGGRAPH '86*, volume 20, pages 151–160. ACM, August 1986.

25. C. Studholme, D. L. G. Hill, and D. J. Hawkes. An overlap invariant entropy measure of 3D medical image alignment. *Pattern Recognition*, 32(1):71–86, 1998.

26. E.Waks, J. L. Prince, and A. S. Douglas. Cardiac motion simulator for tagged MRI. In *Proceedings of the IEEE Workshop on Mathematical Methods in Biomedical Image Analysis*, pages 182–191, June 21–22 1996.

27. A. A. Young, D. L. Kraitchman, L. Dougherty, and L. Axel. Tracking and finite element analysis of stripe deformation in magnetic resonance tagging. *IEEE Transactions on Medical Imaging*, 14(3):413–421, September 1995.

Tracking the Movement of Surgical Tools in a Virtual Temporal Bone Dissection Simulator

Marco Agus, Andrea Giachetti, Enrico Gobbetti,
Gianluigi Zanetti, and Antonio Zorcolo

CRS4, VI Strada Ovest, Z. I. Macchiareddu, I-09010 Uta (CA), Italy
{magus,giach,gobbetti,zag,zarco}@crs4.it
http://www.crs4.it

Abstract. In this paper we present the current state of our research on simulation of temporal bone surgical procedures. We describe the results of tests performed on a virtual surgical training system for middle ear surgery. The work is aimed to demonstrate how expert surgeons and trainees can effectively use the system for training and assessment purposes. Preliminary kinematic and dynamic analysis of simulated mastoidectomy sessions are presented. The simulation system used is characterized by a haptic component exploiting a bone-burr contact and erosion simulation model, a direct volume rendering module as well as a time-critical particle system to simulate secondary visual effects, such as bone debris accumulation, blooding, irrigation, and suction.

1 Introduction

Temporal bone drilling is an extremely delicate task common to several surgical procedures. A successful execution of temporal bone dissection requires a high level of dexterity, experience, and knowledge of the patient anatomy. The current primary teaching tool to acquire these skills is dissection of human cadavers. The physical limitations and decreased availability of the material – as well as its high handling and disposal cost and the risks associated to transmission of diseases – are, however, making this training method increasingly problematic. A VR simulator realistically mimicking a patient-specific operating environment would, therefore, significantly contribute to the improvement of surgical training in this context.

A number of groups are developing virtual reality surgical simulators for bone dissection. Early systems (e.g. [1]) focused on increasing the understanding of the anatomy by providing specialized visualization tools of static models, while following projects such as the VrTool [2] and the VOXEL-MAN system [3,4] mainly concentrate on the accurate visual presentation of free-form volume-sculpting operations. Others systems, such as the Ohio Virtual Temporal Bone Dissection simulator [5,6,7] and IERAPSI simulator [8,9] aim instead at realistically mimicking the visual and haptics effects of a real operation. The IERAPSI system is a visual and haptic surgical simulator, characterized by a physically

N. Ayache and H. Delingette (Eds.): IS4TM 2003, LNCS 2673, pp. 100–107, 2003.

based contact model, the use of patient specific data, and the focus on validating the haptic model with experimental data. References [8,9] provide a general overview of the project, mostly covering pre-operative planning; reference [10] focuses on the human factor analysis; while reference [11] presents an implementation of visual and haptic simulation of bone dissection based on a "first principles" model. The visual and haptic simulation is based upon the use of patient specific digital data acquired from CT scanners and 3D volume models representing the different materials around the temporal bone. It involves also a physical model of the bone-burr interaction that provides impulses to haptic devices(Sensable's Phantoms), a direct volume rendering component and other physically based visual effects. The simulator has been completed and is currently being tested by experienced surgeons and trainees. A complete description of the system architecture and of the algorithms implemented can be found in [11,12].

In this paper, we present the current state of our research on simulation of temporal bone surgical procedures. We report the preliminary results of tests performed on our virtual surgical training system. The data acquisition and analysis involves all the bone-burr interaction dynamic parameters in a series of simulated specific interventions performed by trainees and experienced surgeons. The specialty considered in these sessions is the basic mastoidectomy, that represents the most superficial and common surgery of the temporal bone, and it is undertaken by a wide range of surgeons in everyday practice. The procedure consists in the removal of the air cavities just under the skin behind the ear itself, and it is performed for chronic infection of the mastoid air cells (mastoiditis).

The rest of the paper is organized as follows. Section 2 provides a short description of the virtual surgical training system, while section 3 illustrates our preliminary results with regards to the surgical simulator testing as well as the kinematic and dynamic analysis of the basic mastoidectomy phases.

2   Methods and Tools

Our surgical simulator has been designed following the requirements identified in a human factor analysis[8,9]. The analysis involved a review of existing documentation, training aids, and video recordings, interviews with experienced operators, as well as direct observation of the procedure being performed in theater. The results of our analysis show that the simulator must include burr–bone contact detection, bone erosion, generation of haptic response, and synthesis of secondary visual effects, such as bone debris accumulation, blooding, irrigation, and suction [10]. The human perceptual requirements of a simulator impose very stringent constraints on performance, making bone dissection simulation a technological challenging task.

We harnessed the difference in complexity and frequency requirements of the visual and haptic simulations by modeling the system as a collection of loosely coupled concurrent components. The haptic component exploits a multiresolution representation of the first two moments of the bone density to rapidly compute contact forces and determine bone erosion.

The force estimation is based on a physically based contact and erosion model loosely based on Hertz contact theory. The actual bone erosion is implemented by decreasing the density of the voxels that are in contact with the burr in a manner that is consistent with the predicted local mass flows. The method complexity scales, however, with the cube of the burr tip radius, imposing important limitations on the surgical tool size. A thorough description of the method can be found in [12].

The visual component uses a time-critical particle system evolution method to simulate secondary visual effects, such as bone debris accumulation, blooding, irrigation, and suction. The system runs on two interconnected multiprocessor machines. The data is initially replicated on the two machines. The first is dedicated to the high-frequency tasks: haptic device handling and bone removal simulation, which run at 1 KHz. The second concurrently runs, at about 15–20 Hz, the low-frequency tasks: bone removal, fluid evolution and visual feedback. The

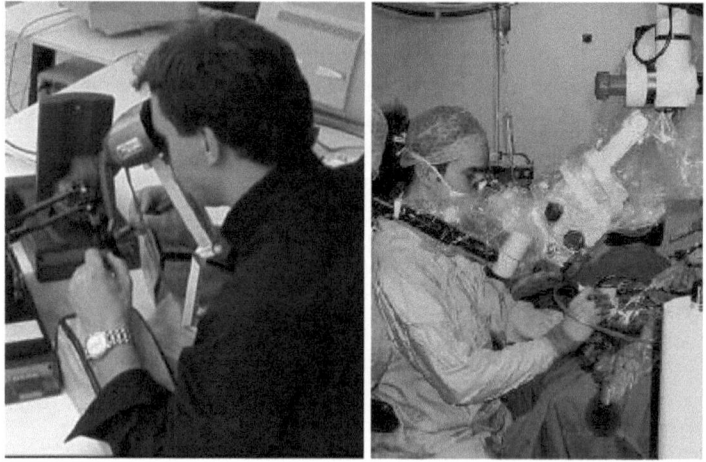

Fig. 1. Surgical simulator setup vs operating room: on the left the virtual surgical training system user interface is composed by two phantom devices that provide force feedback for sucker and burr, as well as an N-vision binocular display that presents images to the user; it simulates the real feelings of the surgeon in the operating room(right)

two machines are synchronized using one-way message passing via the Stanford VRPN library[13]. The Virtual-Reality Peripheral Network (VRPN) system provides a device-independent and network-transparent interface to virtual-reality peripherals. This communication library provides also a suitable mean to record complete traces of the training sessions, which can then be processed off–line by data analysis tools.

3   Results

Our current training system is configured as follows: a single-processor PIV/1500 MHz with 256 MB PC133 RAM for the high-frequency tasks (haptics loop (1KHz) and interprocess communication loop); a dual-processor PIII/800 MHz with 512 MB PC800 RAM and a NVIDIA GeForce 4 Ti 4600 and running a 2.4 linux kernel, for the low frequency tasks(receiving loop, simulator evolution and visual rendering); a Phantom Desktop and a Phantom 1.0 haptic devices, that provide 6DOF tracking and 3DOF force feedback for the burr/irrigator and the sucker; a n-vision VB30 binocular display for presenting images to the user.The performance of the prototype is sufficient to meet timing constraints for display and force-feedback, even though the computational and visualization platform is constructed from affordable and widely accessible components.We are currently using a volume of 256x256x128 cubical voxels (0.3 mm side) to represent the region where the operation takes place.

We are extensively testing the virtual surgical training system in collaboration with surgeons of the Department of NeuroScience of the University of Pisa. In particular, contact model parameters and erosion factors have been tuned according to their indications and there is consensus that they represent a good approximation of reality. Using the tuned system, surgeons can perform complete virtual surgery procedures with satisfactory realism. The possibility of recording dynamic values of a surgical training session provides new opportunities for the analysis and the evaluation of procedures. Different surgical procedure could be recognized by the system and it becomes possible to use the recorded values also to compare the behavior of expert surgeons and trainees in order to evaluate surgical skills.Current available data show consistency between different training sessions of the same user. Average forces exerted by burr are between 0.7 and 1.3 N for the expert surgeon and between 0.8 and 1.1 N for trainees, while average tool velocities are between 8.0 and 12.0 m/sec for the expert surgeon and 10.0 and 17.0 m/sec for trainees.In order to evaluate the possibility of characterizing different procedures according to dynamical parameters computed by the simulator, we recorded all the parameters (i.e. burr and sucker positions and velocities, force vectors, voxels removed) during a series of simulated mastoidectomy procedures.We analyzed four steps of the mastoidectomy procedure. In the first, the surgeon removes the cortex. The drill is applied to the mastoid cortex immediately posterior to the spine of Henle and draws two perpendicular cuts, the first along the temporal line and the second toward the mastoid tip. Then the mastoid cortex is then removed in a systematic fashion of saucerization.

Figure2A shows a snapshot of the scene viewed by the trainee during this step and on the right plots of the force module and of the material removed as a function of time. The second step is the cavity saucerization: before a deeper penetration in the antrum, it is necessary to perform a wide cortical removal and the posterior canal should be thinned so that the shadow of an instrument can be seen through the bone when the canal skin is elevated. Snapshot and plots relative to this step are shown in Figure2B. In the next phase considered there is the identification of the mastoid antrum. It can be identified as a larger

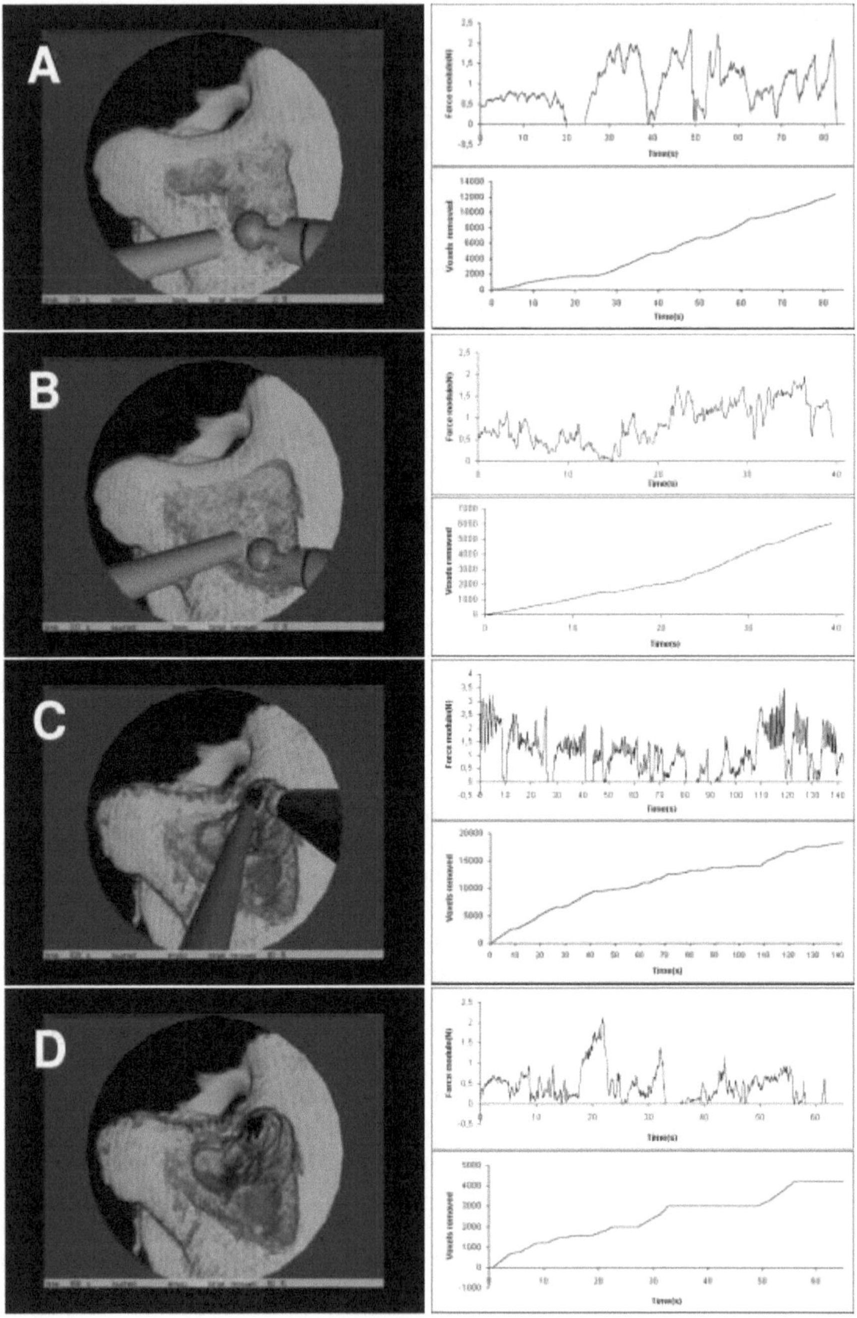

Fig. 2. Snapshot of the simulator (left) and plots of the force modulus and of the bone removal vs time (right) for the four masoidectomy phases considered:A: cortex removal, B: cavity saucerization, C: identification of the mastoid antrum, D: localiztion of the facial recess

air-containing space at whose bottom lies the basic landmark of the smoothly contoured, hard, labyrinthine bone of the horizontal semicircular canal. The localization of this canal allows exposure of the fossa incudis, the epitymphanum anteriorly and superiorly and the external genu of the facial nerve medially and inferiorly. Snapshot and plots relative to this step are shown in Figure2C. The final part of the basic mastoidectomy is represented in Figure2D. During this step several landmarks are identified, and also the facial recess area is discovered. Force and voxel removal plots show that each step in the surgical procedure can be characterized by different actions. In the first step, the force plot presents evident peaks and valleys due to the necessity of creating holes to start the bone removal. In the second step the force is more continuous and not too high. During the mastoid antrum exposure the force is irregular and reaches higher values, up to 3N. The removal rate is similar, about 10.000 voxel removed per second. Finally the last considered phase is characterized by large pauses where there is no voxel removal and even when removal is present its rate is lower than in the previous steps, indicating that critical sites have been reached and consequently burring movements are more careful and accurate.

These facts can be pointed out just taking statistical values relative to the considered steps displayed in figures 3. It is possible, for example, to distinguish two phases with high average values of force and bone removal and two with lower values. The two phases with high bone removal can be distinguished by the average burr velocity: in the mastoid cortex removal, where the user try to start new paths for the bone removal, the velocity is limited, while in the mastoid atrium exposure, where the user removes small quantities of material burr's movements are much faster. The cavity saucerization and the facial nerve identification phases, characterized by lower force values can also be distinguished by correlating with the burr bit movements speed. In fact, in the first phase the burr moves quickly along already determined paths, while in the second it is moved slowly – and carefully – since there is an high risk of damaging the facial nerve.

4   Conclusions and Future Work

This paper was aimed to describe the current state of our research in the field of virtual simulation of temporal bone surgical procedures. We presented preliminary results of the analysis of experimental data acquired during validating session of a novel virtual surgical training system for middle ear surgery. Tests are performed by expert surgeons and trainees and data are acquired in a controlled environment. These data can provide to the surgical community useful information to improve the training methods for critical surgical procedures involving bone dissection. We are currently in the process of acquiring experimental data also to compare the dynamic behavior of real materials, burr tips, and burring velocities with the simulated ones. We are also working on defining metrics appropriate to the quantitative analysis of virtual training session traces.

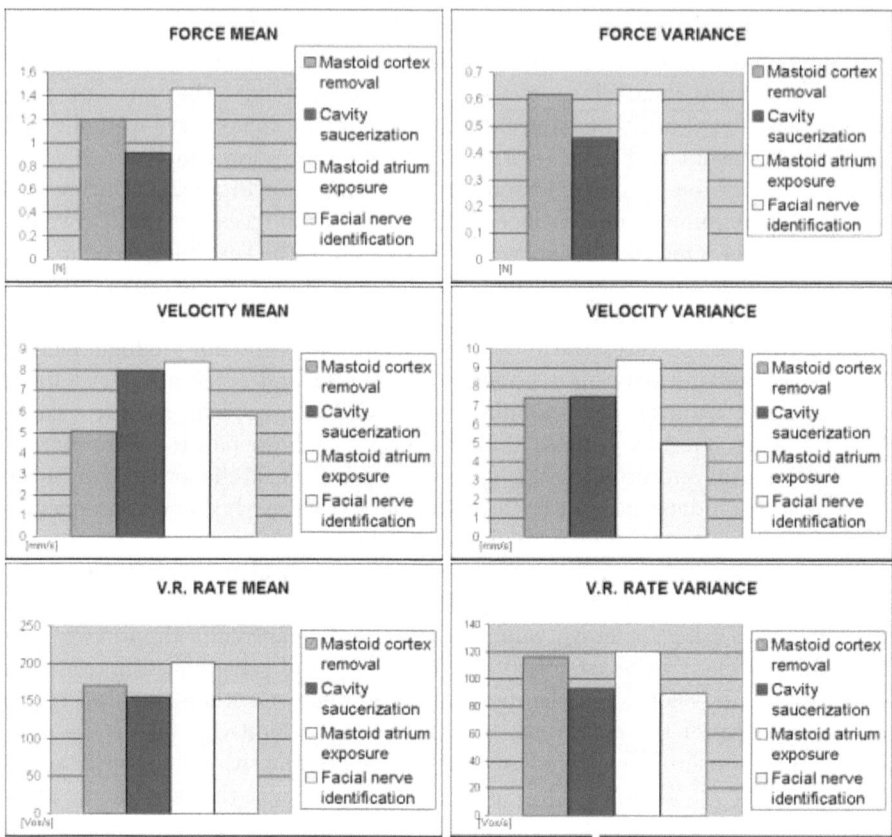

Fig. 3. Average value and variance of the force modulus, velocity and bone voxels removed during the four mastoidectomy phases considered

Acknowledgments

We would like to thank Pietro Ghironi for the precious technical support, Stefano Sellari Franceschini and the surgical residents Michele D'Anteo and Simone Valori for their performances during the training sessions. We also acknowledge the NIH National Research Resource in Molecular Graphics and Microscopy at the University of North Carolina at Chapel Hill for the VRPN library.

References

1. T. Harada, S. Ishii, and N. Tayama, "Three-dimensional reconstruction of the temporal bone from histological sections," *Arch Otolaryngol Head Neck Surg*, vol. 114, pp. 1139–1142, 1988.
2. RB Kuppersmith, R Johnston, D Moreau, RB Loftin, and H Jenkins, "Building a virtual reality temporal bone dissection simulator," in *Medicine Meets Virtual Reality 5*, J. D. Westwood, Ed., Amsterdam, The Netherlands, January 1997, pp. 180–186, IOS Press.

3. Bernhard Pflesser, Andreas Petersik, Ulf Tiede, Karl Heinz Hohne, and Rudolf Leuwer, "Volume based planning and rehearsal of surgical interventions," in *Computer Assisted Radiology and Surgery, Proc. CARS 2000, Excerpta Medica International Congress*, Heinz U. Lemke et al., Ed., Elsevier, Amsterdam, 2000, 1214, pp. 607–612.

4. Bernhard Pflesser, Andreas Petersik, Ulf Tiede, Karl Heinz Hohne, and Rudolf Leuwer, "Haptic volume interaction with anatomic models at sub-voxel resolution," in *10th International Symposium on Haptic Interfaces for Virtual Environment and Teleoperator Systems,Proc. Haptics 2002*, 2002, pp. 66–72.

5. G. Wiet, J. Bryan, D. Sessanna, D. Streadney, P. Schmalbrock, and B. Welling, "Virtual temporal bone dissection simulation," in *Medicine Meets Virtual Reality 2000*, J. D. Westwood, Ed., Amsterdam, The Netherlands, January 2000, pp. 378–384, IOS Press.

6. Jason Bryan, Don Stredney, Greg Wiet, and Dennis Sessanna, "Virtual temporal bone dissection: A case study," in *IEEE Visualization*, 2001, pp. 497–500.

7. D. Stredney, G. Wiet, J. Bryan, D. Sessanna, J. Murakami, O. Schamllbrock, K. Powell, and B. Welling, "Temporal bone dissection simulation – an update," in *Medicine Meets Virtual Reality 2002*, J. D. Westwood, H. M. Hoffmann, G. T. Mogel, and D. Stredney, Eds. Jan. 2002, pp. 507–513, IOS Press.

8. N. W. John, N. Thacker, M. Pokric, A. Jackson, G. Zanetti, E. Gobbetti, A. Giachetti, R. J. Stone, J. Campos, A. Emmen, A. Schwerdtner, E. Neri, S. Sellari Franceschini, and F. Rubio, "An integrated simulator for surgery of the petrous bone," in *Medicine Meets Virtual Reality 2001*, J. D. Westwood, Ed., Amsterdam, The Netherlands, January 2001, pp. 218–224, IOS Press.

9. Marco Agus, Andrea Giachetti, Enrico Gobbetti, Gianluigi Zanetti, and Antonio Zorcolo, "A multiprocessor decoupled system for the simulation of temporal bone surgery," *Computing and Visualization in Science*, vol. 5, no. 1, 2002.

10. Marco Agus, Andrea Giachetti, Enrico Gobbetti, Gianluigi Zanetti, Nigel W. John, and Robert J. Stone, "Mastoidectomy simulation with combined visual and haptic feedback," in *Medicine Meets Virtual Reality 2002*, J. D. Westwood, H. M. Hoffmann, G. T. Mogel, and D. Stredney, Eds. Jan. 2002, pp. 17–23, IOS Press.

11. Marco Agus, Andrea Giachetti, Enrico Gobbetti, Gianluigi Zanetti, and Antonio Zorcolo, "Real–time haptic and visual simulation of bone dissection," in *Proc. of Ieee Virtual Reality 2002*, Orlando, FL, USA, March 2002, pp. 209–216.

12. Marco Agus, Andrea Giachetti, Enrico Gobbetti, Gianluigi Zanetti, and Antonio Zorcolo, "Adaptive techniques for real–time haptic and visual simulation of bone dissection," in *Proc.of IEEE Virtual Reality 2003*, 2003.

13. ll Russell M. Taylor, Thomas C. Hudson, Adam Seeger, Hans Weber, Jeffrey Juliano, and Aron T. Helser, "VRPN: a device-independent, network-transparent vr peripheral system," in *Proceedings of the ACM symposium on Virtual reality software and technology*. 2001, pp. 55–61, ACM Press.

Area-Contact Haptic Simulation

Sang-Youn Kim[1], Jinah Park[2], and Dong-Soo Kwon[1]

[1] Teloperation & Control Laboratory, Dept. of Mechanical Engineering, KAIST,
Guseong-Dong 373-1, Yuseung-gu, Daejeon, 305-701, Korea
{sykim@robot,kwonds@mail}.kaist.ac.kr
[2] Computer Graphics & Visualization Laboratory, School of Engineering, ICU,
Hwaam-Dong 58-4, Yugeong-gu, Daejeon,305-732,Korea
jinah@icu.ac.kr

Abstract. This paper presents the haptic interaction method when the interaction occurs at several points simultaneously. In many virtual training systems that interact with a virtual object, the haptic interface is modeled as a point. However, in the real world, the portion interacting with real material is not a point but rather multiple points, i.e., an area. In this paper, we address an area-based haptic rendering technique that enables the user to distinguish hard regions from softer ones by providing the distributed reflected force and the sensation of rotation at the boundary. We have used a shape retaining chain linked model that is suitable for real-time applications in order to develop a fast area-based volume haptic rendering method for volumetric objects. We experimented with homogeneous and non-homogeneous virtual objects consisting of 421,875 (75x75x75) volume elements.

1 Introduction

In most virtual environments, a deformable object is modeled as a mesh, and the probe of the haptic interface is modeled as a point, which is known as the haptic interface point (HIP). Haptic rendering is a process that generates an interaction force between a virtual object and the haptic interface point [1,2,3]. Until now, most haptic rendering methods have focused on point-based haptic rendering. In point-based haptic rendering, since the probe of the haptic interface is modeled as a point, interaction forces can be calculated easily and quickly.

One of the important factors in haptic rendering is to give a realistic feeling to the user as the user touches a real object. In most cases, when the user interacts with a real object (for example, in palpation) the interaction occurs at several points simultaneously, i.e., area contact. Consider the case when we touch an object with a finger. We have many touch sensors at the fingertip. By integrating the stimulating forces at those sensors, we may perceive the hard portion and soft portion of a non-homogeneous object, simultaneously. Consider that we construct the palpation simulator system. Since the stiffness coefficient of the hard portion is larger than that of the soft portion, the reflected force in the hard portion is larger than that of the soft

N. Ayache and H. Delingette (Eds.): IS4TM 2003, LNCS 2673, pp. 108–120, 2003.

portion. Therefore, we can also perceive the hard portion and soft portion of a non-homogeneous object by using a tactile display.

Consider what would happen if a user explored a virtual volumetric object with a blunt tool like a diagnostic ultrasound system. In this system, since the user explores the internal part of the object not with skin but with a blunt tool, to give the only distributed reflected force (caused by surface tension and normal force) to the user is insufficient for distinguishing the hard portion from the softer one at the same time.

Consider that the user interacts with a non-homogeneous object. Since the stiffness coefficient of the hard portion is different from that of the soft portion, the user is given small reflected force in the soft portion and large reflected force in the hard portion. At this time, torque is generated because the reflected force in the soft portion is different from that in the hard portion. Because of torque, the probe of the haptic device must be rotated in order to go into equilibrium state. We use this torque information to distinguish the inhomogeneity of the object. By giving torque information to the user, the user can perceive the hard portion and soft portion of the non-homogeneous object by the sensation of rotation caused by the difference in stiffness between the hard portion and the softer one.

However, in a point based haptic rendering, because the interaction occurs at only a single point, a user cannot get distributed reflected force and cannot get the sensation of rotation caused by the difference in stiffness between the hard portion and the softer one.

Our proposal is to compute the forces from the contact area simultaneously to render the haptic feeling induced by difference in the stiffness at the border where the hard material meets with soft material. Our area based haptic rendering can apply to the system in which the user touches the virtual object with his or her finger like a palpation simulator. Also area-based haptic rendering is able to apply to the systems in which the user explores the virtual volumetric object with a blunt tool like a diagnostic ultrasound system simulator.

For the realistic haptic rendering of a deformable object, a Mass-Spring model and Finite Element Method (FEM) have been used. Unfortunately, in practice, physics based deformable models are very much limited to surface modeling mainly due to the overwhelming computational requirements. In a haptic simulation of interaction with a deformable object, it is harder to meet the real-time constraint than in a graphic simulation because a virtual object model requires a haptic update rate of 1 kHz (as opposed to the graphics update rate of 30Hz).

We have verified the real-time performance of the 3D volumetric object with a shape retaining chain linked model (or S-chain model) [5,6]. In this paper, we propose an area-based haptic rendering that enables the user to perceive the hard portion and the soft portion of a non-homogeneous object. For real-time simulation, we use the S-chain model for the proposed area-based haptic rendering. This haptic rendering method based on an S-chain model has been simulated with homogeneous and non-homogeneous virtual objects of size 75x75x75, and its performance is shown to be fast and stable.

2   Background

In virtual reality, mainly the visual information has been utilized for the interaction between a user and virtual environments. However, the haptic information has become an important factor since the development of stable haptic devices. In the early stages of its development, haptic rendering focused on displaying the shape of a hard object. Zilles and Salisbury [7] presented a constraint-based method to haptically render a virtual hard object. In [7], they defined a god-object which is a point representing the virtual location of the haptic interface. The god-object is constrained to remain on a particular facet of the object. Ho et al. [3] proposed a ray-based haptic rendering method for displaying 3D virtual objects. In the ray-based haptic rendering, since the probe of the haptic interface is modeled as a line segment, the user can feel reflected forces and interaction torques.

Due to increasing computing power, researches on haptic rendering of deformable objects have progressed rapidly. For haptic rendering, many physics based deformable models have been developed. D'Aulignac et al. [8] proposed a two-layer mass–spring model for the human thigh. The two-layer model is composed of linear springs and a set of non-linear springs orthogonal to the surface. Nedal and Thalmann [9] proposed a method to simulate muscle deformation in real-time using a mass–spring system. To physically simulate deformation, they considered three different forces: elasticity, curvature and constraint forces. To calculate curvature force, they used a new type of spring called an angular spring. Tokumo and Fujita [10] suggested a four-element model to deform a visco-elastic object in real time. Zhu et al. [11] presented a voxel-based biomechanical model for muscle deformation using finite element method (FEM) and volume graphics. In [11], hierarchical voxel meshes were reconstructed from filtered segmented muscle images followed by FEM simulation and volume rendering. For both modeling techniques, however, the use of volumetric models has been limited in applications where realistic and stable real time simulation of interaction is required, mainly due to the computational overhead.

Even with surface models, it is sometimes difficult to increase the update rate of the virtual object model to the haptic update rate. To overcome this problem, Astley and Hayward [12] introduced a multi-layer mesh, where a coarse mesh is applied to the entire body, and a finer mesh is applied to the area of the body to be manipulated. Cavusoglu and Tendick [13] proposed a multi-rate simulation technique. In their work, the full order model has an update rate of over 10Hz, and the area of interest (the low order local model) has an update rate of 1 kHz.

Costa and Balaniuk [14] presented a new modeling method for deformable objects – the Long Element Method (LEM). Their method can calculate reflected forces and display animation in real-time. They filled the interior of the volume with rectangular parallelepiped (Long Elements) to simulate the deformable object, and the equilibrium equation was defined using bulk variables. De et al. [15] introduced a meshless modeling technique for real-time rendering of soft tissues. Nodal points are sprinkled around the surgical tool tip and interpolation is performed by functions that are nonzero only on spheres surrounding the nodes. For the real-time deformation of the volumetric data, Gibson proposed a 3D ChainMail algorithm [16,17]. Since the

3D ChainMail algorithm overcomes the computational burden by rapidly propagating deformation outwards through a volume, it is suitable for the real-time deformation of a volumetric object. Since the behavior of the 3D ChainMail algorithm is like that of dough, it is suitable for highly viscous objects. However, if the 3D ChainMail algorithm is used for elastic objects, it can induce some problems in that it is hard to return the model to its original shape once it is deformed. Therefore, Kim et al. [5,6] propose a shape retaining chain linked model (or S-chain model) for real-time volume haptic rendering based on the 3D ChainMail algorithm. The S-chain model is still as fast as the original 3D ChainMail, and yet displays more realistic deformation of elastic materials than the 3D ChainMail does.

The purpose of area-contact simulation is to get the internal information of an object. In order to get the internal information, we must use volumetric data. However, since most haptic rendering method is for surface data only, area contact haptic simulation with the volumetric data has hardly been studied yet.

3   Area-Based Haptic Rendering

Until now, most haptic simulation method has focused on point-based haptic rendering. In haptic rendering, it is important to provide a realistic feeling to the user as the user touches a real object. In most cases, when the user interacts with a real object, the interaction occurs at several points simultaneously (area contact), not at a single point as shown in Figure 1.

In point-based haptic rendering, since the interaction takes place between the probe of the haptic device and a virtual object, the method is not suitable for applications that interact with several points simultaneously, for example in a palpation simulator. In this section, in order to overcome the limitations of point-based haptic rendering for deformable objects, we propose a new haptic rendering method called area-based haptic rendering. In area-based haptic rendering, we model the probe of the haptic device as not a point but an area as shown in Figure 1.

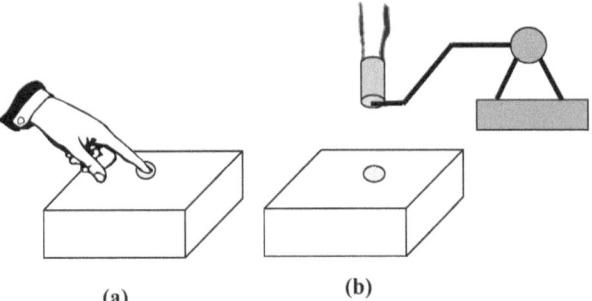

(a)   (b)

Fig. 1. (a) : User can distinguish the hard portion from the softer one by perceiving the distributed reflected force due to skin contact. (b) : User can distinguish the hard portion from the softer one by perceiving the reflected force and sensation of rotation by the differences of the amount of deformation

Consider a 2D deformable object that interacts with a haptic interface at the area (not a point) shown in Figure 2. Since the interaction occurs at the area including nodes n_1, n_2, n_3, and n_4, the user perceives the distributed reflected force due to skin contact.

The reflected forces (f_1, f_2, f_3, and f_4) generated at each node can be calculated with the haptic model. Suppose that the non-homogeneous deformable object is composed of a soft material and hard material as shown in Figure 2. The two left nodes (n_1 and n_2) are included in the soft object and the two right nodes (n_3 and n_4) are included in the hard object. If the user uses a tactile display, the user can perceive the object and can distinguish the hard portion from the softer one because the reflected force at the two right nodes is greater than the reflected force at the left two nodes. This approach is useful for palpation simulators.

Consider again the situation in which the user explores a virtual volumetric object with a blunt tool like a diagnostic ultrasound system simulator. Since the user explore the internal part of the object with not skin but a blunt tool, the user cannot distinguish the hard portion from the softer one at once. In order to distinguish the hard portion from the soft one at once, other information is needed. Therefore, we adopt torque information for distinguishing the inhomogeneity of the object. In Figure 2, since the reflected force at the two right nodes is greater than the reflected force at the left two nodes, the torque occurs about the z-axis and user can feel the boundary of two different materials so that he or she can distinguish the hard portion from the softer one.

We consider a 2D model as shown in Figure 2. Let us consider the movement only in the y-axis for simplicity. Let n_i be the i-th node in the y-axis, and let F_i be the reflected force of the i-th node in the y-axis. The force (F_i) reflected by the movement of the i-th node in the y-axis can be calculated by Newton's third law. Let Δx_i be the distance from each node of the haptic interface area to the axis of rotation. The torque generated by the different forces is given in (1).

$$\sum \tau_z = \sum_i F_i \, l_i \qquad (1)$$

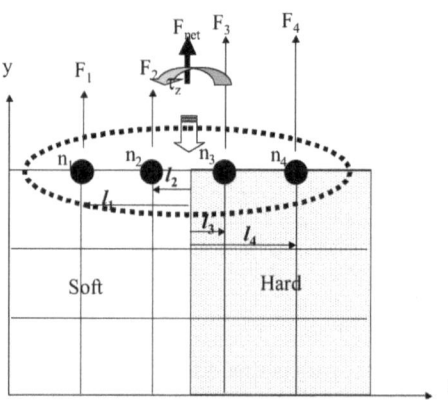

Fig. 2. Non-homogeneous model composed of a soft material and hard material

We now consider a 3D model for interaction with the haptic interface area (HIA) as shown in Figure 3(a). Figure 3(b) shows the configuration of Figure 3(a) from the viewpoint of the z-axis. Figures 3(c) and 3(d) show the configuration of Figure 3(a) from the viewpoints of the x-axis and y-axis, respectively. In Figure 3, each dot represents an individual contact point. When the user interacts with the model in an arbitrary direction, reflected forces generated in each node (i,j) are calculated by the haptic model (mass-spring, FEM) and the torques reflected to the user on each axis are calculated by equation (2),(3),and (4).

$$\tau_z = \sum_i \sum_j (F_{ij})_y \, \Delta x_{ij} + \sum_i \sum_j (F_{ij})_x \, \Delta y_{ij} \quad (2)$$

$$\tau_x = \sum_i \sum_j (F_{ij})_z \, \Delta y_{ij} + \sum_i \sum_j (F_{ij})_y \, \Delta z_{ij} \quad (3)$$

$$\tau_y = \sum_i \sum_j (F_{ij})_x \, \Delta z_{ij} + \sum_i \sum_j (F_{ij})_z \, \Delta x_{ij} \quad (4)$$

where Δx_{ij} , Δy_{ij} , Δz_{ij} are the distances from each node of the haptic interface area to the axis of rotation on the x-axis, y-axis, and z-axis, respectively; and F_{ij} is the reflected force from each node (i,j).

In this section, we have presented our new haptic rendering method (area-based haptic rendering), which enables the user to get the sensation of rotation and to distinguish hard regions from softer ones. When the user interacts with the virtual object with his or her hand, reflected force can be calculated at each node, and the user can perceives the object via our haptic rendering algorithm and tactile display. When the user interacts with the virtual object using a blunt tool, the user can distinguish hard regions from softer ones due to the sensation of rotation and haptic interface.

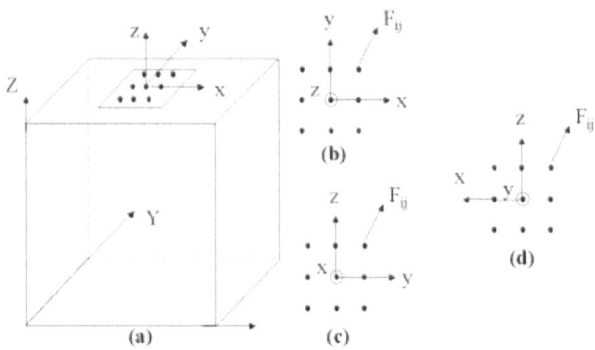

Fig. 3. The 3D model for interaction with HIA

4 Towards a Real-Time Haptic Rendering

In a haptic simulation with a deformable object, it is very hard to meet the real-time constraint because a virtual object model requires a haptic update rate of 1 kHz (as opposed to the graphics update rate of 30Hz). In order to meet computational requirements of the volumetric object, we have previously proposed a shape retaining chain linked model [5, 6]. The S-chain model is suitable for real-time volumetric object deformation and volume haptic rendering because it does not require heavy computation. In this section, we describe how the S-chain model is applied to area-based haptic rendering.

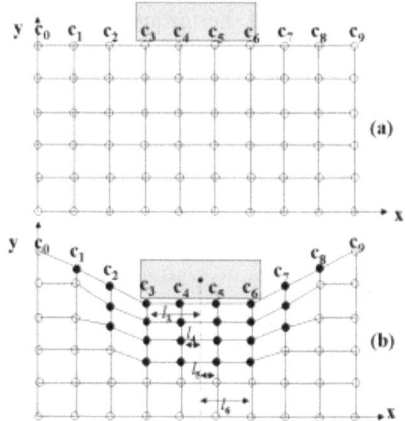

Fig. 4. 2D model of size 6x10

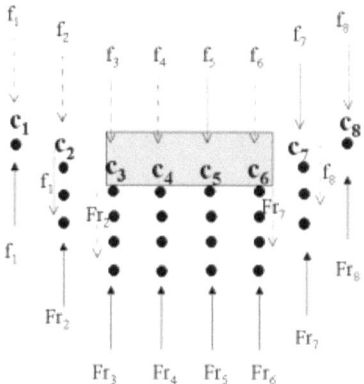

Fig. 5. Force equilibrium of the 2D model

Consider a 2D S-chain model of the size 6x10 as shown in Figure 4. Let us consider a case where compressing along the y-axis only for simplicity. Let c_i be the i-th chain in the y-axis, and let f_i be the force generated by the movement of the i-th chain in the y-axis. Let Fr_i be the reflected force of i-th chain in the y-axis. The force generated by the movement of i-th chain in the y-axis (f_i) can be calculated by the S-chain model. Figure 6(a) shows the initial configuration of the nodes and Figure 6(b) shows the resulting configuration when the user pushes the nodes including c_3, c_4, c_5, and c_6 chain. Since the leftmost chain (c_0) does not move, the reflected force in c_1 becomes $Fr_1 = f_1$. In c_2, there is a reaction force (f_1) generated by c_1, since c_1 is moved by c_2. To satisfy the force equilibrium, the reflected force becomes $Fr_2 = f_1 + f_2$. Since c_3 drags c_1 and c_2 along the direction of its movement, the reflected force becomes $Fr_3 = f_1 + f_2 + f_3$.

The shape of the portion in contact with the haptic interface is the same as that of the HIA. That is, if the shape of the HIA is a round, then the shape of the portion in contact with the haptic interface is a round shape. In this paper, we model the shape of the HIA as a rectangular shape.

Let us suppose that the displacement between a node and a neighboring node in the HIA is less than the shearing limit. Since chains contacting the haptic interface area (HIA) behave independently, the reflected force in c_4 becomes $Fr_4 = f_4$. The expressions in (5)-(8) show how we computed the reflected force components.

$$Fr_3 = f_1 + f_2 + f_3 \qquad (5)$$
$$Fr_4 = f_4 \qquad (6)$$
$$Fr_5 = f_5 \qquad (7)$$
$$Fr_6 = f_6 + f_7 + f_8 \qquad (8)$$

If the user's hand interacts with the nodes including c_3, c_4, c_5, and c_6 chain elements, the reflected force that is calculated in each node is transferred to the user via a tactile display. Therefore, the user can explore the inside of the object. If the user interacts with the virtual object with a blunt tool, torque information for distinguishing the inhomogeneity of the object is needed (refer to section 3).

Let Δx_i be the distance from the axis of rotation to the i-th chain along the x-axis. Then the torque (τ_z) generated by the different forces is

$$\tau_z = \sum_i Fr_i \, \Delta x_i \qquad (9)$$

In this section, in order to simulate a virtual object in real-time, we have presented area-based haptic rendering with an S-chain model, which is suitable for volumetric haptic rendering. The results of the area-based haptic rendering with the S-chain model are discussed in the following section.

5   Result

We have simulated the virtual object with an S-chain model of the size 75x75x75. The virtual object is volumetric in that the interior is also filled with chain elements. The graphical and haptic simulations are carried out by a program written in C++ with OpenGL as a graphics library. The graphic and haptic simulation is conducted with a PHANToM^{TM} device and a PC with dual 800MHz Pentium III processors.

5.1   Area-Contact

Figures 6-9 show the results of the 3D virtual object in interaction with a blunt probe. Figure 6 shows the graphical result of a 3D homogenous virtual object when the position input is given at the middle area (20x20) of the upper part. Figure 6(a) shows the result of the stretching case along the z-direction and Figure 6(b) shows the result of the compression case along the z-direction. In Figure 6, since the probe of the haptic interface is modeled as an area whose shape is rectangular, the interacting portion. Since we assumed that the probe shape of the haptic interface is square in Figure 6, the portion of the object that interacts with the probe is flat in shape.

Figure 7 shows haptic results with a homogeneous 3D model when the position input is given from the point (z = 75) to the point (z=50). Figure 7(a) shows the reflected forces at the left portion of the HIA and at the right portion of the HIA, respectively. Since we suppose that a flat input object compresses the homogeneous object, reflected forces in each node are the same. At this time, torque generated at the HIA is zero as shown in Figure 7(b) because the state is in equilibrium.

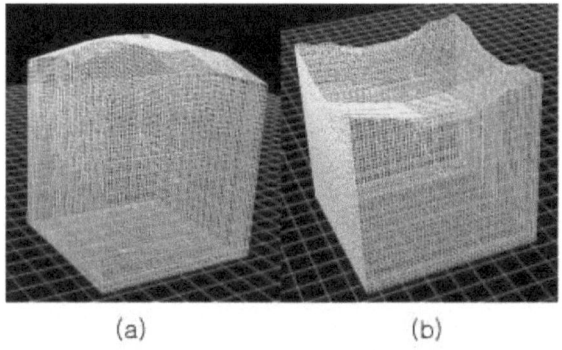

(a)   (b)

Fig. 6. Graphical result of the 3D S-chain Model under the area contact

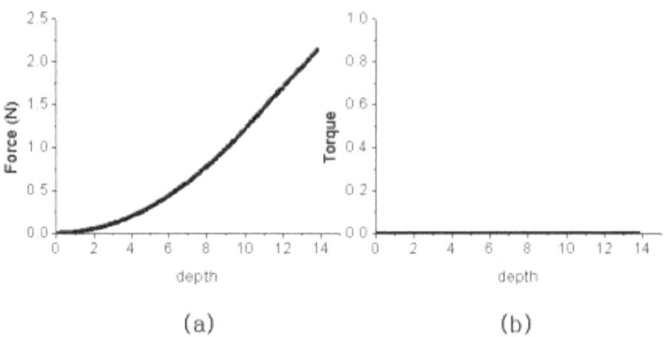

(a)   (b)

Fig. 7. Haptic result with homogeneous 3D model

5.2 Non-homogeneous Volumetric Object

We also simulated the algorithm for non-homogenous objects when the same position input is given as in the case of the homogeneous object. The non-homogenous object is composed of a soft material and a hard material as depicted in Figure 8. Figure 9(a) shows the reflected forces at the soft region and at the hard region, respectively. At the left portion of the object, the reflected force is gradually increased as the depth of penetration is increased. Differently, at the right portion of the object, the reflected force is slowly increased when the haptic interface area (HIA) is in the homogenous region, and significantly increased as the HIA approaches the non-homogeneous region.

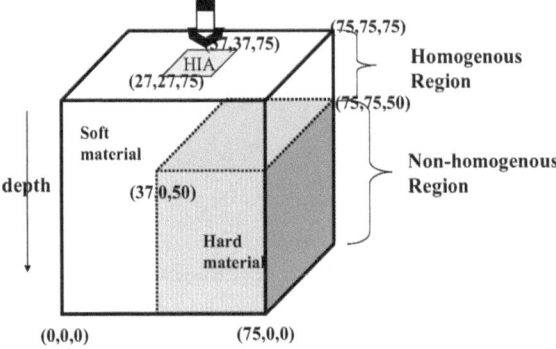

Fig. 8. The non-homogeneous object is composed of soft and hard materials

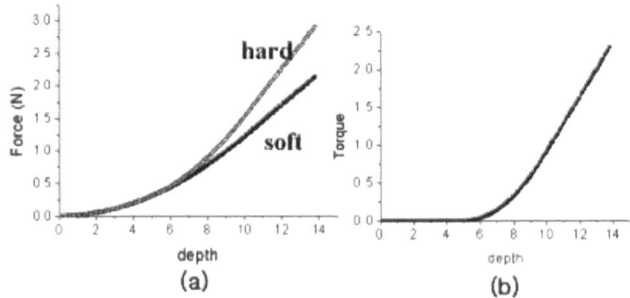

Fig. 9. The reflected force at the soft portion and at the hard portion and the torque generated at the HIA

When the user pushes or pulls the non-homogeneous object as shown in Figure 8, the reflected force at the hard portion is different from the one at the softer portion due to difference in stiffness between the hard portion and the softer portion (Figure 9(a)). By the different forces between the hard portion and the softer portion, torque is generated (Figure 9(b)). Figure 9(b) shows the magnitude of torque generated at the HIA. The torque generated at the HIA is zero when the HIA is in the homogeneous region. However, torque is generated as the HIA approaches the non-homogeneous region. In order to maintain equilibrium state, the user feels the sensation of rotation. If the sensation of rotation is transferred to the user through that gives the torque, the user perceives the hard portion and soft portion of a non-homogeneous object. However, we do not have any haptic device that gives the torque. So, we rotated the HIA until torque is zero. It is because the rotation of HIA makes the soft portion deformed and the reflected force of this soft part gets bigger than the harder part and torque would be zero. Figure 10 shows the graphical result when the generating torque is zero.

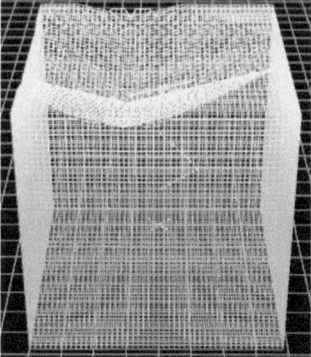

Fig. 10. The HIA is rotated in order to maintain the equilibrium state

From the haptic results (Figure 7, 9), we have verified that the proposed method enables the user to distinguish hard regions from softer ones because of torque feedback. As expected, area-based haptic rendering with an S-chain model can haptically render the homogeneous and non-homogeneous volumetric model in real time.

6   Summary

One of the areas where we can apply our technique is a palpation simulator. By pushing down a virtual volumetric data at the surface level, the user can feel the internal structure of the object. But this cannot be accomplished if we have only point haptic contact. In this paper, we have proposed area-based haptic rendering to overcome the limitations of point-based haptic rendering. Since the real-time performance is also very critical in area-based haptic rendering, we applied the S-chain model to the area-based haptic rendering.

This paper makes the following contributions to area-based haptic rendering: (1) describing a haptic rendering method that gives the distributed forces to the user due to area contact; (2) describing a method that can distinguish hard regions from the soft ones as the rotational feeling (or torque) is given to the user; and (3) implementation of the above in real-time with an S-chain model.

We have performed graphic and haptic simulations with homogeneous and non-homogeneous virtual objects consisting of more than 400000 volume elements and we have been able to verify the real-time haptic interaction. In many applications, especially in the case where we attempt to find a feature within a volume by touching the object, we believe that the area-based haptic rendering algorithm is more effective than the point-based haptic rendering algorithm.

If tactile display for palpation simulator is developed, our algorithm is more and more suitable for palpation simulator. Therefore, we are currently considering creating a tactile display in order to apply our algorithm to palpation simulators.

Acknowledgements

This work is supported by KOSEF through HWRS-ERC at KAIST. The second author was supported by Digital Media Laboratory. We would like to thank Kyubin Lee and Seong-Young Ko for having made useful suggestion.

References

1. C. Basdogan and M.A. Srinivasan, "Haptic Rendering in Virtual Environments", *Handbook of Virtual Environments,* London, Lawrence Earlbaum, Inc: Chapter 6, 2002.
2. C.B. Zilles and J.K. Salisbury, "A Constraint-Based God-Object Method for Haptic Display", *IEEE International Conference on Intelligent Robots and System, Human Robot Interaction, and Co-operative Robots*, IROS, Vol 3, 1995, pp 146-151.
3. C.H Ho, C. Basdogan, and M.A. Srinivasan, "Ray- based haptic rendering: force and torque interactions between a line probe and 3D objects in virtual environments", *International Journal of Robotics Research,* 19(7), 2000, pp668-683.
4. C. Basdogan, C.H Ho, M.A. Srinivasan, S. Small, and S. Dawson, "Force interactions in laparoscopic simulation: haptic rendering of soft tissues", *Medicine Meets Virtual Reality VI Conference*, San Diego, 1998, pp 385-391.
5. J. Park, S.Y. Kim, S.W. Son, and D.S. Kwon, "Shape retaining chain linked model for real-time volume haptic rendering", *IEEE/SIGGRAPH Symposium on Volume Visualization and Graphics*, Boston, 2002, pp 65-72.
6. S.Y. Kim, J. Park, S.W. Son, and D.S. Kwon, "Real-time soft material volume deformation using shape retaining chain linked model", *International Conference on Virtual Systems and MultiMedia (VSMM)*, Gyeongju,Korea,2002, pp 256-263.
7. C.B. Zilles and J.K. Salisbury, "A Constraint-Based God-Object Method for Haptic Display", *IEEE international Conference on Intelligent Robots and System, Human Robot Interaction, and Co-operative Robots,* IROS, 1995, pp146-151.
8. D. d'Aulignac, R. Balaniuk, and C. Laugier, "A Haptic Interface for a Virtual Exam of the Human Thigh", *IEEE International Conference on Robotics and Automation, 2000,* pp 2452-2457.
9. L.P. Nedel, and D. Thalmann, "Real time Muscle Deformations using Mass-Spring Systems", *Computer graphics international (CGI'98),* 1998, pp156-165.
10. S. Tokumoto, Y. Fujita, and S. Hirai, "Deformation Modeling of Viscoelastic Objects for Their Shape Control", *IEEE International Conference on Robotics & Automation,* 1999, pp767-772.
11. Q.H. Zhu, Y. Chen, A.E. Kaufman, "Real-time Biomechanically-based Muscle Volume Deformation using FEM", *Computer Graphics Forum 17(3),* 1998, pp275-284.

12. O. Astley, and V. Hayward, "Multirate haptic simulation achieved by coupling finite element meshes through Norton Equivalents", *IEEE International Conference on Robotics & Automation,* 1998, pp 989-994.
13. M.C. Cavusoglu, and F. Tendick, "Multirate Simulation for High Fidelity haptic Interaction with Deformable Objects in Virtual Environments", *IEEE International Conference on Robotics & Automation.* 2000, pp 2458-2465.
14. I.F. Costa, and R. Balaniuk, "LEM-An approach for real time physically based soft tissue simulation", *IEEE International Conference on Robotics & Automation.* 2001, pp 2337-2343.
15. S. De, J. Kim and M. A. Srinivasan, "A Meshless Numerical Technique for Physically Based Real Time Medical Simulations", *Medicine Meets Virtual Reality*, 2001, pp. 113-118.
16. S. Gibson, "3D ChainMail: A Fast Algorithm for Deforming Volumetric Objects", *Symposium on interactive 3D Graphics,* 1997, pp149-154.
17. S.F. Frisken-Gibson, "Using Linked Volumes to Model Object Collisions, Deformation, Cutting, Carving, and Joining", *IEEE Transactions on Visualization and Computer Graphics 5(4),* 1999, pp333-348.

Integrating Geometric and Biomechanical Models of a Liver Tumour for Cryosurgery Simulation

Alexandra Branzan Albu[1], Jean-Marc Schwartz[1], Denis Laurendeau[1], and Christian Moisan[2]

[1] Computer Vision and Systems Laboratory, Department of Electrical and Computer Engineering, Laval University, Québec (Qc) G1K 7P4, Canada
{Branzan,Schwartz,Laurend}@gel.ulaval.ca
[2] IMRI Unit, Québec City University Hospital, Québec (Qc) G1L 3L5, Canada
{Christian.Moisan}@rad.ulaval.ca

Abstract. In this paper, we present a 3D reconstruction approach of a liver tumour model from a sequence of 2D MR parallel cross-sections, and the integration of this reconstructed 3D model with a mechanical tissue model. The reconstruction algorithm uses shape-based interpolation and extrapolation. While interpolation generates intermediate slices between every pair of adjacent input slices, extrapolation performs a smooth closing of the external surface of the model. Interpolation uses morphological morphing, while extrapolation is based on smoothness surface constraints. Local surface irregularities are further smoothed with Taubin's surface fairing algorithm [5]. Since tumour models are to be used in a planning and simulation system of image-guided cryosurgery, a mechanical model based on a non-linear tensor-mass algorithm was integrated with the tumour geometry. Integration allows the computation of fast deformations and force feedback in the process of cryoprobe insertion.

1 Introduction

Widely used medical imaging systems based on MR, X-rays, positron-emission, or ultrasound scan 3D anatomic structures in a sequence of 2D parallel image slices. In order to visualize, analyze and manipulate this data, one has to deal with the difference between the inter- and intra-slice resolution. Usually, the intra-slice resolution is much higher than the inter-slice resolution, due to technical limitations and/or medical reasons (respiratory motion artefact, limited interval of exposure etc.). This is why interpolation and/or extrapolation techniques are required to estimate the missing slices. While a great variety of interpolation methods are available in the medical imaging literature, extrapolation techniques are rare, probably because of the difficulty in validating the results.

Grey-level interpolation techniques [1][2] consist of direct computation of intensity for every pixel in the interpolated slice. Since medical imaging applications are strongly object-oriented, the main drawback of grey-level interpolation techniques consists in the large amount of input data for further segmentation, and errors occur-

N. Ayache and H. Delingette (Eds.): IS4TM 2003, LNCS 2673, pp. 121–131, 2003.

ring in segmentation due to prior interpolation. Shape-based interpolation techniques are object-oriented and interpolate the binary object cross-section rather than the grey-scale intensity values. A general object reconstruction method based on deformable meshes is proposed in [3]. There is also a rich literature in spline-based interpolation techniques [4]. Mathematical morphology offers a coherent framework for developing effective shape-based interpolation algorithms. The morphological morphing transform in [5] interpolates a new group of slices between each two consecutive input slices, by performing a gradual shape transition. Our proposed scheme is similar to this approach. However, their morphing approach is based on iterative erosion and we observed that, in the case of a non-convex initial shape, iterative erosion may divide the foreground into disjoint regions, thus hindering a smooth shape transition. Instead, we propose a morphing technique based on conditional dilation. Furthermore, we are able to obtain an uniform inter-slice resolution by adjusting the lengths of the morphing sequences.

In this paper, we propose a 3D reconstruction approach using shape-based interpolation and extrapolation. To obtain maximum overlapping between adjacent slices, a *shape alignment* step is necessary prior to morphing. The interpolation process is based on *morphing*, thus performing a smooth transition between every two adjacent input slices. Next, a *closing surface* step is performed using an extrapolation technique. The 3D reconstructed model integrates the "closing" and "morphing" sequences in a coherent manner, featuring an adjustable uniform inter-slice resolution. Figure 1 presents the diagram of the proposed reconstruction process.

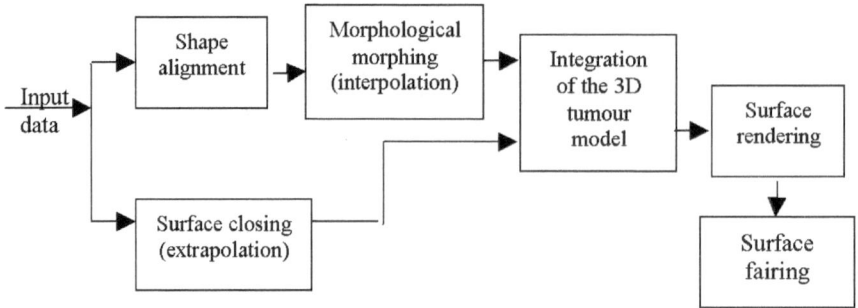

Fig. 1. The diagram of the proposed reconstruction process.

The organization of this paper is as follows. Section 2 presents the proposed 3D reconstruction approach. Section 3 shows and validates our reconstruction results, while section 4 describes the integration of the geometric model with a mechanical tissue model. Finally, we draw the conclusions and describe future work.

2   Reconstruction Approach

For every patient, a serial sequence of 2D MR segmented images of transversal liver slices is provided. The foreground of each segmented image represents a cross-

section through the targeted tumour. The used segmentation method has been described in [6]. Respiratory movements prevent the slice thickness from being reduced
below 10 mm in the abdominal MR image acquisition process. Small liver tumours of
5 mm in diameter (the standard threshold for significant lesions) are therefore visible
in only one image. Thus, it is impossible to create a 3D model of a smaller tumour
using only MR images of transversal slices. For tumours of medium size, the number
of contributing slices is usually three, but in some cases it may be up to four or five.
Sequences of three segmented MR slices are considered as input data for the reconstruction approach, as it is the most frequent situation. However, our approach can be
easily adapted for longer input sequences. In any input cross-section, the interior of
the tumour does not contain any holes and is represented by a single compact region.
The following sub-sections present the main steps of the reconstruction process.

2.1 Shape Alignment

A morphing process is impossible between two planar (xy) shapes that do not overlap, when viewed in the z–direction, thus in the general case shape alignment is necessary prior to morphing. In the particular case of liver tumours, their nodular appearance always results in a partial overlap between adjacent slices. Thus, shape alignment is not absolutely necessary for liver tumours, but plays nevertheless an important role in the design of our reconstruction approach because the morphing process
obtains best results when the common area shared by the two input shapes is maximal. To align two shapes we use a simple translation-based method. Only one shape
is translated, while the other one remains immobile and is considered as reference.
The search of maximal overlap may result in more than one possible translation. In
order to provide a unique solution, we minimize the Haussdorff's distance between
the contours of the translated and the reference shape.

Figure 2 presents the results of shape alignment for two configurations of the input
data. Two displacements are computed, t_1 and t_3 for objects *obj*1 and *obj*3 respectively, considering *obj*2 as a reference object.

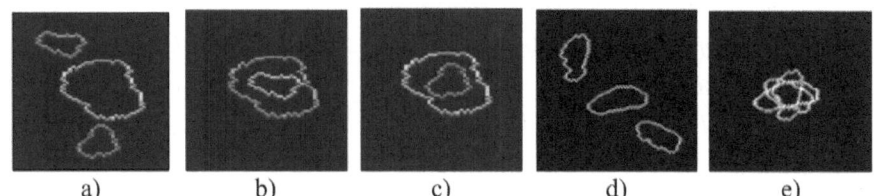

| a) | b) | c) | d) | e) |

Fig. 2. a) An example of input configuration, containing the initial relative position of
obj1(top), *obj2*(middle) and *obj3* (bottom) ; b) alignment of *obj1* with respect to *obj2*; c) alignment of *obj3* with respect to *obj2*; d) another example of input configuration, containing *obj1*,
obj2 and *obj3*; e) shape alignment of *obj1* and *obj3* with respect to *obj2*.

The aligned sequence of binary objects *obj1t*, *obj2* and *obj3t* represents the input
data for the next step, that is the morphological morphing.

2.2 Morphological Morphing Based on Conditional Dilation

We propose a new morphing technique based on conditional dilation.

Definition : Let A and B be two sets, such that $B \subset A$. The conditional dilation of set B using the structuring element K with respect to the reference set A is expressed as :

$$B \oplus K|_A = \left(\bigcup \left\{ B_k | k \in K \right\} \right) \cap A \tag{1}$$

The input data for this morphing technique consists in an initial binary object *objA* and a final binary object *objB*, located in adjacent slices. The only constraint imposed to the input configuration is $objA \cap objB \neq \emptyset$, which is always satisfied after shape alignment. Let $objint = objA \cap objB$. The result of this morphing technique is a sequence of intermediate binary objects gradually changing their shape from *objA* towards *objB*. Figure 3a) contains the contours of *objA* and *objB* respectively, while figure 3b) highlights the common area of *objA* and *objB*.

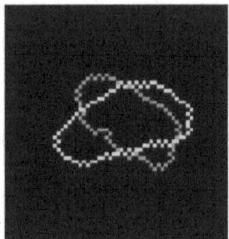 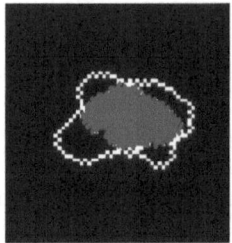

Fig. 3. a) the contours of *objA* and *objB* ; b) the common area shared by *objA* and *objB*.

In order to gradually transform *objA* into *objB* we perform two parallel morphing processes based on conditional dilation. These processes transform *objint* into *objA* and into *objB* using k_1 and k_2 iterations respectively. We name $obj_1(i)$ and $obj_2(i)$ the objects generated after i conditional dilations of *objint* with respect to *objA* ($i<k_1$) and, *objB* ($i<k_2$) respectively. The morphing process transforms *objA* into *objB* by generating a sequence of intermediate objects *objAB* as follows :

$$objAB(i) = \begin{cases} obj_1(k_1 - i) \cup obj_2(i) & \text{if } i < k_1 \leq k_2 \text{ or } i < k_2 \leq k_1 \\ obj_2(i) & \text{if } k_1 \leq i \leq k_2 \\ obj_1(k_1 - i) \cup obj_2(k_2) & \text{if } k_2 \leq i \leq k_1 \end{cases} \tag{2}$$

$i = 1..\max(k_1, k_2)$.

The length of the morphing sequence is equal to the largest number among k_1 and k_2.

For input data containing three parallel equidistant tumour slices, two morphing sequences are to be integrated in the 3D tumour model : the sequence obj_{12} of length L_{12}, which gradually transforms obj_1 into obj_2 and the sequence obj_{23} of length L_{23} which gradually transforms obj_2 into obj_3. The intermediate objects in the morphing sequences are to be located in equidistant planes. Since the lengths L_{12} and L_{23} are usually different, we eliminate $|L_{13} - L_{23}|$ intermediate objects from the longer sequence. Due to the anisotropy of the conditional dilation, it is possible to encounter very slow shape variations between adjacent intermediate objects at some instances of the morphing process. In order to achieve a quasi-uniform rate of shape change in the sequence of intermediate shapes, we eliminate "redundant" intermediate objects by using a *distance measure*, defined as $d(obj_1, obj_2) = card(obj_1 \Delta obj_2)$ where obj_1, obj_2 are adjacent objects in the morphing sequence and Δ stands for the symmetric difference. At iteration k, a *redundancy coefficient* is assigned to every intermediate object of the input sequence :

$$R(obj(i)) = min\left(dist(obj(i), obj(i-1))^{-1}, dist(obj(i), obj(i+1))^{-1}\right) \qquad (3)$$

$i = 1..L(k)$, where $L(k)$ is the length of the input sequence at iteration k.

The first and the last objects of the input sequence cannot be eliminated, since they represent input slices in the reconstruction process. The object with the highest redundancy coefficient is eliminated at the current iteration, and the resulting sequence represents the input for the next iteration. The redundancy coefficients are updated at each iteration.

The equal-length constraint for the two morphing sequences results in an uniform inter-slice resolution of the 3D reconstructed model. Furthermore, we use the method of redundancy coefficients for varying the common length of the morphing sequences. Both equal-length sequences can be shortened by eliminating a given number of intermediate objects. Thus, we are able to generate 3D tumour models of variable size and adjustable inter-slice resolution.

Once the two morphing sequences are adjusted to the same length, the 3D interpolated tumour model is obtained by a simple concatenation of the two sequences. Furthermore, to obtain a tumour model consistent with the input data, we have to reverse the shape alignment process. Since translation is reversible, we replace *obj1t* and *obj3t* at their initial locations. The objects belonging to the morphing sequences are also translated, in order to perform a smooth transition from obj_1 to obj_2 and from obj_2 to obj_3 respectively.

2.3 Surface Closing

Due to the finite inter-slice distance, the acquisition process does not offer any information about the tumour's extremities. However, we cannot accept flat-endings in the 3D reconstructed model. To close the surface, we perform a shape-based extrapolation which respects the surface smoothness constraint. We assume that the first and the last horizontal cross-sections of the real tumour are one pixel-sized objects, which is a reasonable assumption for liver tumours.

We create two sequences of closing objects located in horizontal slices. These sequences gradually shrink obj_1 and obj_3 towards pixel P_1 and P_3 respectively. When viewed in the z-direction, pixels P_1 and P_3 are located inside obj_1 and obj_3 respectively, as shown in Figure 4. The length of a closing sequence is set to $(N/2)$, where N is the even-valued length of the morphing sequences.

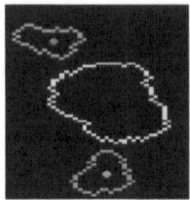

Fig. 4. Pixel $P_1 \bullet$ is located inside obj_1 (top), while pixel $P_3 \bullet$ is located inside obj_3 (bottom).

To obtain smooth closing, pixels P_1 and P_3 are chosen to be the centroids of obj_1 and obj_3 respectively. This choice is justified by the fact that real liver tumours are egg-shaped. The centroid of an object is defined as the pixel inside the object which generates this object in a minimum number of conditional dilations.

To generate a closing sequence from an initial 2D object and its centroid, the distances from the centroid to each pixel in the object's boundary are computed. We use a parametric representation for the contour of the object, which allows the storage of the contour pixels in a 1D array. Therefore, the distances from the centroid to the contour pixels are as follows :

$$DIST(k) = \sqrt{(x - X(k))^2 + (y - Y(k))^2} \qquad K = 1..\,length(X) \qquad (4)$$

where (x, y) are the coordinates of the centroid and $X(k)$, $Y(k)$ are the coordinates of the kth element in the contour parametrisation. These distances are to decrease gradually towards 0 in $N/2$ iterations, where N is the length of the morphing sequences. More explicitly, we generate $N/2$ intermediate closing contours which shrink gradually towards the centroid. Choosing a linear decreasing pattern leads to an angular, disturbing appearance of the closing parts of the object. Instead, we set the difference between the distances computed at two successive iterations to be proportional to the index of the last iteration :

$$DIST(k)_i - DIST(k)_{i-1} = i \qquad (\forall)k = 1..\,length(X) \qquad (5)$$

where i is the index of the iteration, $i = 1..\,L/2$.

Next, the closing sequences are concatenated at the corresponding extremities of the interpolated 3D tumour model. Furthermore, a surface rendering technique is used to generate a triangular mesh on the external surface of the reconstructed tumour model.

2.4 Surface Fairing

The previously described morphological morphing and surface closing processes should result in a 3D object with a smooth surface. However, local irregularities may occur. Some possible reasons for their presence are : a) the shape of the elementary structuring element in the 2D discrete space used in conditional dilation; b) the constraint of integer horizontal displacements in the translation of intermediate cross-sections; c) the successive elimination of intermediate objects with high-valued redundancy coefficients; d) the fixed length of the closing sequences.

We consider Taubin's surface fairing algorithm [7] for its linear complexity and for the fact that it moves the vertices of the mesh without changing the connectivity of the faces. The fairing process conserves the number of vertices and faces, thus allowing us to compare and measure the smoothness of the faired surface and of the original surface.

3 Reconstruction Results and Geometric Evaluation

The geometric validation of the 3D liver tumour model does not compare this model to the real tumour, since there is a big gap between the amount of input information (3 serial tumour cross-sections) and the amount of output data ($2L+1$ object cross-sections, $L \geq 6$). As a consequence of undersampling, no technique can guarantee to reconstruct the actual anatomy from any set of cross-sections. However, since it contains the three input cross-sections at the original z-levels as specified in the acquisition process, the model is coherent with the input data. Taubin's surface fairing algorithm [5] smoothes the shape of the 2D cross-sections corresponding to the input data, but it performs no shrinking or expanding.

Since the proposed 3D reconstruction approach aims towards a smooth transition between adjacent input shapes and towards a smooth 3D surface closing, we propose a measure of surface smoothness for result evaluation. For each vertex P, the normals to every triangular face containing P are computed, using the classical parametric equations.

Among the k normals corresponding to P we arbitrarily choose a reference direction (l_0, m_0, n_0), and compute the cosine of the angle between every normal in the set and the reference direction. The average value $\overline{\cos\alpha}|_P$ of $\cos\alpha_i$, $i = \overline{1,k}$-1 represents a local measure of smoothness at vertex P. The local smoothness at P increases when $\overline{\cos\alpha}|_P$ approaches 1. A global smoothness measure is represented by the histogram of local smoothness measures computed over the entire surface. The histogram of a smooth surface presents a peak value near 1, and low values elsewhere.

The input sequence in Figure 5a) was interpolated using morphing based on conditional dilation. Shape-based extrapolation for surface closing was performed afterwards. The surface fairing process consisted in two iterations of Taubin's algorithm [5]. The results and evaluation of the reconstructed 3D model are shown in Figure 5.

The reconstructed 3D tumour model presented in Figure 5 shows a reasonable quality of surface smoothness even before the fairing process. However, the surface fairing considerably improves the surface smoothness without changing the global appearance of the object.

4 Mechanical Model

Reconstructed 3D tumour models are to be integrated into a complete system for the simulation of cryosurgery of liver cancer. A mechanical model has been developed for this simulator and was presented in [8]. In this section we present the integration of this mechanical model with a 3D tumour model reconstructed by the algorithm described in section 2.

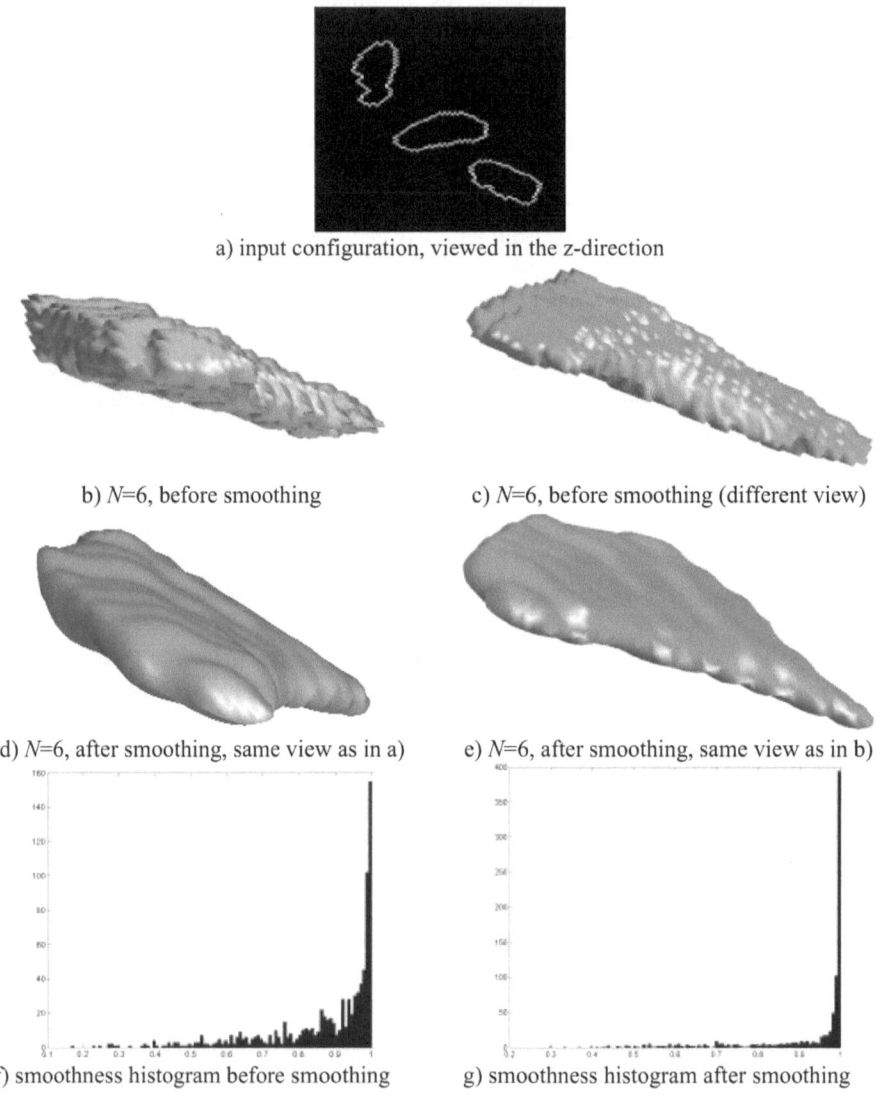

a) input configuration, viewed in the z-direction

b) $N=6$, before smoothing   c) $N=6$, before smoothing (different view)

d) $N=6$, after smoothing, same view as in a)   e) $N=6$, after smoothing, same view as in b)

f) smoothness histogram before smoothing   g) smoothness histogram after smoothing

Fig. 5. Results and evaluation for the reconstruction approach using morphing based on conditional dilation and shape-based extrapolation; N is the length of the morphing sequence.

Our mechanical model uses on the finite element based tensor-mass algorithm [9], which computes forces from a combination of local stiffness tensors attached to every mesh element. These tensors depend only on the mesh geometry at rest, and on the mechanical properties of the tissue. Therefore they can be computed in a preliminary step, while computation in the actual simulation is limited to a linear combination of stiffness matrices and displacement vectors, meeting real-time constraints.

We have previously shown that it is possible to extend the linear tensor-mass model in order to simulate different types of non-linear and visco-elastic mechanical properties [8][10]. Previous results were obtained using meshes consisting of a regular assembly of cubic elements divided into tetrahedrons. We show in this section that the same mechanical model can be applied on a non-uniform mesh derived from an reconstructed 3D tumour model. The mechanical tissue properties used for testing were obtained from experimental *in vitro* compression on deer liver membrane by a biopsy needle [8], since no *in vivo* mechanical data of liver tumour could be measured so far.

The faired triangular surface mesh obtained in section 3 was first transformed into a tetrahedral volume mesh using the Geompack package [11]. Next, compression of this mesh was simulated using the mechanical parameters measured experimentally. Figure 6a) presents five independent experimental force measurements, as well as the simulated force on the reference mesh used to fit our model parameters, and the simulated force computed on the non-uniform tumour mesh.

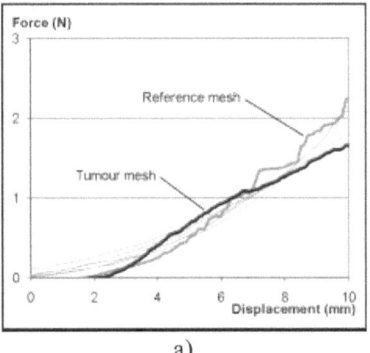 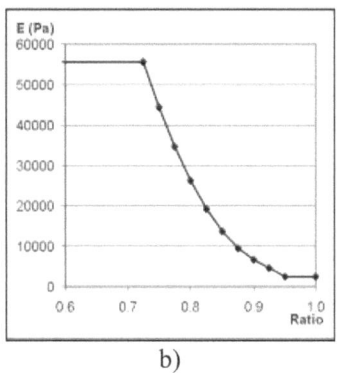

a)   b)

Fig. 6. a) Five independent experimental measurements of forces in compression of a deer liver sample by a biopsy needle (light grey curves), and simulated forces computed by a non-linear tensor-mass algorithm on the reference mesh and on a reconstructed 3D tumour model. Compression speed was in all cases constant at 10 mm/s. b) Non-linear function introduced into the tensor-mass algorithm to obtain the simulated curves in a). Non-linearity is expressed as a function of a value quantifying local deformation, which is the ratio of the current tetrahedron mean ratio on the tetrahedron mean ratio at rest.

Although differences can be observed between the two simulations due to the different mesh geometries, accordance between experimental data and both simulations can be considered satisfactory. Due to the thin the tumour geometry (approximately 8 mm in thickness), forces on the tumour mesh tend to increase more slowly at higher deformations, as it becomes almost entirely pierced. This comparison shows that the proposed mechanical model can be successfully applied to variable geometries. The accurate mechanical parameters of liver tumours remain yet to be determined.

The tissue model derived from these measurements was highly non-linear. Figure 6b) shows the non-linear function introduced into the tensor-mass model to account for these properties. For low deformations the Young modulus was $E = 3600$ Pa, and the Poisson coefficient was kept constant at $v = 0.4$.

Figure 7 shows a few deformed mesh configurations at different time steps. The tumour mesh contained 1135 vertices and 3535 tetrahedrons, and a computation rate of 50 iterations per second was achieved on a 2 GHz Pentium III computer.

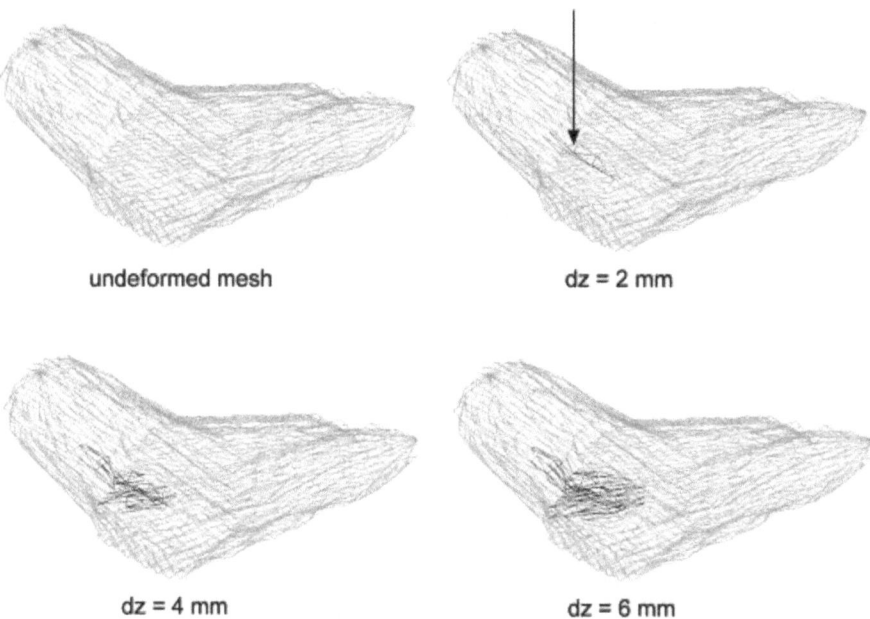

undeformed mesh   dz = 2 mm

dz = 4 mm   dz = 6 mm

Fig. 7. Deformation of a reconstructed 3D mesh under simulated compression by a biopsy needle. The arrow on the second frame shows the position of the needle. Values indicate compression depth, and deformed mesh elements are highlighted.

5   Conclusion

In this paper, we proposed a new 3D reconstruction technique integrating morphological morphing between adjacent slices and shape-based extrapolation of extremity slices. The presented reconstruction algorithm is appropriate for modelling anatomical structures and was successfully integrated with a biomechanical model allowing fast computation of deformations and force-feedback. However, the accurate mechanical properties of liver tumours *in vivo* remain to be measured. Future work will focus on the 3D reconstruction of the entire liver and its hepatovascular system from 2D MR cross-sections. Furthermore, tumour and liver geometries are to be integrated with their specific mechanical and thermal models into a complete planning and

References

1. Goldwasser, S.M., Reynolds, R.A., Talton, D.A., and Walsh, E.S.: Techniques for the rapid display and manipulation of 3-D biomedical data. Comput. Med. Imag. Graphics, Vol.12 (1998), 1-24
2. Goshtasby, A., Turner, D.A., and Ackermann, L.V.: Matching of tomographic slices for interpolation. IEEE Trans. Med. Imag., Vol.12 (1993), 366-379
3. Delingette, H. : General Object Reconstruction based on Simplex Meshes. International Journal of Computer Vision, Vol.32 (1999), 111-146
4. Herman, G., T., Zheng, J., and Buchholtz, C., A.: Shape-Based Interpolation. IEEE Computer Graphics and Applicat., Vol.12 (1992), 69-79
5. Bors, A., Kechagias, L., and Pitas, I.: Binary morphological shape-based interpolation applied to 3-D tooth reconstruction. IEEE Trans. Med. Imag., Vol.21 (2002), 100-108
6. Branzan-Albu, A., Moisan, C., and Laurendeau, D.: Tumour detection in MR liver images by integrating edge and region information. Proc. MS4CMS '02 (Modelling and Simulation for Computer-aided Medicine and Surgery, Rocquencourt, France, 12-15 November 2002)
7. Taubin, G.: A signal processing approach to fair surface design, Proc. SIGGRAPH'95 (Los Angeles, CA, 6-11 August 1995), 351-358
8. Schwartz, J.-M., Dellinger, M., Rancourt, D., Moisan, C., and Laurendeau, D.: Modelling liver tissue properties using a non-linear viscoelastic model for surgery simulation. Proc. MS4CMS '02 (Modelling and Simulation for Computer-aided Medicine and Surgery, Rocquencourt, France, 12-15 November 2002)
9. Cotin, S., Delingette, H., and Ayache, N.: A hybrid elastic model for real-time cutting, deformations, and force feedback for surgery training and simulation. Visual Computer, Vol. 16 (2000), 437-452
10. Schwartz, J.-M., Langelier, È., Moisan, C., and Laurendeau, D.: Non-linear soft tissue deformations for the simulation of percutaneous surgeries. Proc. MICCAI 2001 (Medical Image Computing and Computer-Assisted Intervention, Utrecht, The Netherlands, 14-17 October 2001), Lecture Notes on Computer Science, Vol. 2208 (2001), 1271-72
11. Joe, B.: GEOMPACK - a software package for the generation of meshes using geometric algorithms. Adv. Eng. Software, Vol. 13 (1991), 325-331

3D Reconstruction of Large Tubular Geometries from CT Data

Andrea Giachetti and Gianluigi Zanetti

CRS4, VI Strada Ovest, Z. I. Macchiareddu, I-09010 Uta (CA), Italy
{giach,zag,}@crs4.it
http://www.crs4.it

Abstract. In several medical applications it is necessary to have a good reconstruction of approximately tubular structures – mainly blood vessels but also intestine or bones – providing a description of both the internal lumen (usually a triangulated surface) and its networked structure (skeleton). This description should be such that it allows lengths and diameters estimation. Several methods have been proposed for these tasks, each one with advantages and drawbacks and, typically, specialized to a particular application. We focused our attention on methods making as few assumptions as possible on the structure to be determined in order to capture also anomalous features like bulges and bifurcations. We looked for a method able to obtain surfaces that are smooth, with a limited number of triangles but accurate and skeletons that are continuously connected and centered. The results of our work is the use of customized deformable surface and multi-scale regularized voxel coding centerlines to obtain geometries and skeletons with the desired properties. The algorithms are being tested for real clinical analysis and results are promising.

Introduction

The extraction of approximately tubular structures from medical images is a common task in medical image analysis. CT scans can give a sufficiently good representation of the interested structures, usually blood vessels or intestine. Modern image processing techniques are able to extract from these images accurate 3D representations of vessels and other organs. These techniques have many applications in the diagnostic and surgical activities: vessel repair surgical planning [1], blood flow simulation [2], virtual colonoscopy [18,20], liver surgery planning [23].

For all these applications, the reconstruction algorithms should not be limited to a simple voxel classification after smoothing, skeletonization or cylinder approximation. An accurate reconstruction of the lumen, adaptable to tubular geometries with rapidly changing radius and not too sensitive to image noise has to be found. Skeletonization is, however, still necessary, in order to be able to evaluate topology of the structure, to describe paths along it and to estimate distances and sections. For this reasons we analyzed surface extraction methods and algorithms for the skeletonization of the extracted volume and we have

N. Ayache and H. Delingette (Eds.): IS4TM 2003, LNCS 2673, pp. 132–144, 2003.

developed improved algorithms for the two tasks. The paper is structured as follows: in Section 1 we present a short review of existing methods, in Section 2 our solutions are described and in Section 3 experimental results are shown.

1   Existing Approaches

Several different methods have been proposed for the detection of approximately tubular structures from CT or MRI 3D datasets. Some approaches first search the vessel skeletons from ridges at different scales, and then analyze the local geometry of the lumen [14,7,8]. In these approaches, surfaces are extracted later from local directions with generalized cylinders or deformable models. These methods can give good results if images are well contrasted, but are extremely time consuming due to the necessity of filtering large datasets at different scales.

Being interested to the reconstruction of large and irregular vessels not easy to detect with ridge search, we are interested in methods to recover the vessel lumen with no a priori information.

1.1   Vessel Lumen Extraction

Methods starting from the vessel volume classification or lumen surface extraction can use several data processing tools. The first one is the standard iso-surface extraction with marching cubes or similar algorithms [11], like in [9]. The extraction is done usually on classified or pre-processed data. The surfaces extracted require post processing for mesh simplification and smoothing. An alternative method is based on region growing, front propagation or fast marching algorithms that can give a good detection of well contrasted vessels [10,9]. Deformable models [12] can give smoother and simpler results, even if they have to be initialized correctly and parameters controlling their evolution must be carefully tuned. If the vessel is sufficiently contrasted in the original images and the acquisition resolution is not too low, the extracted surface has usually a good correspondence to the correct boundaries.

1.2   The 3D Skeletonization Problem

If we assume that we are segmenting an approximately tubular structure and if the method does not start with the recovery of its "skeleton", in order to capture the local direction of the "tube" and the networked structure of the organ it is necessary to evaluate this structure later from the lumen extracted. What we need to compute is a local vessel direction and a path inside the lumen, possibly "centered" inside it. This problem is clearly ill-posed. In 3D things are much more complicated than in 2D, where skeletons can be easily extracted with a medial axis transform [21]. The "medial axis", intuitively is the locus of points center of a N dimensional ball tangent to the lumen surface in at least two points. While in 2D this locus is a line, in 3D it is a 2D surface, unless the geometry has spherical symmetry (0D centerline) or a cylindrical (1D centerline). Furthermore, in 3D it is extremely complicated to have an estimation of the medial axis (i.e. of the surface). Methods to extract this surface from have been proposed in [4,5]. They are based on geometrical approximations or on the Voronoi diagram of the boundary points, a computationally heavy task. To extract 1D skeletal curves from boundary points an algorithm has been proposed by Verroust and Lazarus [6]. It is based on a cylindrical decomposition of the surface started from a source point. Most authors, however, extract skeletons from voxelized geometries. Int this case the skeleton is defined as a set of one voxel thick lines with particular properties, i.e. they are centered, they are connected (using 6, 18 or 26 neighborhoods), they are smooth.

It is clear that this definition is not at all satisfactory. What does, for example,"centered" mean? The lack of a "correct" definition, leads to ambiguities. Where is the centerline of the vessel with the section represented in Fig. 2? If we are asked to answer this question, we would say probably near the cross, but if we use the "maximum ball" criteria we find two circles. Furthermore, if we try to join points that are maximum ball centers in the 3D structure we could get discontinuous lines.

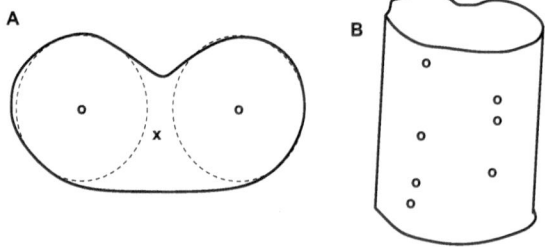

Fig. 1. A: The detection of centerline as local directional maximum of a distance from border function or "maximum ball" may be ambiguous. In this case, local maxima of the distance from centerline does not form a continuous line (B).

Two main techniques have been proposed to extract centerlines from binarized volumes: topological thinning based on "simple points" removal [3,20] or voxel coding based [16,18,19]. The first class of methods, is based on a progressive removal of voxel called "simple", that does not change the topology, i.e. the connectivity of the voxel set. This can be done by defining correct rules that can lead to very high complexity. The use of large look-up tables can somewhat overcome computational problems. Furthermore, it is necessary to preserve endpoints. This cannot be done automatically. Results depends on the search strategy for removable points. Furthermore, the method fails in the case of a vessel with irregular sections, like the banjo-shaped section described by Bitter et al [19]. The error is due to the fact that the erosion is not done at a constant speed from the borders to the center. When somewhere a 1 voxel section is reached, the erosion is stopped due to connectivity preserving rules. In the example of Fig. 2

Fig. 2. Example, of voxel removal failure. If we use this technique on a vessel with an irregular section, the resulting centerline may be even composed by border voxels.

we see that, if we suppose to have an high voxel column with a section like in A, doing a standard raster removal of the voxels that does not cause connectivity violation leads to a centerline that is not "centered" at all. Another drawback of this technique is that it does not provide a hierarchical branch structure, useful, for example, to reconstruct an arterial tree.

Voxel coding algorithms have been recently introduced by Zhou and Toga and other authors [16,19]. Considering that a distance map from the border (BSC, boundary seeded code) in the 3D case is not sufficient to extract one voxel thick skeletons as in the 2D case, the idea of these methods is to compute first paths inside the volume, and then center them using the distance map. To extract the paths another voxel coding is defined, called "Single Seeded Code" (SSC) or "Distance from Seed" (DFS), measuring the distance of volume points from one seed voxel. Taking as starting points local maxima of the SSC with high values, paths are extracted searching for voxels with lower SSC in the neighborhood. When the voxel has been found, it is added to the centerline and then a new voxel with lower SSC is searched around it. The procedure is stopped when the seed is reached or when the line is close to a previously extracted one. The extracted lines are approximately "shortest" path joining the starting point and the seed. Paths are depending on the metric used to compute the SSC and on the search strategy. The skeleton defined in this way have some nice properties: it is composed by lines, i.e by lists of connected point, it capture the network structure of the vessel and its points are inside the volume.

Two desired properties can be, however, still missing: centering and smoothness. The usual approach presented by authors to center the lines is derived by Zhou and Toga and consists of the following steps:

- For each point of the skeleton, find the cluster of voxels with the same SSC, connected with that point.
- Find the voxel of the cluster with maximum BSC
- Move the centerline point to the position of that voxel

Results are not always satisfactory. This procedure can give reasonable skeletons in the case, for example, of vascular structures with approximately constant radius, but are strongly dependent on the position of the seed and on the shape of the object to be skeletonized. This fact can be made evident with a simple example. Consider the shape of the 2D structure of Fig. 3. It can be considered as a 2D section of an aorta with an aneurysm. The gray level encodes the distance from a seed placed near the center of the bulge. In the example in the left, the seed is near the center, int the example on the right, it is at one extremal point. If we start the detection of a line from the farthest point and we apply the

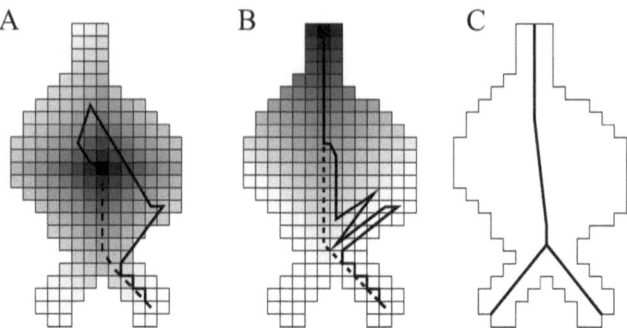

Fig. 3. 2D aneurysm example showing failure of standard Voxel coding algorithm. While the Shortest path (dashed line) seems reasonable, the centered one (thick line) is discontinuous, irregular and far from the expected position. In A paths toward a seed in the center of the aneurysm are shown. In B paths toward a seed in the top of the vessel are shown. In C the intuitive centerline is shown.

shortest path algorithm, we obtain the dashed lines. Then, if we take clusters and find the center of 26-connected regions, we obtain the continuous line and the result is not satisfactorily at all. Points are moved far from each other, connection is lost and not all the points are moved to a point near the expected centerline. Maxima of cluster with close SSC values can be distant due to the complex shape of the object. Some authors have introduced methods to handle this problem introducing penalized distances [19], or using iterative shift of the contour of limited value.

2   Solutions Proposed

2.1   Lumen Reconstruction

For lumen reconstruction, we were looking for an algorithm able to extract smooth surfaces well adapted to the organ boundary even in the case of bulges, and bifurcations, and that does not change topology, defining an unique connected region. We developed therefore a surface reconstruction method that is a specialized surface expansion that improves the performances of usual deformable models techniques. The method here used is based on the simplex mesh geometry introduced by Delingette [13]. As defined in this paper, the generic Simplex Mesh is a N dimensional mesh with N+1 connectivity (Fig. 4).

A 2D simplex mesh in the 3D space is therefore a closed surface mesh composed by nodes each one connected with three neighbors. This geometry makes simple to evaluate local curvature and give regularization rules as well as to compute local properties of the surface. We built on this structure a deformable models defining a closed simplex surface and making each node move with a Newtonian law of motion:

$$m_i \frac{d^2 \boldsymbol{X}(i)}{dt^2} = -\gamma \frac{d \boldsymbol{X}(i)}{dt} + \boldsymbol{F_i}(\boldsymbol{i}) + \boldsymbol{F_e}(i) \qquad (1)$$

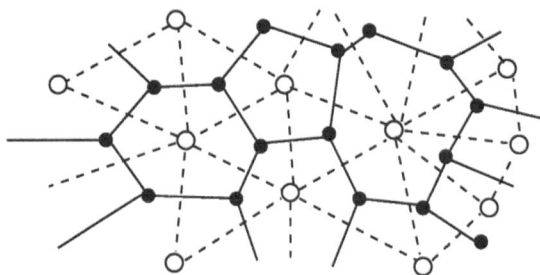

Fig. 4. A 2D simplex mesh (Black dots) and its dual triangulation (white dots).

where m_i is the mass, $\boldsymbol{F}_i$ the internal force, $\boldsymbol{F}_e(i)$ the external force. $\boldsymbol{F}_e$ is the sum of an inflating force directed along the surface normal, and two image forces: a deflating force directed against the surface normal and compensating the inflating one; its modulus is relevant where the local average of the gray level differs from the internal value more than a fixed threshold and an edge attraction moving nodes toward the maximum of the gray level gradient modulus in the neighboring. $\boldsymbol{F}_i(\boldsymbol{i})$ an elastic smoothing force; we give the possibility of using two forces described in [13] ("surface orientation continuity constraint", averaging the local normal vector and "simplex angle constraint" averaging the mean curvature at neighboring nodes).

After a fixed number of iterations of the node evolution algorithm, faces are resized in order to keep their size approximately in a fixed range. This procedure is a bit complex, due to the necessity of preserving the simplex structure.

We customized the algorithm not only through the image forces; but also making it adaptive and optimized for the extraction of tubular structures: the maximum face size is not a global value, but is proportional to the local curvature. Furthermore, the computation is made faster by labeling as fixed the nodes that already reached the desired border (i.e. an edge with the gray level which differs more than a fixed thresholds from the internal value). In this way, the surface is well adapted to complex structures and the computation is fast because at each step only the free "front" of the surface in the tubular structure is moved. The final mesh is converted in the "Dual" form (see Fig. 4 (i.e. a new mesh with nodes in the center of the faces and connections corresponding to the simplex edges) to have a smooth triangulation. The simplex mesh, in fact, is, in general, composed by polygons that are not necessarily planar and cannot be easily rendered.

The surface is usually initialized as a small sphere inside the lumen and is inflated until the surface is not blocked by edges or changed image value. The user can control the maximum number of iterations to be performed, force parameters and the maximum and minimum size of the polygons.

The initialization of the surface can be made as a small sphere inside the lumen (the center is selected clicking on the interface on a seed point) or can also be made from a close isosurface extracted after a data filtering and binarization.

The simplex conversion allow an immediate polygon reduction and smoothing and can be refined as desired through the image based forces.

2.2   Centerline Extraction

For the centerline extraction we propose an improved voxel coding method using a multi-scale approach and a snake-based regularization. Given a seed point, we compute the single seeded code inside the lumen with a region growing procedure at a reduced resolution. If the starting data is a surface, voxel classification as inner is made immediately with a region growing from the seed while computing the single seeded code. The boundary seeded code is then computed in the inner region with an iterative rule keeping into account also a possible non-uniform quantization (i.e. a different voxel size in the three directions). BSC is initialized equal to zero in the boundaries and put at infinity in the other regions. Then until for each iteration all the voxel with a finite BSC, a search in a 26-point neighborhood is performed and if the sum of the central BSC and an integer approximation of the distance between the neighbor and the center is less than the current value, the value is replaced with that sum. The iterations are stopped when the map does no longer changes. The algorithm then works like the other voxel coding method. Starting from the boundary point that is farest from the seed, the shortest path to the seed is evaluated, each point is then moved at most of a few voxel towards a larger BSC, clusters of connected voxel with the same SSC of the shortest path point are removed from the search space, then the procedure is repeated to extract all the skeleton branches. Finally the high resolution correction and a snake based centering and regularization is performed: for each only points at extrema are kept fixed and the others are moved with an "image" force driving them toward the maximum of the distance from boundaries and an internal force keeping the curve smooth.

The final algorithm works as follows:

- Find the internal region on the full dataset by region growing inside the previously extracted surface.
- Create a low resolution binarized dataset labeling as internal all the voxel at the low resolution including an internal voxel at the high resolution. The user interaction is therefore only in the choice of the desired initial resolution and in the selection of the starting point.
- Compute the centerline at the low resolution, i.e. compute the BSC and the SSC at the low resolution, and find the tree structure.
- Compute the boundary seeded code map at the high resolution, called BSCH
- Move each skeleton point to the high resolution voxel location corresponding to the maximum of the BSCH inside the low resolution voxel.
- Resample and regularize the line with the snake-based algorithm.
- Go on finding the other branches, joining their last point to the closest point of the previous lines. Lines shorter than a fixed threshold are removed.

The multi-scale approach makes the algorithm faster and less influenced by local structure. The snake based regularization driven by the distance from border map acting as an energy field and elastic forces, makes the lines centered and smooth. With our method we obtain results compliant with our requirements: i. e. continuous curves connected in a tree structure and locally centered in the volume. Note that the voxelization of the volume is now correspondent to the true voxel dimensions. The algorithm is, however, independent and in the next release the user will put the desired resolution as an input value and the SSC and BSC maps will be computed at that resolution (subsampled in the multiresolution part).

3   Experimental Results

We applied our algorithms to detect several structures of surgical interest. A quantitative analysis of the segmentation results is also ongoing in a trial surgical planning application for the EU funded project AQUATICS [1]. The trial is aimed at the reconstruction of measurable models of the aorta for surgical planning of endovascular procedures. The aortic geometry in this case is a very good testbed for automatic skeleton algorithms, because large bulges in the aorta may create some of the problems we have analyzed to standard algorithms. Fig. 6 shows the centerlines extracted from a patient specific acquisition (courtesy of Radiology dept, Univ. of Innsbruck). An example of geometry growth is shown in fig 5.

The algorithm is also sufficiently fast, the only problem is that in the case of small vessels with noise it is necessary to tune carefully the simplex parameters (forces and face size) in order to prevent surface bending. For this reason the standard procedure for the reconstruction must be user controlled to change parameters during the evolution if necessary. We plan to reduce these problems by improving the elastic force model and/or introduce a collision detection scheme for geometry nodes.

For all the geometries extracted, we have taken the voxelized region inside the surface as the binary data to be skeletonized and computed the skeletal lines. An example of the procedure is shown in Fig 6. Starting from a resolution reduced by a factor 4 in x and y direction, we obtained the shortest path (black). The refined centered and smoothed result is shown in white. Fig 7 shows the same result obtained on the scan of a phantom model with a very large bulge that causes the failure of some commercial centerline algorithms.

Another application where centerline extraction is fundamental is an emerging visualization modality used for polyp screening called virtual colonoscopy. Here the goal is to navigate inside a model of the colon lumen or a volume rendering of the colon after a CT scan on the patient, following the path corresponding to the skeleton of the colon. Fig. 8 shows an example of the reconstructed geometry of a colon with the three skeletons computed: the shortest path at lower resolution, the centered one and the snake smoothed one. Finally, we present the results obtained on the vascularization of liver from a CT scan (Fig. 9). This application can also become relevant for the planning of transplants. Also in this case the structure is well extracted, even if it the parameter tuning must be much more careful due to the complex structure and the quality of the contrast, not good as in the aneurysm scans.

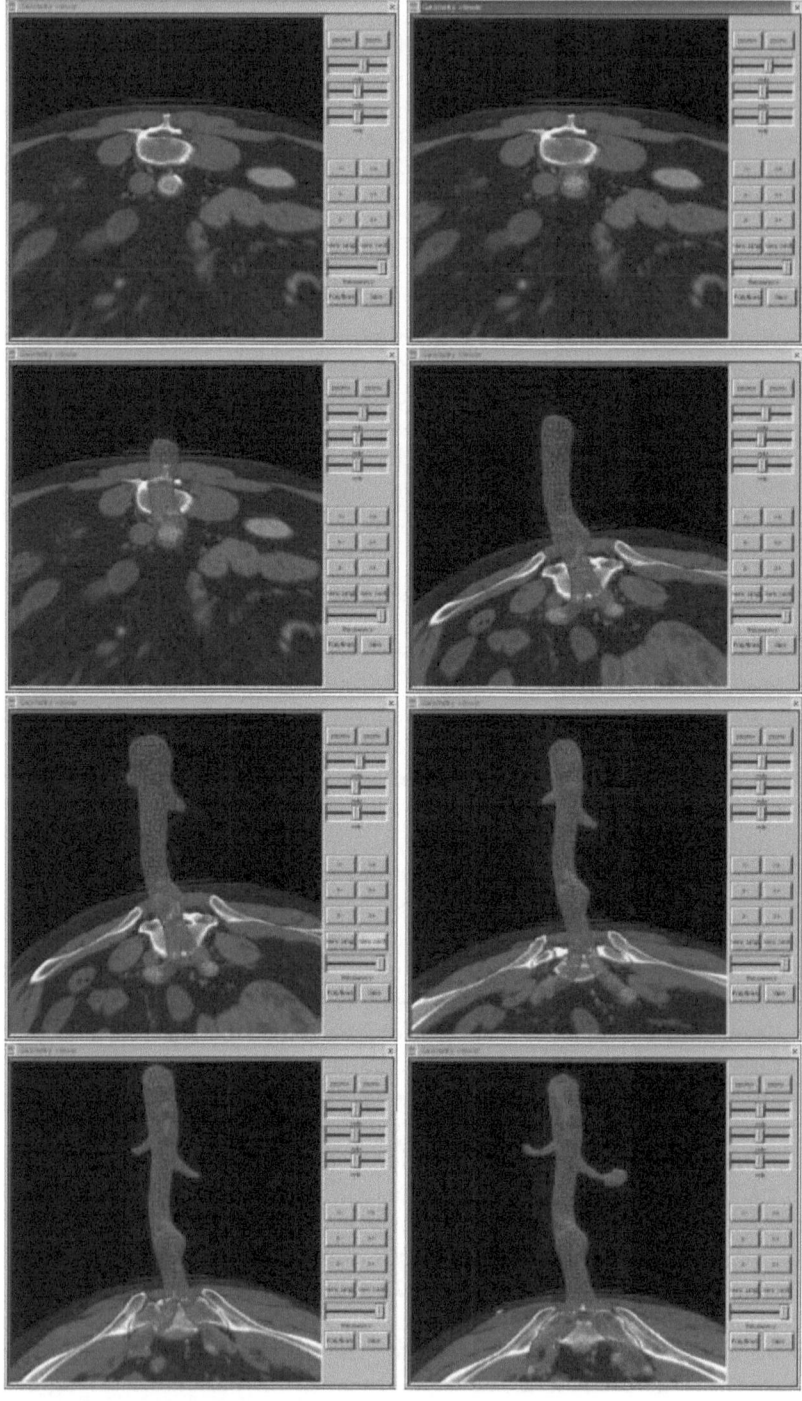

Fig. 5. Mesh growing inside the lumen of the aorta.

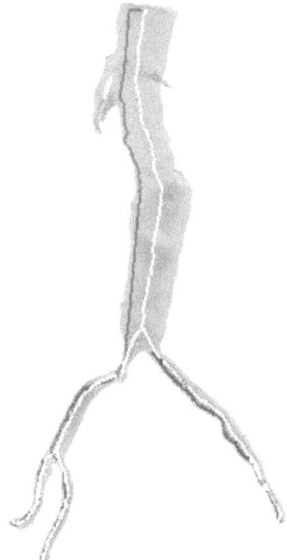

Fig. 6. Lumen and centerline computed on a CT scan of an abdominal aortic aneurysm of a real patient (courtesy of Radiology dept, Univ. of Innsbruck). The black line is the shortest path, the white one the final skeletal line extracted.

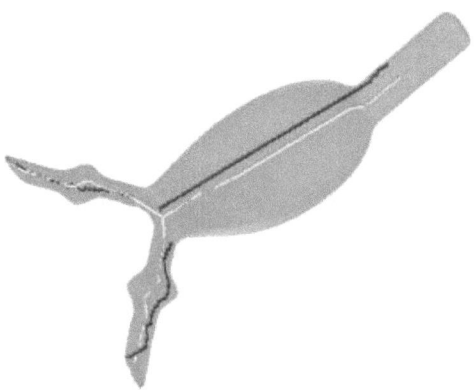

Fig. 7. Lumen and centerline computed on a CT scan of a plastic phantom (courtesy of Radiology dept, Univ. of Innsbruck). The black line is the shortest path, the white one the final skeletal line extracted.

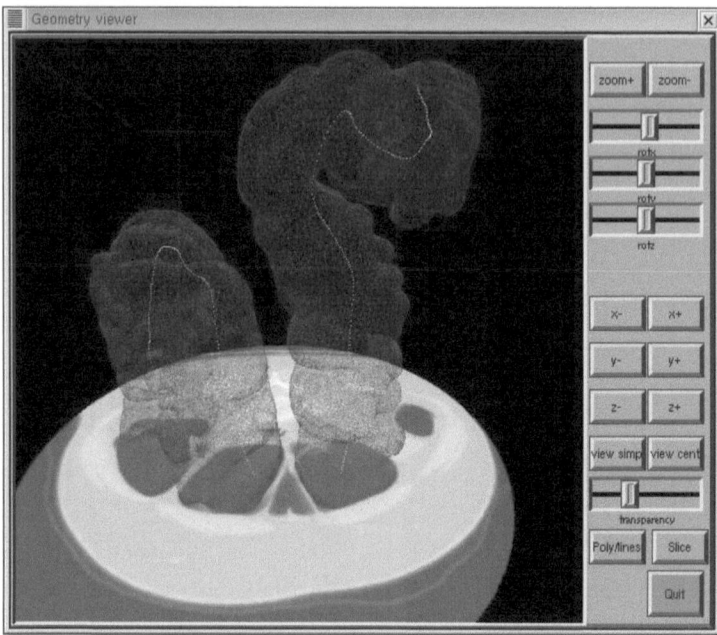

Fig. 8. Lumen and centerline computed from a CT scan of a air-inflated colon (courtesy of Radiology dept, Univ. Pisa.

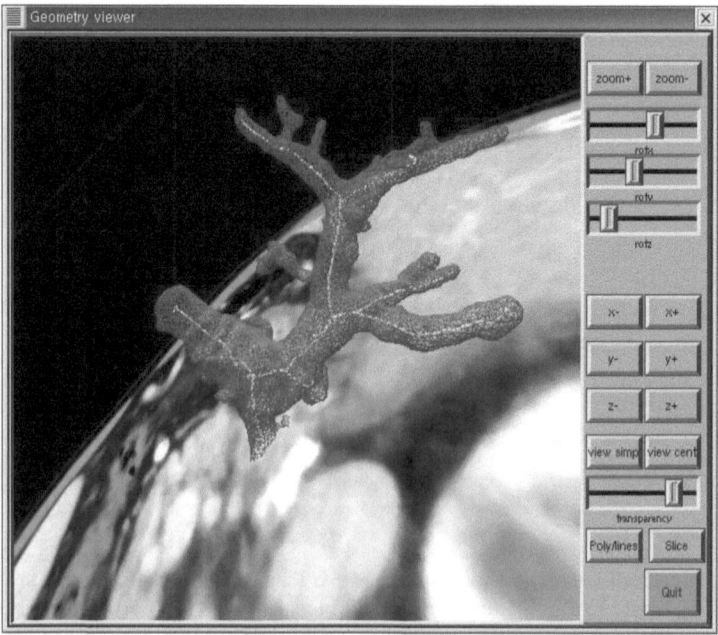

Fig. 9. Vessel surface and skeleton extracted from a liver CT scan (Courtesy of Hospital G. Brotzu, Cagliari.

4   Conclusions

We proposed a method to extract surfaces and skeletal lines of 3D structure with an approximately tubular structure. The method provides a reconstruction precise and reliable, but also flexible and suitable for different applications. These results were obtained within the framework of the European Union AQUATICS project, part of the Eutist-M cluster (EU-IST-1999-20226).

References

1. AQUATICS project web site: http://aquatics.crs4.it.
2. G.Abdulaev et. al.,"ViVa: The Virtual Vascular Project" IEEE trans. on Information Technology in medicine, 14: 1 34–48 (1998)
3. G. Bertrand and G. Malandain, "A new characterization of three dimensional simple points" Pattern Rec.Lett, 15,2 169–175 (1994)
4. D. Attali and A. Montanvert, "Computing and Simplifying 2D and 3D Semicontinuous Skeletons of 2D and 3D shapes" CVIU, Vol. 67, No. 3, September, pages 261–273, (1997)
5. T. K. Dey and W. Zhao, "Approximate medial axis as a Voronoi subcomplex" Proc. 7th ACM Sympos. Solid Modeling Applications, 356–366 (2002)
6. A. Verroust and F. Lazarus, "Extracting Skeletal Curves from 3D Scattered Data" In The Visual Computer . Vol 16. No 1, pp 15-25. 2000.
7. S.R. Aylward,"Initialization, Noise, Singularities and Scale in Height Ridge Traversal for Tubular Object Centerline Extraction" IEEE Trans. on Medical Imaging, 21,2 61–75 (2002)
8. L. Florez-Valencia, J. Montagnat, M. Orkisz, "3D Graphical models for vascular-stent pose simulation" proc. ICCVG 2002 Zakopane
9. A.F. Frangi et, al., "Quantitative Analysis of Vascular Morphology from 3D MR angiograms: in Vitro and In Vivo Results" Magn Res. in Med. 45,311-322 (2001)
10. F.K.H. Quek and C. Kirbas, "Vessel extraction in Medical Images by Wave Propagation and Traceback" IEEE Trans. on Medical Imaging, 2, 20 pp. 117–131 (2002)
11. W. E. Lorensen and H. E. Cline, "Marching cubes: A high resolution 3D surface construction algorithm," Proc. ACM Conference on Computer Graphics (SIGGRAPH'87), Anaheim, USA, pp. 163–169, July 1987.
12. T. Mc Inrey and D. Terzopulos, "Deformable models in medical images analysis:a survey" Medical Image Analysis, 1(2):91–108, 1996.
13. H. Delingette, "Simplex meshes: a general representation for 3d shape reconstruction", in CVPR94, pp. 856–859, 1994.
14. K. Krissian et al. "Model Based Multiscale Detection and Reconstruction of 3D vessels" INRIA Sophie Antipolis Report n.3442 (1998)
15. Montagnat, J, Delingette, H, and Ayache, N, "A review of deformable surfaces: topology, geometry and deformation," Image and Vision Computing, vol. 19, pp. 1023-1040, 2001.
16. Yong Zhou, Arthur W. Toga, Efficient Skeletonization of Volumetric Objects IEEE Transactions on Visualization and Computer Graphics 5:3 July-September 1999
17. M. Wan, F. Dachille, and A. Kaufman (2001) "Distance-Field Based Skeletons for Virtual Navigation," Visualization 2001, San Diego, CA, October 2001
18. D. Chen et al, "A Novel Approach to Extract Colon Lumen from CT Images for Virtual Colonoscopy," IEEE Transactions on Medical Imaging, Vol. 19, No. 12, December 2000, pp. 1220-1226.

19. I. Bitter, A. Kaufman and M. Sato(2001) "Penalized-Distance Volumetric Skeleton Algorithm," IEEE Transactions on Visualization and Computer Graphics, Vol. 7, No. 3, July-Sept. 2001, pp. 195-206
20. R.J.T. Sadleir and P.F. Whelan, "Colon Centerline Calculation for CT Colonography using Optimised 3D topological thinning" Proc. IEEE 3DPVT, pp.800-803 (2002)
21. H. Blum, "A tranformation for extracting new descriptors of shape, "Proc. Symp. Models for the Perception of Speech and Visual Form (W.W. Dunn, Ed.), MIT Press, Cambridge, MA, pp. 362-380, 1967
22. A. Kass, A. Witkin and D. Terzopoulos, "Snakes: Active contour models," Int. J. of Comp. Vision **1,** 321–331 (1988).
23. H. Bourquain et al., "HepaVision2 - a software assistant for preoperative planning in LRLT and oncologic liver surgery", Computer Assisted Radiology and Surgery (CARS), June 2002, pp. 341-346

Tetrahedral Mass Spring Model
for Fast Soft Tissue Deformation

Wouter Mollemans, Filip Schutyser,
Johan Van Cleynenbreugel, and Paul Suetens

Medical Image Computing (Radiology and ESAT/PSI),
Faculties of Medicine and Engineering, University Hospital GasthuisBerg,
Herestraat 49, B-3000 Leuven, Belgium

Abstract. Maxillofacial surgery treats abnormalities of the skeleton of
the head. Skull remodelling implies osteotomies, bone fragment repositioning, restoration of bone defects, inserting implants, ... Recently,
the use of 3D image-based surgery planning systems is more and more
accepted in this field. Although the bone-related planning concepts and
methods are maturing, prediction of soft tissue deformation needs further
fundamental research. In this paper we present a tetrahedral soft tissue
model that can be used in a surgery planning system to predict soft tissue changes due to skeletal changes. Our model consists of mass points
connected by springs. We propose a way to directly calculate the deformation of the model due to external changes. To achieve fast calculations
we take advantage of the fact that most deformations are local and we
compare our results with pre-computed reference models, to prove the
accuracy of our model.

1 Introduction

Maxillofacial surgery is an extremely challenging area of research combining
medical imagery, computer graphics and mathematical modelling. Since the human face plays a key role in interpersonal relationships, people are very sensitive
to changes to their outlook. Therefore planning of the operation and reliable prediction of the facial changes are very important. Various approaches have been
proposed for simulating deformable soft tissue.

The Finite Element Method (FEM) [1,2] is a common and accurate way to
compute complex deformations of soft tissue, but conventional FEM has a high
computational cost and large memory usage. This makes FEM models inap-
propriate for realtime simulation. Hybrid models based on global parameterized
deformations and local deformations based on FEM, have been introduced to
solve this problem [3,4]. Most of these methods, however, are only applicable
to linear deformations and valid for small displacements. Furthermore they rely
on pre-computing the complete matrix system and are therefore unable to cope
with topological changes when these occur during simulation.

Mass Spring systems (MSS) [5,6,7] are widely used to model deformable
objects. They are applied to a variety of problems, such as cloth modelling,

N. Ayache and H. Delingette (Eds.): IS4TM 2003, LNCS 2673, pp. 145–154, 2003.

facial animation or real-time deformation. All approaches use models consisting of mass points and springs. The dynamic behavior of these models is simulated by considering forces at mass points.

A mass-spring model assumes a discretization of the object into n points x_i with mass m_i. These points are linked by springs and dampers. The relation between position, velocity and acceleration for point x_i at time t is described by

$$m_i \frac{d^2 x_i(t)}{dt^2} + \gamma \frac{dx_i(t)}{dt} + F_i^{int}(t) = -F_i^{ext}(t) \tag{1}$$

with γ denoting a damping factor, $F_i^{int}(t)$ denoting the resulting internal elastic force caused by strains of adjacent springs and $F_i^{ext}(t)$ denoting the sum of external forces, such as gravity or collision reaction forces. In order to solve (1), it is reduced to two coupled first-order differential equations for each object point $x_i(t)$

$$\frac{dx_i(t)}{dt} = v_i(t) \tag{2}$$

$$\frac{dv_i(t)}{dt} = a_i(t, x_i(t), v_i(t)) = \frac{-F_i^{ext}(t) - F_i^{int}(t) - \gamma v_i(t)}{m_i} \tag{3}$$

whereas the new variable $v_i(t)$ corresponds to the velocity of point $x_i(t)$.

Given initial values for $x_i(t$, $v_i(t)$, $F_i^{int}(t)$ and $F_i^{ext}(t)$ at time t, various methods are commonly applied to numerically integrate through time.

Excellent results can be achieved by applying these numerical integration methods in order to animate deformable models. However, due to numerical problems and slow convergence, these approaches are not very well suited to estimate the rest position of mass spring systems. In maxillofacial surgery this rest position is more important than the exact animation. In [5], Teschner proposes an other approach to directly estimate the deformation, without calculating the animation.

Teschner uses a layered based model. This model consists of several layers with each the same topology: mass points connected by springs to form triangular structures. The successive layers are connected by a set of vertical and diagonal inter-layer springs. Nevertheless the good results that can be achieved with this model, obtaining such a layered model starting from volumetric data or surface meshes, is not straightforward and topology errors can occur.

In this paper we propose a new method that uses a tetrahedral based extended mass spring model and direct computation of the rest position. Due to the flexibility of the tetrahedral build up, our model can be used for almost any kind of shape.

2   Our Model

2.1   Model Components

The geometric model is a discretization of the tissue into n points that are connected to form a tetrahedral 3D mesh. The model is extracted from CT data by the Amira software[1]. In contrast to most mass spring models our extended model consists of three types of elements: points, springs and tetrahedra.

The tetrahedra are directly extracted from the tetrahedral 3D mesh. Mass points are set on the mesh nodes and linear springs on the mesh edges.

In our model we distinguish two types of points: joint points and free points. Free points are points of which the movement is completely determined by the resulting force active in the point. Joint points are the points of the soft tissue model that will be connected to the bone structure and are not able to move in free space; the position of the joint points is dependent on the position of the bone structure.

To make our model more realistic, we add a mass to each soft tissue point, whereby the effect of gravity can be simulated. A certain initial strain is also assigned to the springs that represent the skin surface. The purpose of this initial strain is to model the so-called skin turgor that is responsible for gaps in the tissue as a result of tissue cuts.

2.2   Direct Computation of Deformation

Due to the existence of free and joint points, the displacement of some of the joint points, while keeping the others fixed, will result in object deformation. Transformation and deformation can be described as a displacement of soft tissue points. For each soft tissue point $p \in \Re^3$ an new position $p' \in \Re^3$ is found. The displacement d can be found by minimizing the force in each mass point.

The force at soft tissue point p_i due to all forces caused by springs connected to this node, is referred to as internal force F_i^{int}:

$$F_i^{int} = \sum_{\forall p_j \in S_i} k_j * (\mid p_j - p_i \mid -L_j) * \frac{p_j - p_i}{\mid p_j - p_i \mid} \tag{4}$$

with S_i denoting the set of all springs connected to p_i, L_j and k_j denote the initial length and the spring constant of spring j, respectively.

Additional forces at soft tissue point p_i caused by external pressure due to for example surgical instruments are referred to as external force F_i^{ext}.

The overall force F_i at soft tissue point p_i consists of internal components F_i^{int} and external components F_i^{ext}:

$$F_i = F_i^{int} + F_i^{ext} \tag{5}$$

Consider P_{free} as the collection of the coordinates of all free soft tissue points and $f_{force}(P_{free})$ as the sum of all magnitudes of the forces at these points. This function is parameterized by the positions of all free soft tissue points. The new position of the free points after displacement of some of the joint points, will be given by

$$P'_{free} = \text{argmin} f_{force}(P_{free}) \tag{6}$$

---

[1] http://www.amiravis.com

Minimizing $f_{force}(P_{free})$ corresponds to varying the positions of the free soft tissue points in order to neutralize internal forces F_i^{int} and external forces F_i^{ext} in each free point. Joint points are bound to bone patches and are not considered in the optimization process. Therefore joint points are not displaced by the optimization process and forces will not be neutralized in joint points.

To solve the optimization problem, best results were achieved with following algorithm: Suppose that k_{tot}^{min} is the minimal spring constant of all spring constants, then:

1. Acquire the maximal force over all free points.
2. Repeat until $F^{max} < k_{tot}^{min} * 0.1$:
   (a) Determine the force in point i.
   (b) Update position of point i: $x_i^{new} \leftarrow x_i^{old} + \frac{f_i^{tot}}{k_i^{mean} * N_i}$ with k_i^{mean} the average spring constant of the springs connected to point i and N_i the number of springs connected to point i.
   (c) Define the new maximum force

The calculation speed can be seriously improved by implementing a dynamic cut-out. In each iteration there are a lot of points in which the resulting force is very small. More over the forces of point i at the start of iteration j can only have been change when point i or a neighboring point has been moved during iteration $j - 1$. This is why the following to improvements are made. First of all, only points with a resulting force larger than one tenth of maximum force are considered in each iteration. Second, only points which are neighbor of a point that is moved in last iteration have to be evaluated in next iteration.

2.3   Volume Conservation

Simulating volume conservation with a layered model or a classical mass spring system is certainly not straightforward. With a classical mass spring system only forces in the direction of springs can be simulated, while maintaining a constant volume rather requires radial forces. By extension of the standard mass spring system with tetrahedral elements, we can simulate the effect of volume conservation.

To ensure this volume conservation we define a *volume force* in each point. These forces are related to the change in volume of the tetrahedra containing the point. The direction of the volume force must be somehow radial. Like suggested in [6] we let the volume force point to the barycenter of the tetrahedron. Let us define p_{Bi} as the barycenter of tetrahedron i, then the volume force in point j is given by,

$$F_j^{vol} = \sum_{\forall i \in \Omega_j} (V_i^{cur} - V_i^{init}) \frac{p_j - p_{Bi}}{|p_j - p_{Bi}|} \tag{7}$$

$$p_{Bi} = \frac{1}{4} \sum_{\forall j \in T_i : j = 1}^{4} x_j \tag{8}$$

with Ω_j the collection of all tetrahedra containing point j, V_i^{cur} and V_i^{init} the current volume and initial volume of tetrahedra i respectively. Total force in point j will now be given by

$$F_j = F_j^{int} + F_j^{ext} + \alpha * F_j^{vol} \tag{9}$$

The value $\alpha \in \Re$ weights the influence of the elastic deformation part and the volume conservation part. As will be shown good results can be achieved by application of these volume forces.

2.4   Model Initialization

Modelling a Given Stress-Strain Relationship. The parametrization of spring constants of a soft tissue model provides a certain stress-strain relationship. In order to meet a given stress-strain relationship this parametrization has to consider the geometry of the model, the number of mass points and their interconnection.

In [8], Van Gelder suggested a formula to compute spring stiffness for a 3D mesh that is closest to an elastic continuous representation. Let E_i be the local material elastic modulus, then spring constant of spring i is given by,

$$k_i = \frac{E_i * \sum_{\forall j \in \Omega_i} V_j}{l_{0i}^2} \tag{10}$$

with Ω_i the collection of all tetrahedra containing spring i and l_{0i} rest length of spring i.

Mass Initialization. Calculation of masses in each model point can been done by methods based on Voronoi zones [9]. To speed up our program we use a simplified method that presumes that the mass of a tetrahedron is equally divided among his vertices. The mass m_i of model point i is thus estimated as:

$$m_i = \sum_{\forall j \in \Omega_i} \frac{1}{4} \rho_j V_j \tag{11}$$

with Ω_i the union of all tetrahedra that contain point i, ρ_j the local density of the tissue and V_j the volume of tetrahedron j.

Skin Turgor. If soft tissue is cut, it commonly deform and a small gap is formed at the cutting line. If all springs of the soft tissue model were relaxed in the initial state, cutting the model would cause nothing to happen.

In order to simulate the skin turgor, a certain initial strain is applied to springs that represent the skin surface [5]. The initial strain is realized by setting the initial length of all springs i that represent the skin surface to $\lambda_{turgor} * l_{0i}$ with $0 < \lambda_{turgor} < 1$.

Initial Spring Length Adaptation due to Mass and Skin Turgor. Since mass and skin turgor are added to the initial geometry, there are forces at each soft tissue point which can deform this initial model. In order to prevent this, these forces have to be neutralized. This is done by determining an initial strain for each spring for which the position of all free soft tissue points doesn't change. Initial strain is adapted by varying the initial length of the springs.

We define $f_{turgor}(l_{00}, l_{01}, l_{02),...}$ as the function that calculates the sum of all internal and external forces of all free tissue points and that is parameterized by initial spring lengths l_{0i}. We then find the new initial spring lengths by minimizing f_{turgor}.

$$(l_{00}, l_{01}, l_{02}, \ldots) = \mathrm{argmin}(f_{turgor}(l_{00}, l_{01}, l_{02}, \ldots)) \qquad (12)$$

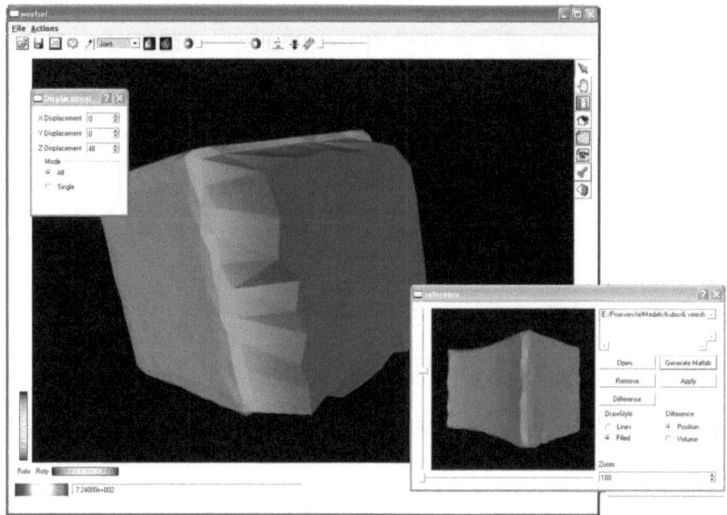

Fig. 1. The simulation environment showing a cubic representation of soft tissue. Points are defined as free or joint by using a drawing tool. Users can displace the joint points and calculate the deformed cube. The result can be compared with a FEM reference model shown in the lower right corner.

3   The Simulator

A basic core of functionality is currently available in the simulator. It is under continuous development. In this section we give an overview of this functionality. The main viewer is based on the Open Inventor[2] libraries. The simulator is shown in figure 1 where a cubic piece of soft tissue is loaded.

---

* http://oss.sgi.com/projects/inventor

Table 1. Test results for different deformations, showing the displacement vector, the average error and maximum error compared to the FEM reference model, the number of iterations and time needed to achieve this rest position.

Displ. vector	Av. error (mm)	Max. error (mm)	Num. Iter.	Time (s)
$(1,0,0)$	0.04	0.19	186	113
$(2,0,0)$	0.06	0.44	227	164
$(3,0,0)$	0.06	0.34	295	183
$(4,0,0)$	0.08	0.37	277	179
$(5,0,0)$	0.10	1.0	259	173
$(6,0,0)$	0.13	1.2	262	182
$(4,4,0)$	0.15	0.71	488	326
$(6,6,0)$	0.21	0.91	540	378
$(0,4,0)$	0.12	0.81	238	167
$(0,5,0)$	0.11	1.2	329	187
$(0,0,-3)$	0.12	0.65	329	187

3.1   Point Definition

As mentioned in 2.1 we distinguish two types of points: free and joint points.

At startup all points are set to be free points. By using a simple drawing tool, the user can interactively determine which points have to be free and which have to be joint points. The drawing tool is illustrated in figure 1. Light grey represents free points, while dark grey points are defined as joint points.

3.2   Deformation

Now the model is completely initialized and deformations of the model can be calculated. A deformation is realized by adding external forces or by movement of some joint points. The new rest position of the model can be calculated.

3.3   Performance Evaluation

The simulator runs on a 2GHz processor of a Windows XP workstation with 1 GB Ram. To quantitatively evaluate our result we compare the deformed object with a reference model that is the result of a FEM based calculation of the same deformation. The FEM algorithm is known to be accurate and correct. Because FEM and our MSS model starts from the same geometric model, we know the point correspondence between both models. This gives us the ability to precisely estimate the accuracy of our computed deformation. Two main tools are available for evaluation of the result. The program can generate a matlab file, with quantitative data about the difference between the calculated and the reference model. Second we can visually inspect the differences between both models by use of a color code.

For evaluation we created a rectangular mesh which consists of 1402 nodes, 6387 tetrahedra and 13374 faces. The cube has dimensions $20mm*20mm*20mm$. Like listed in table 1 we applied a number of various deformations to the cube.

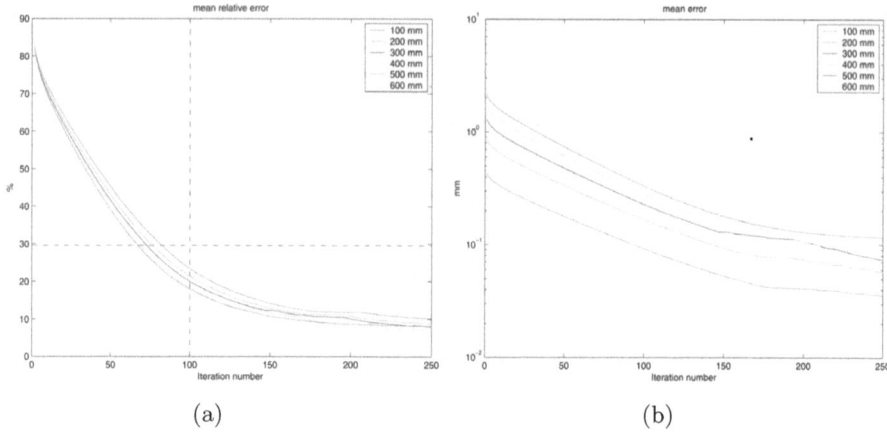

(a)   (b)

Fig. 2. (a) Average relative error and (b) absolute error for different elongations in direction the of x-axis.

For each deformation we displaced one side of the cube with one of the vectors listed in table 1. The table shows average and maximum error in relation to the FEM reference model, number of iterations and time needed to achieve this rest position.

Figure 2 shows average error and average relative error in respect to the number of iterations for different elongations in direction of the x-axis. We see that errors increase with increasing elongation, but after large number of iterations average error can be made sufficient small. Both average relative error and average error have a clear logarithmic slope. The logarithmic slope has as advantage that we can get a pretty good idea of what the displacement of all points will be in a relative small number of iteration. If the deformation is satisfactory we can get a more accurate estimation. As can be seen in figure 2 the average relative error can be reduced to 30% in less than 100 iterations.

As shown in table 1 for a big deformation the maximum error can exceed $1mm$. But out of the visual inspection we could clearly state that differences between FEM and MSS only appears at the edges and vertices of the cube. In the main bulk all differences were smaller than $0.3mm$. This can be seen in figure 3 where the result of elongation with $5mm$ in x direction is shown. Bright points represent the largest differences between the FEM reference model and our computed result, dark points represent places with smallest distance between both models.

The effect of the volume forces can clearly been seen in figure 4 At the left we see the result after elongation with $5mm$ in y-direction without volume forces at the right with volume forces. In respect to the initial cube we recorded a volume difference of 14.5% when no volume forces were used, while with volume forces the volume difference was reduced to 7.3%. When we compared the volumes of the cubes computed with FEM or MSS, less than 0.5% volume difference was seen between both methods. Even when we applied a displacement of the joint points so that the total volume changed with more than 100%, the volume

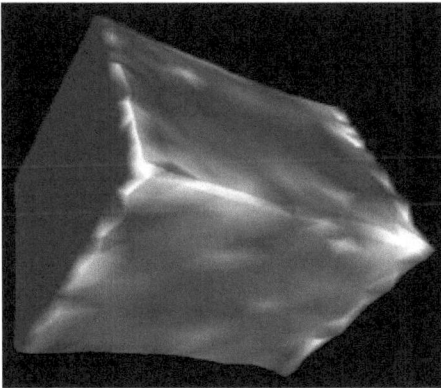

Fig. 3. Visual inspection of the difference between our calculated result and the FEM reference model. Bright points represent the largest differences between FEM and our result, dark points represent places with smallest distance between both models.

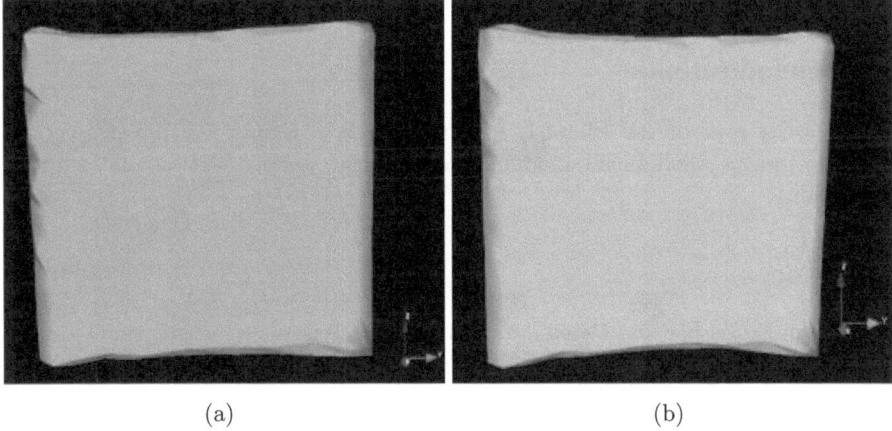

(a)   (b)

Fig. 4. Result of elongation with 5 mm in y direction. (a) Calculated without volume forces, (b) calculated with volume force.

difference between the computed deformation and the initial cube was less than 20%. Application of the volume forces, doesn't only reduce volume differences but also reduce the mean error between both methods. For all deformation a reduction of the mean error with 50% was achieved.

4   Conclusion

At the moment we have set the first step in creating a simulation environment for soft tissue changes. We have proposed a new extended mass spring model based on a tetrahedral geometry, with direct computation of soft tissue changes. In our application field these deformations are more important than the correct animations. Skin turgor and mass are added to increase the realism of our model. We implemented a method to simulate the effect of volume conservation.

Based on this new model we have created a simulation environment to predict soft tissue deformations due to surgical intervention. User interactivity and the speed of calculations are key issues for this simulator.

Comparison between our results and FEM reference models have shown that our method is accurate and gives a close approximation of the FEM based result. Volume conservation was achieved for different deformations, with a clear visible effect. Because of the iterative character of our method, a global idea of what the deformation will be, can be shown in a small amount of time. When the result is satisfactory a more accurate result can be calculated.

Now we are able to simulate the deformations of basic geometric shapes, we will start to work on patient specific data. Once we can simulate these soft tissue deformations caused by surgical interventions, we will set up an evaluation procedure for our model. Based on a set of pre-operative and post-operative patient data, we will evaluate our predictions. These results will provide further information to improve our model.

Acknowledgments

This work is part of the Flemish government IWT GBOU 020195 project on Realistic image-based facial modeling for forensic reconstruction and surgery simulation.

References

1. Gladilin, E., Zachow, S., Deuflhard, P., Hege, H.: Towards a realistic simulation of individual facial mimincs. In: SPIE Medical Imaging Conference, San Diego (2002)
2. Nebel, J.C.: Soft tissue modelling form 3d scanned data. (2001)
3. Ramanathan, R.: Dynamic deformable models for enhanced haptic rendering in virtual environments. In: IEEE Virtual Reality. (2000) 31–35
4. Muller, M., Dorsey, J., McMillan, L., Jagnow, R., Cutler, B.: Stable real-time deformations. In: ACM SIGGRAPH Symposium on Computer Animation (SCA). (2002) 49–54
5. Teschner, M.: Direct Computation of Soft-Tissue Deformation in Craniofacial Surgery Simulation. PhD thesis, Der Technischen Fakultat der Friedrich-Alexander-Universitat, Erlangen (2000)
6. Bourguignon, D., Cani, M.P.: Controlling anisotropy in mass-spring systems. In: Proceedings of the 11th Eurographics Workshop on Animation and Simulation, Springer-Verlag (2000) 113–123
7. Brown, J., Sorkin, S., Bruyns, C., Latombe, J.: Real-time simulation of deformable objects: Tools and application. In: Proc. Computer Animation, Korea (2001)
8. Gelder, A.V.: Approximate simulation of elastic membranes by triangulated spring meshes. Journal of Graphics Tools **3** (1998) 21–41
9. Deussen, O., Kobbelt, L., Tucke, P.: Using simulated annealing to obtain good nodal approximations of deformable objects. In: Proceedings of the 6th Eurographics Workshop on Animation and Simulation, Springer-Verlag (1995)

Surgery Simulation System
with Haptic Sensation and Modeling of Elastic Organ
That Reflect the Patients' Anatomy

Naoki Suzuki and Shigeyuki Suzuki

Institute for High Dimensional Medical Imaging, The Jikei University School of Medicine,
4-11-1 Izumihoncho Komae-shi, Tokyo, 201-8601, Japan
{nsuzuki,sshige}@jikei.ac.jp

Abstract. Surgery simulation is one of the largest applications of Medical Virtual Reality. In order to perform the real-time simulation, we constructed an elastic organ model known as a sphere-filled model. This proposed organ model allows us to perform surgical maneuvers such as pushing, pinching and incising and show the deformation of the inner structures such as blood vessels on our system. In addition, we tried to obtain haptic sensation with the patients' organs in a surgical simulation. Developed system made it possible to handle elastic organs with two force feedback devices attached to the both of users' hands. At the same time, we have been developing a VR cockpit suited for virtual surgery and tele-surgery. Using our VR cockpit, our system allows us to provide the users with an environment closely resembling the open surgery situation.

1 Preface

Three dimensional images reconstructed from MRI or CT images are currently used in various applications not only for diagnosis but also treatment [1]-[3]. The application of virtual reality (VR) techniques to 3D images has a large potential to provide future medical treatments through tele-diagnosis and tele-medicine. Surgical simulation systems also have useful applications in the medical field [4]-[12]. By using such a surgical simulator, surgeons are able to go through various approaches to urgent operations and carefully plan the procedures beforehand. It is also especially useful for experiencing operations for deformities or those with some expected difficulties. Such systems also allow the user to repeatedly perform virtual surgery until a suitable procedure is established. However, in general, it is difficult to manipulate, deform and incise volumetric organ models as one would on elastic objects in real-time, and surgeons require more realistic simulation involving tactile sensation. In this paper we would like to describe our course of the development of a surgical simulation system that has resolved such difficulties.

2 Elastic Model for Surgical Simulation

We have been developing a virtual surgery simulation system so that doctors can conduct trial runs of operations or to master surgical techniques [7]-[10]. With this

N. Ayache and H. Delingette (Eds.): IS4TM 2003, LNCS 2673, pp. 155–164, 2003.

system, we have aimed to develop a virtual surgery system capable of performing surgical maneuvers on elastic organs, the structure of which having been obtained from a patient. Specifically, we constructed an elastic organ model known as a sphere-filled model for real-time simulation. The sphere-filled model consists of a group of small spheres inside and polyhedrons at the surface. If an external force is applied to this model, it is deformed by the movement of each of the spherical elements. Each sphere elements is pressed and displaced in turn by measuring its distance from the neighboring sphere elements. In addition to that, the gravity parameter and the relationship between spheres relative to each other also determine the displacement of the spheres. And the polyhedrons at the surface move because they are linked to the internal spheres. In short, the sphere-filled method expresses the organ deformation according to the sphere's behavior. Fig.1 shows the organ model's deformation at three points as if being pushed by three fingers. In this image, three white spheres show the thumb, forefinger and middle finger. Thus, This simple method is suited for simple deformations in which amputations of internal structures don't occur, such as with pushing and pinching because these deformations express each sphere's reciprocal force.

This simple method is problematic for the complicated incisions that cause amputation of internal structures. To counteract this problem, we modified the model by using a revised algorithm for organs. The incision procedure for the modified model has four steps roughly. First, the model creates the incision surface by getting the position of the scalpel in virtual space. The force feedback device decides its position. Then the incision surface divides internal spheres to the left and right of it. In short, an incision divides the internal spheres into two regions. We called the process "Internal Division". Then, an incision causes the spheres to change their combination and displace around the opening. The displaced spheres in turn act upon the adjacent spheres. Finally, the organ model is distorted by the spheres' movement. In this method, the opening shape of the incision surface depends on the number of spheres displaced on the surface. This incision procedure allows the model to deform for a lot of incisions by repeating "Internal Division". Fig.2 shows an incision deformation. Fig.2a shows the deformation of the surface while Fig.2b shows the internal spheres. The spheres shown in color are the divided spheres by the "Internal Division". Fig.3a1 and b1also shows the deformation of an incision. Fig3a1 shows the condition of the surface while Fig3b1 shows the internal spheres. Fig.3a2 and b2 shows the model creates the resection surface after making some incisions. Fig.3a3 and b3 shows the grasping of the resection region. In this resection, the volume of the organ remained of 71.2 percent. Thus, the sphere-filled model allows our system to enable the execution of various incisions in addition to basic maneuvers such as pressing.

3   Deformation of Volumetric Data

Up to now the main type of elastic organ model for surgical simulation has been a surface model that keeps the surface data but cannot hold the internal structure data due to the decreased quantity of calculation. We have tried to deal with volumetric data itself acquired with MRI or CT as an elastic body and to realize volumetric data deformation in real-time like our surface organ model [12]. By assigning the voxel group to internal sphere it follows the motion of each sphere, and as a result the

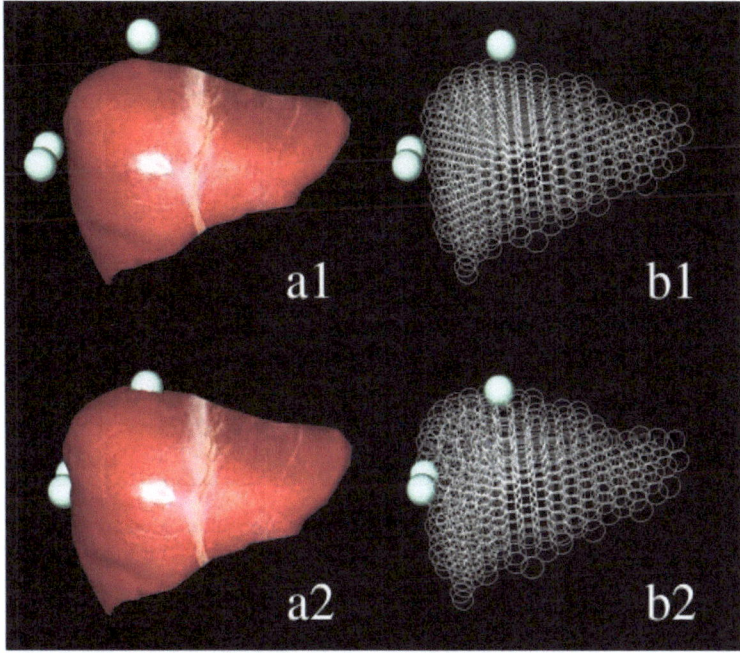

Fig. 1. Deformation of the liver model as if being pushed by three fingers. Three white spheres shows thumb, forefinger and middle finger.

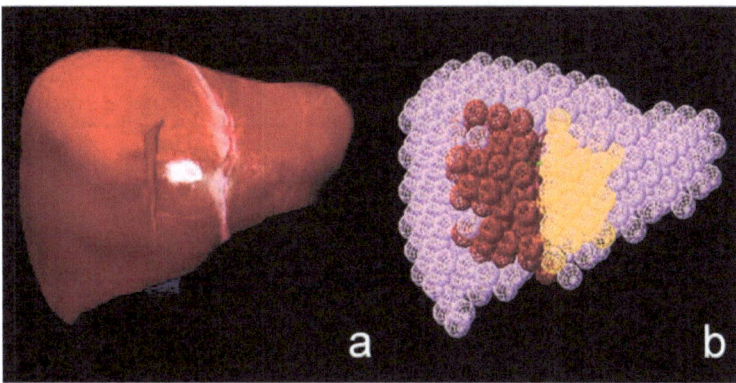

Fig. 2. Incision deformation of the liver model. a. The deformation of the surface b. The condition of the internal spheres. The spheres shown in color are the divided spheres.

volumetric data is deformed. We divided volumetric data among elements and made an elastic organ model by using the abdominal data set (512x512x176) obtained by MRI with a 1mm slice/pitch. After that, we assigned the voxel group to the spheres. Regarding the formation of the voxel group, it is adopted as a rhombus dodecahedron. The reason is that the formation does not overlap spatially with others and as a result the internal sphere in connection with each voxel is decided automatically. Fig.4 shows the relative location between the internal sphere and the rhombus dodecahe-

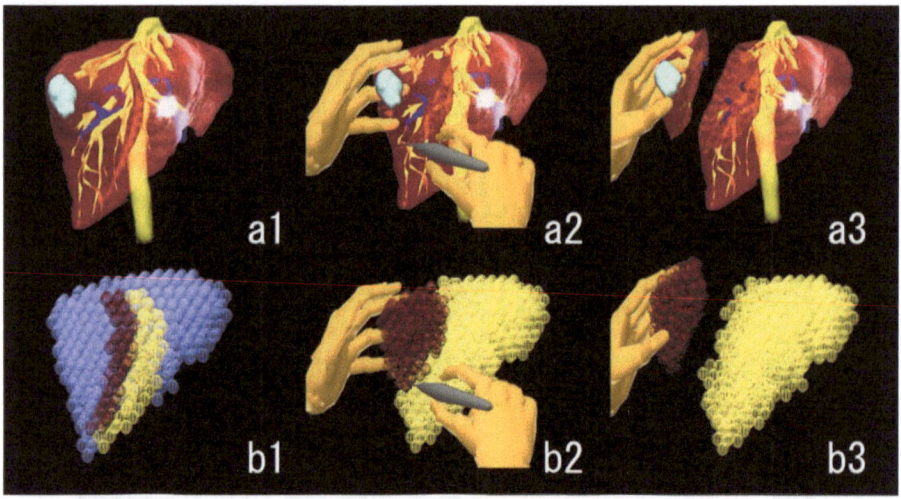

Fig. 3. The deformation of our organ model. a1 and b1 shows an incision deformation while a2 and b2 shows the generation of the resection surface. a3 and b3 shows the grasping of the resection region using user's left hand.

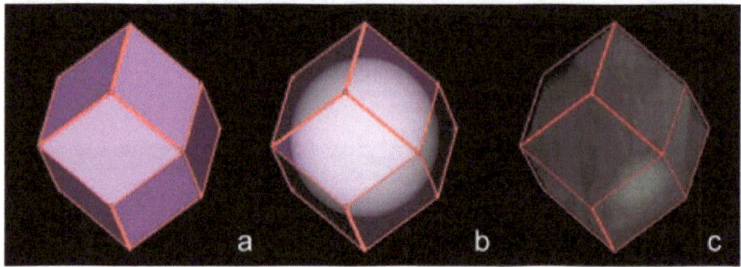

Fig. 4. The formation of voxel group assigned to internal spheres. a The rhombus dodecahedron. b. The transparent rhombus dodecahedron with a sphere. c. A voxel group in the shape of the rhombus dodecahedron.

dron. The formation makes the processing of the deformation easier because the shape of the voxel group is entirely the same. If the radius of the sphere is 8mm, about 20,000 voxels are assigned to the spheres.

Fig.5 shows the volumetric liver model being deformed by basic surgical maneuvers such as pushing and incising. Fig.5a shows the deformation of the pushing while Fig.5b shows the deformation of the incision. We have established that the frame rate of the deformation of the volumetric liver model is about 5 to 10 frame/sec on a graphic workstation (Onyx3400, Silicon Graphics Inc.). This frame rate was realized in all the surgical maneuvers we have developed with the force feedback device.

4   Haptic Devices

We have developed a haptic device which allows the user to experience tactile sensations [9]-[11]. We also have been developing a force feedback device which possesses 16 degrees-of-freedom (DOF) for manual interactions with virtual environments. The features of the device manufactured for our virtual surgery system can be summarized as follows.

1) The force feedback system is composed of two types of manipulators: a force control manipulator and a motion control manipulator.
2) Three force control manipulators for each hand are attached to the end of the motion control manipulator.
3) Both ends of each force control manipulator are attached to the thumb, forefinger, and middle finger of the operator.
4) The force control manipulator has a joint structure with minimal inertia and little friction.
5) The motion control manipulator has a degree of mechanical stiffness.

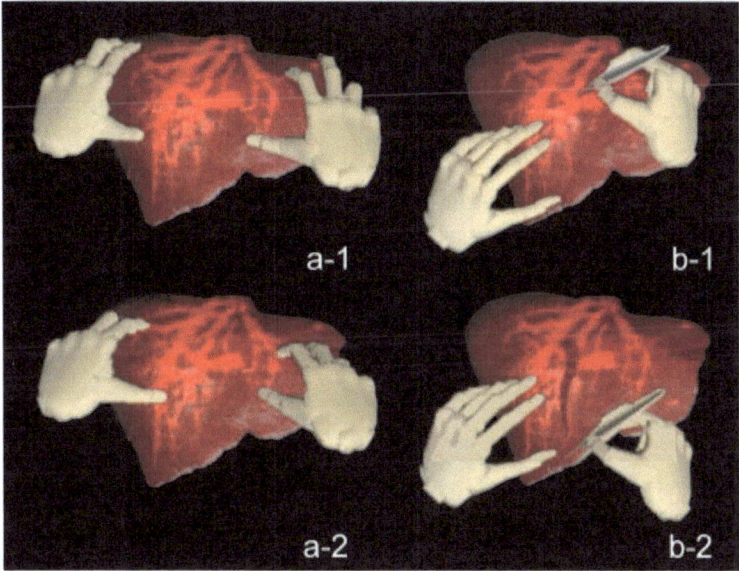

Fig. 5. The deformation of volumetric liver model by basic surgical maneuvers a pushing, b. incising.

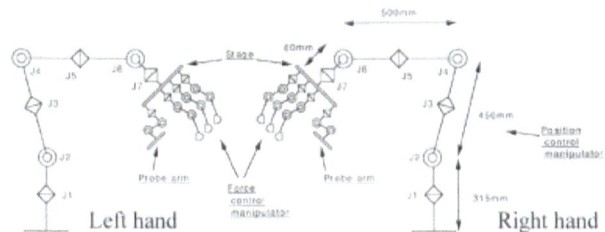

Fig. 6. Block diagram of the dual force feedback devices

Fig.6 shows the block diagram of the force feedback device. Both the right and left force feedback devices have the same internal structure and functions. The force feedback device for the right hand is a mirror image of the left one. On both hands,

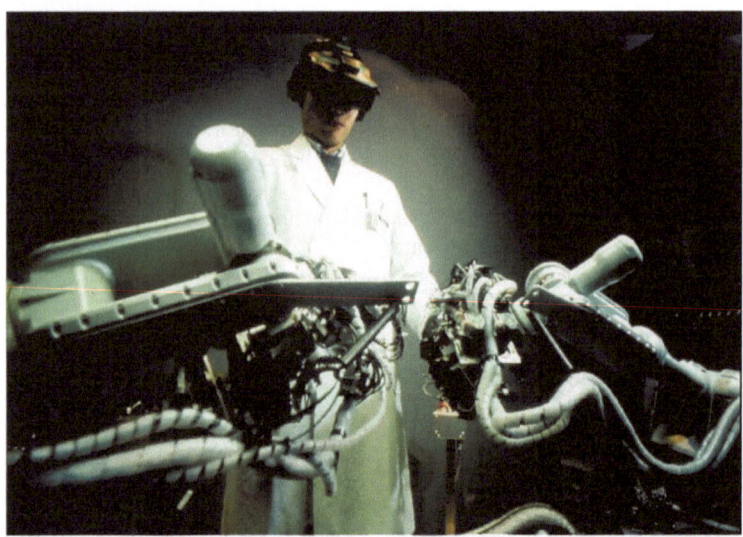

Fig. 7. A demonstration of the force feedback devices for the right and left hands.

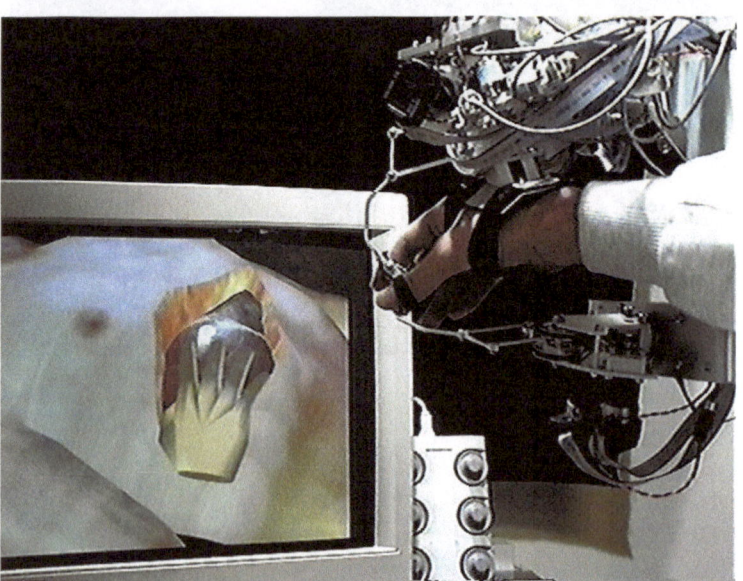

Fig. 8. An experiment to touch the surface of the liver using the manipulator with the 3D image.

the force control manipulator producing haptic sensation is attached to the thumbs, forefingers and middle fingers of the right and left hands respectively. Fig.7 shows a user and the devices attached to both hands. These devices communicate data (finger location etc.) with the surgical simulation system through a LAN. When an interaction occurs between the user's fingers and a 3D object in virtual space, a force parameter of tactile sensations calculated by the surgical simulation system is transferred to these devices. This allows the user to experience tactile sensations in each finger. Fig.8 shows the linkage between the device and the real time image. The user's hand is perceived as a 3D image in order to identify its location as it comes in contact with the liver surface. When the liver surface was compressed on the image, the user's finger corresponding experienced tactile sensations. Fig.9 shows sequential images of a situation in which the surgeon pinches the edge of the elastic incision plane on the liver.

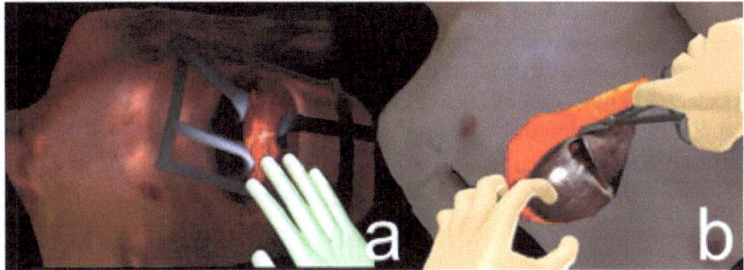

Fig. 9. Sequential images of the surgeon pinching the edge of the elastic incision plane.

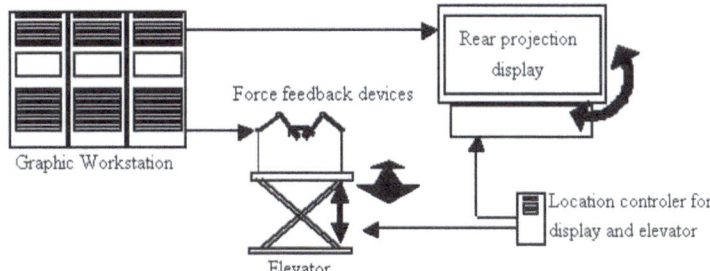

Fig. 10. The system component of our VR cockpit. The projection display's angle of elevation and the elevator position is controlled by the location controller.

5 VR Cockpit

In the case of an ordinary simulator, the user cannot immerse himself into virtual surgical space because of the method of presentation of the simulation results using conventional computer displays. To perform a surgical simulation and tele-surgery in the same environment as open surgery relative to the user's viewpoint and the operation, we have been developing the VR cockpit [13]. Fig.10 shows the system component. Our VR cockpit is composed of a large projection display with a high resolution and a large field of view, an elevator on which the user stands, force feedback devices (CyberForce, Virtual Technologies Inc.) and a graphic workstation (Onyx3400 Infinity Reality3 3-pipes, Silicon Graphics Inc.). To perform the surgical simulation in the same environment as open surgery reflecting the surgeon's viewpoint and the operation, an angle of elevation of this display can be changed from 0 degrees to 45 degrees, and the elevator that is equipped with force feedback devices can be shifted up and down or back and forth. Furthermore, by enabling the stereo view, this cockpit has the function in which users are provided with the spatial conditions of the target organ and the field of operation includes the 3D shape and the 4D movements. This cockpit is also equipped with two PDP in order to encourage communication between surgeons and to display the condition of the patient, such as vital data, for the tele-surgery. Fig.11 shows a general view of this cockpit. Fig.11a shows the situation of the open surgery simulation. Fig.11b shows the initial position of the screen and the elevator while Fig.11c shows their maximum positions. A surgeon attaches the force feedback device to both of his hands and stands on an elevator as shown in Fig.11d.

Using our cockpit, we can perform surgical simulations in an environment that includes the whole field of vision and uses highest quality graphics. Furthermore, the user can select an environment resembling open surgery because they can change the viewpoint and the position of both hands easily.

6 Conclusions

Here we have presented our surgical simulation system that can perform real-time and realistic simulation with tactile sensations including deformation of the inner structures such as blood vessels and tumors. By utilizing the advantages of the surface model and volumetric model, it is possible to simulate the target organ in detail. In addition, using our VR cockpit, we can perform the simulation in an environment that includes the whole field of vision and uses the highest quality graphics.

In the future, we will be able to simulate in accordance with actual surgical techniques and apply the simulation to the clinical situation. Moreover, we will be able to realize the development of tele-surgery system and robotic surgery system.

References

[1] JM. Rosen, H.Soltanian, RJ.Redett, D.R.Luab, "Evolution of Virtual Reality," IEEE Engineering in Medicine and Biology, 1996, pp.16-22.
[2] Robb RA, Camerson B. "Virtual Reality Assisted Surgery Program," Interactive Technology and the New Paradigm for Healthcare, Eds. R.Satava, et al., Vol.18, 1995, pp.309-321.
[3] A.Norton, G.Turk, B.Bacon, J.Gerth, and P.Sweeney. "Animation of fracture by physical modeling," The Visual Computer, Vol.7, 1991, pp.210-219.
[4] Robb RA, Hanson DP. "The ANALYZE software system for visualization and analysis in sugery simulation," Computer Integrated Surgery, Eds. Steve Lavalle, Russ Taylor, Greg Burdea and Ralph Mosges, MIT Press, 1995, pp.175-190.
[5] H.Delingette, "Simplex Meshes: a General Representation for 3D Shape Reconstruction," Technical Report 2214, INRIA, Sophia-Antipolis, France, 1994.
[6] S.Cotin, H.Delingette, J.Marescaux, "Geometric and Physical Representations for a Simulator of Hepatic Surgery," MMVR4, 1996, pp.139-151.
[7] T.Ezumi, N.Suzuki, A.Takatsu, T.Kumano, A.Ikemoto, Y.Adachi, A.Uchiyama. "An Elastic Organ Model for Force Feedback Manipulation and Real-time Surgical Simulation," International conference on artificial reality and telexistence, 1997, pp.115-121.

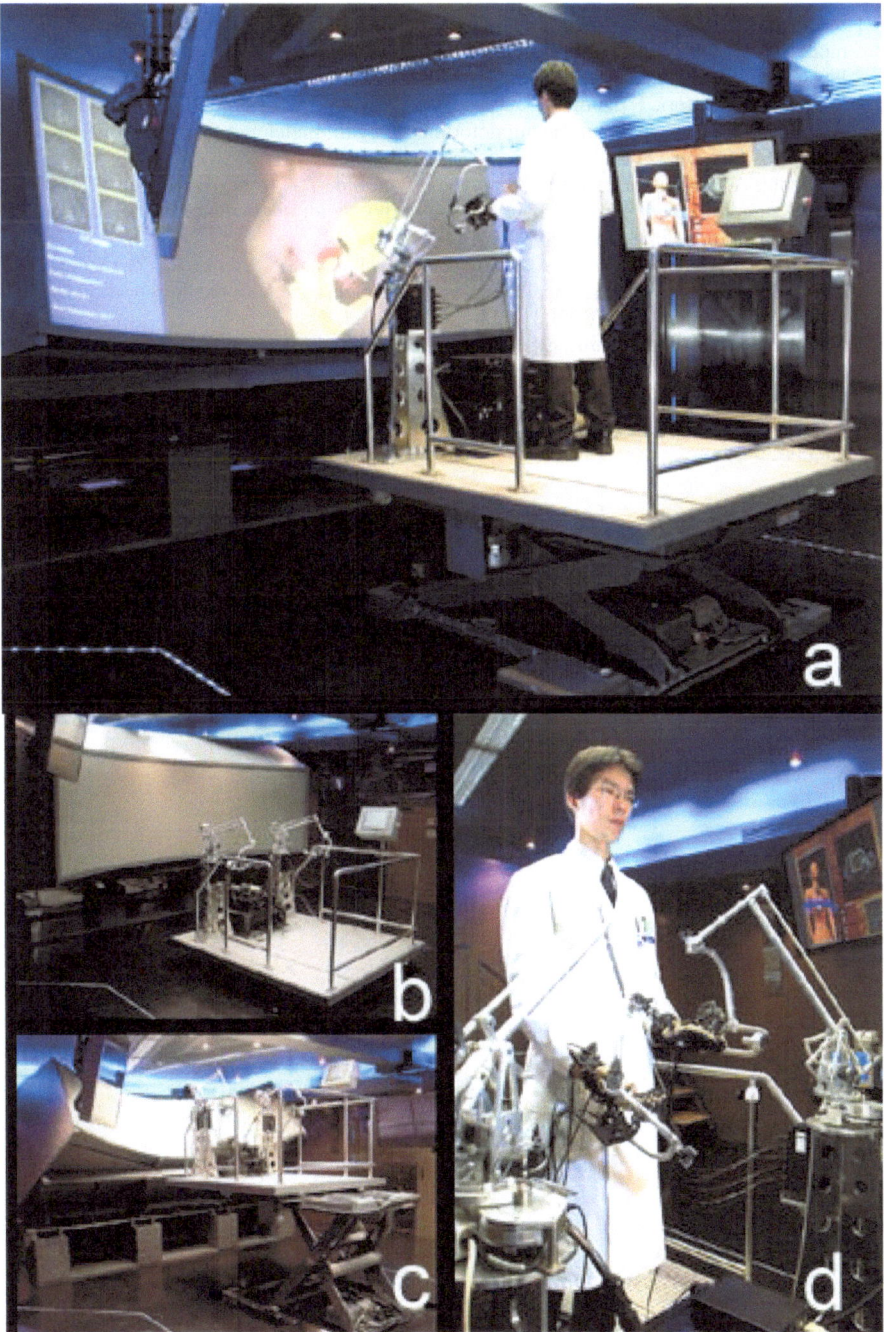

Fig. 11. A general view of our VR cockpit. To attain the same position and field of view as in open surgery, a surgeon can change the position of the screen and the elevator (b,c). A surgeon simulating surgery attaches the force feedback devices to his hands and stands on an elevator (d).

[8] S.Suzuki, N.Suzuki, A.Hattori, K.Sumiyama, S.Wakai, A.Uchiyama. "Deformable organ model using the sphere-filled method for virtual surgery," World Congress on Medical Physics and Biomedical Engineering, 2000, 27(6), CD-ROM 5368-91825.pdf.

[9] N.Suzuki, A.Hattori, T.Ezumi, A.Uchiyama, T.Kumano, A.Ikemoto, Y.Adachi, A.Takatsu. "Simulator for virtual surgery using deformable organ model and force feedback system," MMVR6 1998, pp.227-233.

[10] N.Suzuki, A.Hattori, S.Suzuki, T.Kumano, A.Ikemoto, Y.Adachi, A.Takatsu. "Performing virtual surgery with a force feedback system," The eighth international conference on artificial reality and telexistence, 1998, pp.182-187.

[11] N.Suzuki, A.Hattori, S.Suzuki, K.Sumiyama, S.Kobayashi, Y.Yamazaki, Y.Adachi, "Collaborated surgical works (Surgical planning) in virtual space with tactile sensation between Japan and Germany," MMVR2001, 2001, pp.479-484.

[12] S.Wakai, N.Suzuki, A.Hattori, S.Suzuki, A.Uchiyama. "Real-time volumetric Deformation for Surgical Simulation using Force Feedback Device," MMVR11, 2003, pp.386-388.

[13] S.Suzuki, N.Suzuki, A.Hattori, A.Uchiyama. "Dynamic Deformation of Elastic Organ Model and the VR Cockpit for Virtual Surgery and Tele-surgery," MMVR11, 2003, pp.354-356.

How to Add Force Feedback to a Surgery Simulator

Heiko Maass, Benjamin B.A. Chantier, Hüseyin K. Çakmak,
and Uwe G. Kühnapfel

Forschungszentrum Karlsruhe, Institut für Angewandte Informatik, Postfach 3640, 76021
Karlsruhe, Germany
{maass,benjamin.chantier,cakmak,kuehnapfel}@iai.fzk.de

Abstract. Methods and fundamental considerations for adding force feedback
to a surgery simulator are presented. As an example, the virtual endoscopic surgery trainer "VS-One", developed at the Forschungszentrum Karlsruhe, is
taken. An overview on building a general interface to force feedback applications is given. Implementations and satellite modules of the simulation software
KISMET are presented as the results. The concept and the implementations are
found to be flexible, stable and for universal use. Conclusions are drawn from
the results regarding actual and further developments.

Introduction

At the Forschungszentrum Karlsruhe (FZK), a virtual endoscopic surgery training
(VEST) system has been developed, based on the virtual reality, real-time simulation
software KISMET [1] [2]. The trainer produces an endoscopic view of a surgical
training task. Surgical instruments are located in an input box and can be moved like
in a real surgical operation. According to the operators' movements the real-time
simulation calculates interactions with the virtual environment. Adding force feedback to this system enables the human operator to touch and feel the virtual scenario
by using his haptic senses.

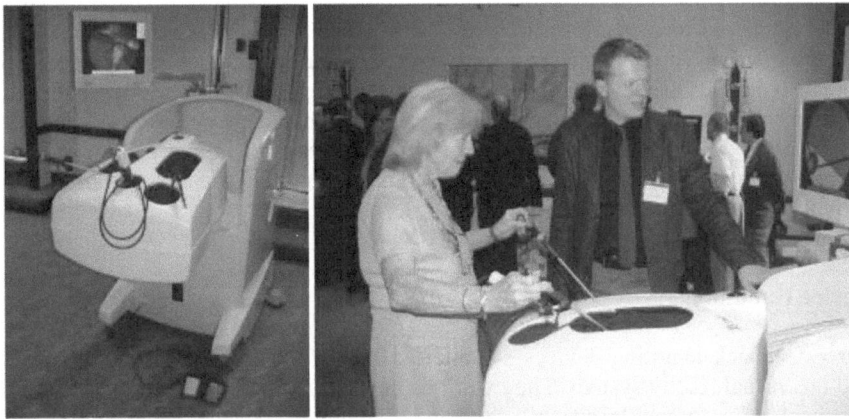

Fig. 1. "VS-One" system overview and operation

N. Ayache and H. Delingette (Eds.): IS4TM 2003, LNCS 2673, pp. 165–174, 2003.

In close cooperation with our commercial partner Select-IT VEST Systems AG we built the laparoscopic surgery trainer "VS-One". The goal is to create a system that meets the demands for surgeons' education and rehearsal. Recent developments and cost reduction in 3D Graphics and computational hardware provide platforms to incorporate affordable working systems. Fig. 1 shows the "VS-One" system.

A fundamental demand for interactive human tissue simulation is real-time computation performance. Elastodynamic object deformation is essential for close-to-reality, high-fidelity simulations. The realism is dependent on the set of parameters, the solving algorithms, and the data conversion and transfer structure. Especially when employing force feedback, the intuitive impression strongly depends on the model preferences.

A high-fidelity simulation system requires methods for collision detection (touching objects), interaction handling (e.g. grasping) and model modification (e.g. cutting). Photo realistic graphical computer techniques further immerse the user into the simulation environment. The data conversion and transmission from and to the haptic display has to consist of continuative treatment of derived forces, adequate filtering, active control algorithms, and reliable error handling. Considerations of the real-time condition and of the authentic perception of physiological tissue behavior are important for all modules.

This paper focuses on the surgery simulation for minimally invasive surgery. The aim is to create a laparoscopic training application using the "Laparoscopic Impulse Engines" by Immersion Corp. We present the KISMET force feedback module (KFF) of the "VS-One" system. Addressing issues like stability, elastodynamic real-time deformation or interaction handling goes beyond the scope of this paper. Some of the force feedback applications using the simulation software, KISMET and KFF, are presented in the results chapter.

Materials and Methods

Haptic devices are used more and more as force displays for computer games and manipulators. Improvements in robotic device design over the last years are largely due to advances in dynamics efficiency, and higher strength-to-weight ratio. Larger actuators and drive mechanisms lead to more inertia and friction in haptic devices. Natural dynamics and inertia of the haptic device might prohibit finer force discrimination for dexterous tasks, even if the force output is adequate. Abrupt changes in mechanical resistance during interaction is typical for surgical interventions. Current generation materials and drive components reduced mass inertia and friction dramatically. [3] [4]

Device Handling

Force feedback handling devices are interfaces between the human operator's hand and the virtual reality system. They work as input devices for movements and as output display for forces. In general, the haptic equipment consists of a handle for the human operator, a set of movable mechanical linkages attached to sensors and actuators coupled by drive transmissions. Counterbalancing reduces gravitational forces and torques within the device. The position and orientation of the gripper can be determined by using the sensors to read the joint angles of the linkages, and calculating forward kinematics. Force data returned by the simulation, is mapped to a set of torques to be produced by the motors by using a Jacobian. This signal is converted into desired currents for amplifiers driving the motors.

Input devices for laparoscopic, minimally invasive surgery require a special mechanical set up. The surgical operation is carried out by using a long instrument through an invariant incision point. The haptical display has to serve this condition.

The "Laparoscopic Impulse Engine" consists of 5 DOF, 3 of which are force reflecting joints actuators. A gripper is mounted to a shaft that can be rotated through three main rotational axes and one translational linkage in the shaft direction. The incision point is held in a spatially constant position, analogous to a three-dimensional pivot. The 5th joint is the gripper's opening/closing mechanism. Amplifiers and drivers are included in the Immersion Corp. package. Except the 5th joint, the rotational joint angles are resolved by encoders, providing incremental values. Hence, the definition of a reference-position is necessary, in which the instruments have to be put, each time the interface starts up.

Control Scheme

In this paper we define force feedback as the display of calculated forces to the operating user. In some other publications the phrase force feedback is defined as the part of a closed loop system. In that case the sensed forces from the input device are returned to the active control system.

Open-loop control is used because there is no redirection of measured operationforces from the device to the controller possible to regulate the output. The force depends upon the device controller assuming negligible dynamics of the mechanics e.g. time-dependent disturbances, friction or inertial forces and torques. It has to be accepted that unpredictable inaccuracies are sensed by the user. We used an openloop control without sensors in order to reduce cost and size of the assembly. The haptic quality in a open-loop control is susceptible to environment model errors and faulty divergences in the pre-computation of forces. Therefore, flexible and model dependent algorithms have to be implemented. Time-independent values like gravity and friction can be reduced by using feed forward methods. A more robust design can be achieved by mounting force-torque or acceleration sensors to the input handle and using the data in the active control. For each model a working range of stable behavior has to be achieved using appropriate parameters. [5] [6] [7] [8]

A block diagram of the control scheme of a surgery simulator is shown in Fig. 2. The simulator components correspond in the following way: the simulation software generates a visual display from the anatomical model. The human operator interacts with the visualization by moving the haptic devices. The controller interface processes this data and passes positional information to the interaction handling algorithms of the simulation. The computed interaction forces are returned to the haptic device using the controller interface. The transmission of data through this controller interface is parameterized by the model.

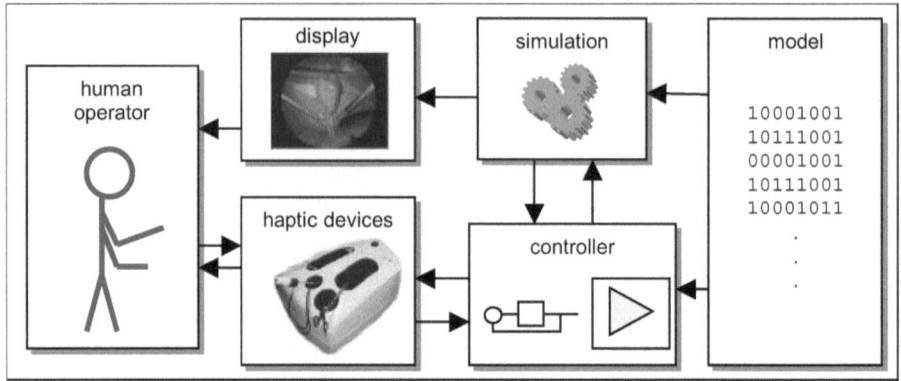

Fig. 2. Simulator control scheme

The cognition of the human operator is included in the loop. The physician interacts by positioning the haptic interface. The desired values for forces and torques are
calculated from the anatomical model, based upon the physical properties of the anatomical objects and the sensed position of the tool. Morphological updates are sent to
the visual display and the new haptic information is transmitted to the force display.
As the operating human is included in the system loop this creates an element of uncertainty which has to be taken into account. The range of man-in-the-loop stability is
dependent on the "stiffness" of the operator.

Impedance Feedback Control

Two basic control methods are known: impedance and admittance control. The impedance control receives positions and controls the forces. Admittance control
measures the operator's force and controls the displacement. Generally, direct forward impedance control techniques are used for human, soft tissue surgical simulations. Admittance control is more suitable for rigid environments. For laparoscopic
surgery simulation we decide to use the impedance control method, mainly because
no force sensing system is present at the "Laparoscopic Impulse Engines".

In Fig. 3, the model dependent controller structure is described. By "model", we
mean all values for a surgical exercise like anatomical morphology, elastodynamic
properties, and filter/control settings. The deformations and forces are calculated in
real-time in the elastodynamics module. The interactions of the instruments with the
elastodynamic model is calculated via the collision handling. World-coordinate impedance values are generated from the displacement and the forces affecting the instruments by the impedance converter. These values are processed by the impedance
controller. A cascaded control and filtering structure for forces, DOF-values and impedances allows flexibility in parameterizing model-dependent control.

The impedance controller measures the difference between the computed position
deflection and the actual position deflection from the haptic device. Using the deflection values, calculated model force, and the actual device force, the impedance function can be derived. This structure is called model-feed-forward open-loop impedance
control. A new force command is generated by interpolating between discrete impe-

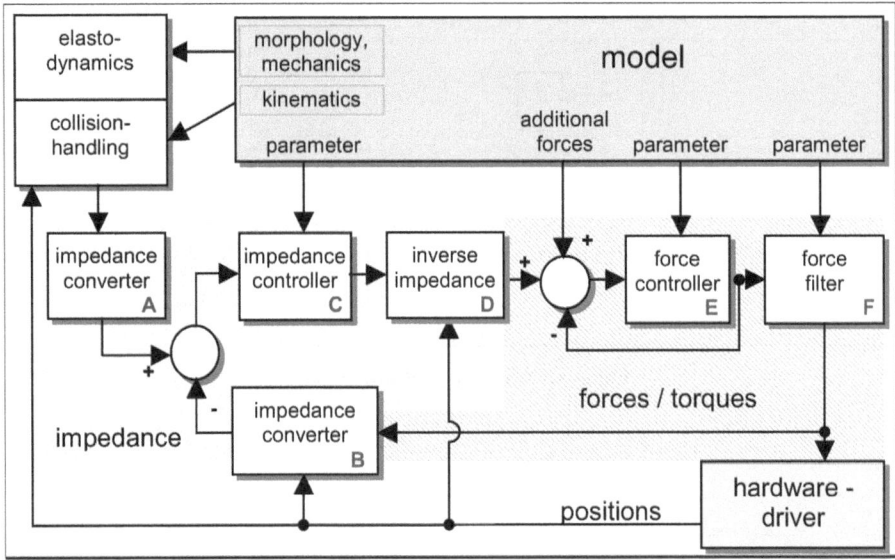

Fig. 3. Model dependent controller structure

dance updates. Values are interpolated by using superposition of filters and automatic control. The closed-loop haptic impedance is the relationship between the positional deviation of the haptic interface and the corresponding force applied by the user. Strictly speaking, the impedance is the relationship between velocity and force. We take the linearized proportion between deflections and force as the impedance by assuming the time intervals to be constant.

Without employing force/torque and six-axis acceleration sensors we have to be aware that we disregard mass inertia and acceleration components in a open loop control. Actually, the loop will be closed by the human operator.

Control Modules

The impedance converter (A, see Fig. 3) reduces the spatial deflections and forces derived from the elastodynamics and the collision handling module to time-dependent stiffness vectors. The difference between the simulation impedances and the current device impedance (B) is processed by an impedance controller (C). The forces and the torques are regained in the inverse impedance converter (D) by combining the impedance values with the positions from the hardware. This conversion indicates the transition between the low-rate calculation and the high-rate driver control.

The forces are superpositioned by additional forces (e.g. for gravity compensation) and provide the command values for a force controller (E). In some configurations an additional force filter (F) enlarges the stability range of the overall control behaviour. The controller structures (C) and (E) are implemented as free-parameterizable PID-modules. We are using low-pass force filters (F) in order to increase stability conditions in some cases. However, by applying filters simulation dynamics can be reduced.

170   Heiko Maass et al.

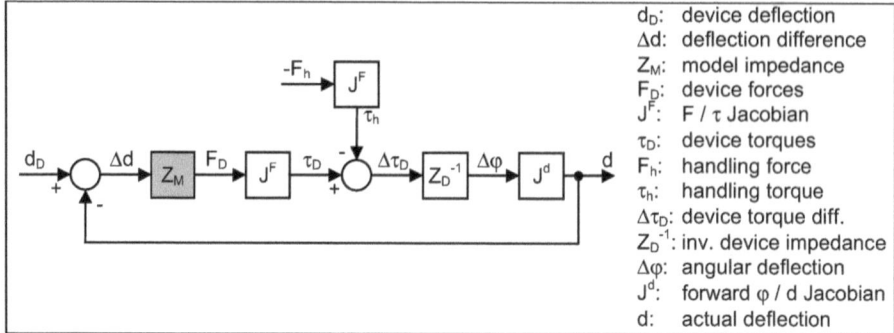

Fig. 4. Closed loop controller structure

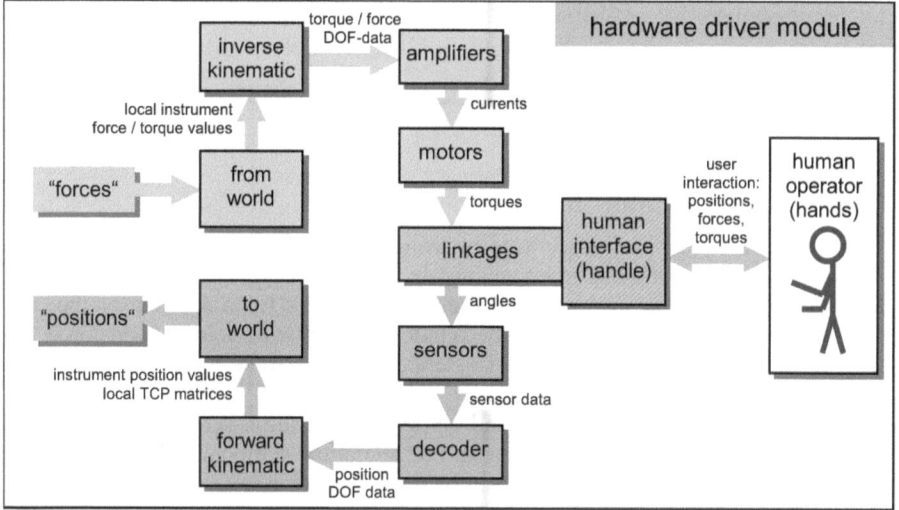

Fig. 5. Data flow in the hardware driver module

The schematic drawing in Fig. 4 depicts an equivalent closed-loop controller structure. The device deflection d_D is the difference between a starting position x_0 and the actual position x caused by the operating human. The individual blocks represent physical and computational elements. The highlighted module Z_M (model impedance) corresponds to Fig. 3 except the hardware driver module, which is equivalent to the other blocks.

Theoretically, if we disregard mass, energy storing elements and backlash and if we linearize the transformations at an operating point, then the closed loop impedance can be calculated and the model impedance can be adjusted for stable closed loop control. Practically, the neglect and linearization of characteristics does not meet physical behavior. Minor modeling errors in an open-loop controller lead to unpredictable output divergences.

A more detailed data flow block diagram of the hardware driver module is shown in Fig. 5. The desired forces arriving at the hardware driver are converted into local coordinates by a "from world" transformation. Through the inverse kinematics calculation torque and force data are sent to their respective amplifiers. The driven motors transfer torques to the mechanical device and transmit forces and torques via linkages to the user interface (handle-hand interface to the human operator). The positions are obtained by sensing the mechanical angular position values of the linkages using appropriate sensors. These data are converted to DOF-values and are transformed to Cartesian positions by using the forward kinematics calculation. The "to world" transformation takes the physical placement of the instruments into account and generates world-coordinate TCP-matrices and positions.

Control Parameters

Finding the appropriate control parameters for each model is a very optimization-intensive task. The resolution, sensitiveness and dynamics have to be maximized while keeping a wide range of stability and a high accuracy of the displayed forces. Stable control has to be guaranteed while assuming a large variety of human-operator behavior.

In a typical implementation the haptic device must be updated at a fairly fast rate (300 - 1000Hz) to ensure stability and a responsive interface. The visual computation is updated at a much lower rate (5 - 75Hz) depending on the complexity of the model. This results in interpolation time constants in the order of 10ms up to 200ms.

Results

We used heavily modified "Laparoscopic Impulse Engines". We mounted small security plates to prevent the force transmitting cables from slipping off the driving wheels. We ordered a custom design with a shaft length of 25cm and with a dis-assembled supporting unit in order to obtain a larger working space.

KISMET Force Feedback (KFF) modules can be distinguished as device-dependent and hardware-independent modules. We designed all structures and algorithms for universal, multifunctional use. Current software modules are implemented on a WIN32 platform and are running on Windows-NT and Windows2000 workstations. The functional structure of the components is depicted in Fig. 6.

KFF_run is a hardware-independent, stand-alone software and works as an interface between KISMET, the hardware specific driver and the administrative interventions (e.g. for calibration purposes) using shared memory technology. The module starts automatically when executing a KISMET model that includes force feedback interface calls. It requests the operating user to place the instruments in a reference position once, in order to set original values for the instruments' position. This service-like module keeps running in the background so that the encoder values are not re-referenced each time a new KISMET model starts.

KFF_dll is implemented as a link library, unique for each hardware set up supported. Modules for multiple "Laparoscopic Impulse Engines" and the SensAble PHANToM™ device (GHOST-Library 3.0 and 4.0) are available and have been tested extensively (different KFF_dll libraries). A fast rate timer guarantees the data transmission cycle between KISMET and the hardware interface. Depending on the calibration information, retrieved by a settings file (KFF.ini), device specific transformations are applied. Up to 15 haptic input devices are supported, each of them provides control for up to 7 DOF with or without force feedback. Filter and control settings for each single DOF can be adjusted, as well as those for the automatic regulator control for world-coordinate impedance and elastodynamic pre-calculation. The respective library is linked to KFF_run.

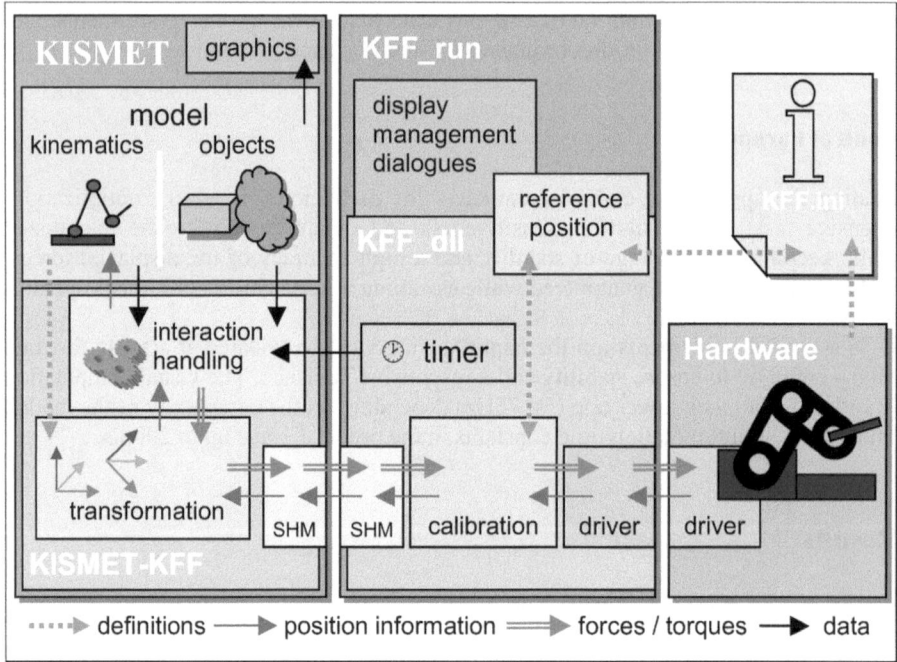

Fig. 6. Functional structure of implemented modules

The main software package KISMET provides the following features:

- retrieves the simulation model information
- calculates the kinematics of the instruments from the joint rotations / translations of the haptic devices
- computes the real-time elastodynamic deformation of objects
- evaluates the instrument collision with objects
- performs the impedance control
- handles instrument interactions: grasping, cutting, placing clips, coagulation, irrigation, suction, needle insertion
- recreates physiological effects: bleeding and pulsation of vessels, or cyclic body/organ movements.

KISMET generates the graphical representation of the scenario. Important parameters for each model enclose kinematical instrument structures, morphologic, elastic and dynamic object definitions, and data control specifications. Parameter optimizations for the best dynamics and stability have to be regarded, which are dependent on the computing hardware and the model complexity.

Two main approaches are traced. One is the rigid model, direct forward impedance control for simple tasks. The second is the elastodynamic model cascaded control, used for high-fidelity models, enclosing two different cycle loops. The low-rate elastodynamics and collision detection loop combines with a high rate impedance control and force-feed-forward loop.

The "VS-One" system includes a variety of models for education. Simple basic training tasks (BTT) are implemented to instruct the trainee to learn precise, spatial instrument orientation, coordination of multiple instruments and the usage of a 30° camera optics. Simple anatomical models are designed to teach the user in the handling of different instruments and operation techniques. Several high-fidelity models, like the laparoscopic removal of the gall bladder (Cholecystectomy), are intended to teach surgery operation techniques including complex procedures.

Conclusions

The model-feed-forward impedance control concept was found to be well-suited for soft tissue simulation. The implemented, cascaded control and filter methods allow a flexible model-dependent parameter design. Stability can be achieved in a wide range of the closed-loop impedances. The quality of the force feedback is particularly dependent on precise and optimized modeling. For a more stable control the measurement of the operator's handling forces, torques and accelerations is necessary. However, this will result in increasing cost and assembly complexity.

It can be concluded from the experiences using the "Laparoscopic Impulse Engines" that this product is appropriate for basic applications. The limitation to only 3 linkages exerting force feedback results in reduced perceptibility. The invariant pivot point of the devices is located inside the mechanical unit. In order to build a close-to-reality laparoscopic surgery simulator the instrument incision point has to be placed on the patient body surface. The mechanical setup is space consuming, the components indicate low operating reliability and not more than 2-3 devices can be installed in a system, because the number of extension-bus slots provided by conventional computer systems is limited.

Within the scope of a project called "HapticIO" new haptic devices for laparoscopic surgery simulators are currently in the designing process at the Forschungszentrum Karlsruhe. These devices will provide an enlarged operating space, improved mechanical specifications and operating reliability and will facilitate a more precise control. The project "HapticIO" is funded by the German Federal Ministry of Education and Research.

References

1. Kühnapfel U., Çakmak H.K., Maass H.: Endoscopic Surgery Training using Virtual Reality and deformable Tissue Simulation. Computers & Graphics 24, 671-682, Elsevier (2000)
2. Kühnapfel U., Çakmak H.K., Maass H., Waldhausen, S.: Models for simulating instrument-tissue interactions. 9th Medicine Meets Virtual Reality 2001 (MMVR 2001), Newport Beach, CA, USA, Jan. 23-27, 2001 (2001)

3. Burdea, G.: Force and Touch Feedback for Virtual Reality. John Wiley and Sons (1996), ISBN: 0-471-02141-5
4. Massie, T.H., Salisbury, J.K.: The PHANTOM Haptic Interface: A Device for Probing Virtual Objects. Proceedings of the ASME Winter Annual Meeting, Symposium on Haptic Interfaces for Virtual Environment and Teleoperator Systems, Chicago, IL. (1994)
5. Carignan, C.R. Cleary, K.R.: Closed-Loop Force Control for Haptic Simulation of Virtual Environments. Haptics-e Vol 1 No.2 (2000)
6. Basdogan, C., Srinivasan, M.A.: Haptic rendering in virtual environments. Virtual Environments HandBook, K. Stanney, Ed., (2001)
7. Hannaford, B., Ryu, J.H, Kim, Y.S.: Stable Control of Haptics. In: Touch in Virtual Environments: Proceedings USC Workshop on Haptic Interfaces, Feb 23 2001. Margret McLaughlin, Ed., Prentice Hall, (2001)
8. Picinbono, G., Lombardo, J.C.: Extrapolation: a Solution for Force Feedback? In: International Scientic Workshop on Virtual Reality and Prototyping, pages 117-125, Laval, France, June (1999)

Tissue Cutting Using Finite Elements and Force Feedback

Cesar Mendoza and Christian Laugier

INRIA* Rhône-Alpes & GRAVIR, ZIRST 655 av. de l'Europe,
38330 Montbonnot Saint Martin, France
{Cesar.Mendoza-Serrano,Christian.Laugier}@inrialpes.fr
http://www.inrialpes.fr/sharp

Abstract. This paper presents a methodology to simulate 3D cuts in deformable objects. It uses an *explicit* finite element approach to simulate deformations in real-time. We use a non-linear strain tensor formulation (Green tensor) to allow large displacements. Haptics is used to allow touching sensation of the cutting procedure.

1   Introduction

Cutting is an essential component in medical simulators. Simulating cutting of biological tissues requires real-time interactions, accurate geometric models of anatomical structures and realistic dynamic models. Previous works address cutting by *removing* [1] from the simulation the elements that collide with the cutting tool or by *subdividing* [2,3] the colliding elements. Removing elements destroys the material from the virtual organ. In some cases, this is not realistic since the mass of the organ is not preserved. To increase realism, the number of simulated elements is incremented. This might cause a slow down in the simulation. On the other hand, subdivision is more realistic, but the number of simulated elements increases and therefore the simulation is slowed down as well.

In our previous works [7], we have started a new approach: *separating the elements* instead of removing or dividing them. The approach does not increment the number of elements during the simulations and preserves the mass of the organ. We implemented it in a 2D mass-spring model. Later, Nienhuys et. al. [4] has used the same idea to approach 3D cutting.

This paper is organized as follows: first we describe the physical deformable model used in section 2. Next, we present the cutting algorithm in section 3. Section 4 presents some of our results. Finally, we conclude and present future research in section 5.

2   Physical Model

We used an explicit formulation of finite element methods [1,5,8] to simulate the dynamics of the biological tissue. The model is similar to the mass-spring deformable model, since each node of the tetrahedral mesh uses Newtonian laws of

---

* Inst. Nat. de Rech. en Informatique et en Automatique.

N. Ayache and H. Delingette (Eds.): IS4TM 2003, LNCS 2673, pp. 175–182, 2003.

motion. However, the finite element method has a stronger physical and mathematical foundation and therefore, we have chosen it for our simulations. Finite element methods (FEM) partition the object into sub-elements on which the physical equations are expressed. Instead of merging all these equations in a large matrix system, an *explicit* FEM solves each element independently. The idea is to write the elastic energy of a tetrahedron as a function of the displacement of its 4 vertices. It uses the balance equation of each element to obtain the force at each node in function of the displacement of neighbor nodes. Then, instead of obtaining the equilibrium position by solving a large matrix system, we only integrate the force at each node to obtain the new position for the node. We use a non-linear Green strain tensor, ϵ, allowing *large displacements*. The Green strain is expressed by a 3 x 3 matrix. Its (i, j) coefficient is:

$$\epsilon_{ij} = (\frac{\partial \boldsymbol{x}}{\partial u_i} \frac{\partial \boldsymbol{x}}{\partial u_j} - \delta_{ij}) \tag{1}$$

where the Kronecker delta is $\delta_{ij} = 1$ if $i = j$ or zero otherwise. We assume that our material is isotropic and consider linear elasticity to link stress and strain, as follows:

$$\sigma_{ij}^{(\epsilon)} = \sum_{k=1}^{3} \lambda \epsilon_{kk} \delta_{ij} + 2\mu \epsilon_{ij}. \tag{2}$$

The material's rigidity is determined by the value of μ, and the resistance to changes in volume (dilation) is controlled by λ. The total internal force that a tetrahedron exerts on a node is [5]:

$$\mathbf{f}_{[i]}^{el} = -\frac{vol}{2} \sum_{j=1}^{4} \mathbf{P}_{[j]} \sum_{k=1}^{3} \sum_{l=1}^{3} \beta_{jl} \beta_{ik} \sigma_{kl} \tag{3}$$

where *vol* is the volume of the tetrahedron, **p** the position of the nodes of the tetrahedron in the world coordinates and β, the inverse barycentric matrix that links the world positions to the material coordinates. The total internal force acting on the node is obtained by summing the forces exerted by all elements that are attached to the node. Finally, we use a modified-Euler scheme to integrate the dynamics of each node.

3   3D Volumetric Cutting Algorithm

Once a collision detection has been detected between the cutting tool and the object, we follow the next steps to carry out cutting phenomena: (1) a geometric and physical criteria, (2) select and separate tetrahedrons, (3) local remeshing and (4) force feedback.

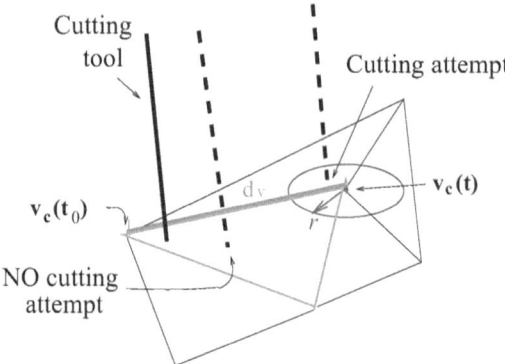

Fig. 1. Determining cutting attempts.

3.1 Geometric and Physical Criteria

Geometric criteria: We first determine if the user displacements on the surface of the object corresponds to a *cutting attempt* or not.

Let $C_p(t)$ be the colliding point at time t between the virtual tool and a facet on the surface of the object. Let $v_c(t)$ be the closest vertex to $C_p(t)$ and t_0 the moment of the first contact. Define a neighborhood ⊌ around the vertex $v_c(t)$. We consider a *cut attempt* if:

1. $v_c(t) \neq v_c(t_0)$
2. $Cp(t) \in$ ⊌

The first condition states that the closest vertex to the colliding points at times t and t_0 must be different. The second condition avoids degenerated cuts due to small movements (e.g. a very small displacement of the tool in the middle of the facet may satisfy the first condition). To cut, the user is forced to execute larger displacements by constraining the tool to lie on the neighborhood ⊌, see figure 1. For simplicity, we have chosen the neighborhood to be a circular region with radius r. The value of r determines the size of ⊌. Since the facets of a mesh are, in general, of different sizes and forms, the value of r must be computed as a function of the size of the current colliding facet. Thus,

$$r = \alpha d_v \tag{4}$$

where d_v is the distance between $v_c(t)$ and $v_c(t_0)$, see figure 1. This distance changes depending on the colliding facet. The parameter α determines the size of the neighborhood.

Physical criteria. A cut attempt is not enough to break apart the object. Some physical aspects, that take into account the physical interaction between the object and the cutting tool, have to considered. To do that, we consider the internal behavior of the object when it is subjected to external loads produced by the tool. According to fracture mechanics, an object may be broken due to two different type of failures: (a) *Tensile failure:* This corresponds to loading *normal*

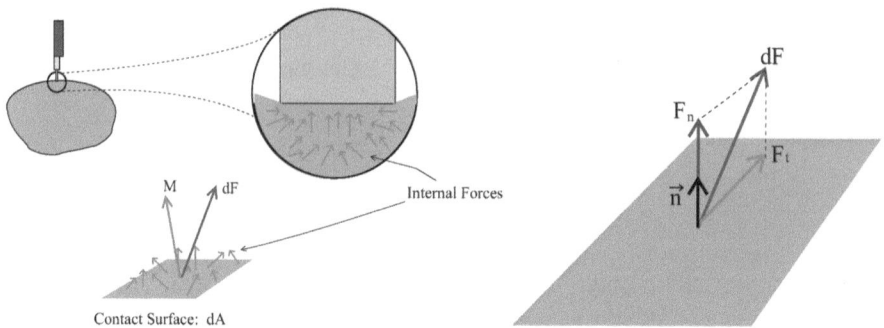

Fig. 2. Internal forces during contact between cutting tool and object.

to the failure surface. If the failure is produced by pushing rather than pulling then we can have a *compressive failure*. (b) *Shear failure:* This corresponds to loading *tangential* to the failure surface. We analyze the forces that cause these failures.

First, note that during the contact, the internal forces equilibrate the external load produced by the tool. The internal force distribution can be represented by an equivalent set of resultants, F, and moments, M, see figure 2. From classical mechanics [9], the traction, Tr, provides a measure of the direction and intensity of the loading at a given point and it is defined as:

$$Tr = \lim_{dA \to 0} \frac{dF}{dA}. \tag{5}$$

Decomposing the force into normal and tangential components, see figure 2, and introducing a sharpness factor, κ, for the cutting tool, we have a *cutting traction vector*:

$$Tc = \frac{1}{\kappa} \left[(\lim_{dA \to 0} \frac{|F_n|}{dA}) \boldsymbol{n}_1 + (\lim_{dA \to 0} \frac{|F_t|}{dA}) \boldsymbol{n}_2 \right] \tag{6}$$

where n_1 is the normal to the plane and n_2 is the tangent to that same plane,(i.e. a normal in another perpendicular plane). Most of the measurable parameters available in the literature are given using the *fracture toughness*, K_I, of the material which is the critical stress intensity required to produce a failure in a material. Therefore, we put the *cutting traction vector*, T_c, in terms of the stress:

$$T_c = \frac{1}{\kappa} (\sigma \boldsymbol{n}_1 + \tau \boldsymbol{n}_2). \tag{7}$$

where σ is the *normal* stress and τ the *shear* stress. In the 3D case, T_c takes the following form:

$$\begin{bmatrix} t_{e_x} \\ t_{e_y} \\ t_{e_z} \end{bmatrix} = \frac{1}{\kappa} \begin{bmatrix} \sigma_{e_x x} & \tau_{e_x y} & \tau_{e_x z} \\ \tau_{e_y x} & \sigma_{e_y y} & \tau_{e_y z} \\ \tau_{e_z x} & \tau_{e_z y} & \sigma_{e_z z} \end{bmatrix} \begin{bmatrix} n_{e_x} & n_{e_y} & n_{e_z} \end{bmatrix}. \tag{8}$$

where $\boldsymbol{n}_i$ is the normal of each plane of the infinitesimal cube. For simplicity, take Γ as the set of normal and shearing stresses. The object is broken when

the maximum stress takes a value greater than the material toughness, K_I. From classical mechanics, the maximum shearing stress is computed using the eigenvalues, σ_1, σ_2 and σ_3 of Γ.

$$\tau_{max} = \frac{1}{2}max\{|\sigma_1 - \sigma_2|, |\sigma_1 - \sigma_3|, |\sigma_2 - \sigma_3|\}. \tag{9}$$

and the maximum normal stress, σ_{max} is the greatest eigenvalue of Γ. Finally, we define our *cutting stress*, σ_c as:

$$\sigma_c = \frac{1}{\zeta} \min{(\sigma_{max}, \tau_{max})}. \tag{10}$$

where $\zeta \in [0.1\ 1]$ is a parameter representing the *damage* in the cutting area. Finally, a cut is produced if a *cutting attempt* has occurred and if

$$\sigma_c \geq K_I \tag{11}$$

where K_I is the material toughness of the object.

3.2   Select and Separate Tetrahedrons

To select the tetrahedrons which need to be separated to broken the object we consider a *cutting line* on the surface of the object. This cutting line is given by the set of vertices selected as follows: it starts at $v_0 = v_c(t_0)$, the closest vertex to the previous colliding point at t_0, it continues to $v_1 = v_c(t)$, the closest vertex to the current colliding point, such that $v_c(t_0) \neq v_c(t)$. The ending vertex, v_2 of the cutting line is the one that best fits the profile of the cut. We consider that the *cut attempt* is executed in the direction, s_1, from $C_p(t_0)$ to $C_p(t)$ and that the cut is as straight as possible. We define s_i as the vectors from the possible projected vertices to $C_p(t)$; v_2 will be chosen as the vertex v_i whose vector s_i is minimum with respect to s_1:

$$v_2 = v_i \quad \text{such that} \quad \min\{\angle(s_1, s_i)\}. \tag{12}$$

Let s_1 be the vector from v_0 to v_1 and n the normal to the facet as shown in figure 3. Define P as the plane spanned by n and s_1. Set T as the set of tetrahedrons, e^T, sharing the vertex v_1. Then, from e^T, we separate the tetrahedrons, that are in one side of the plane from those that belongs to the other side of the plane, by only splitting v_1. When a tetrahedron, e^T, is divided by the plane, P, the tetrahedron will lie in the side where its furthest vertex lies.

Separating tetrahedrons may create singularities or zero area joints (e.g. tetrahedrons connected only by one vertex). Our data structure, based in a *abstract simplicial complex K*, let us identify these singularities efficiently.

3.3   Local Remeshing

To reflect the cut, we reposition the vertices of the cutting line by translating v_0 and v_1 to $C_p(t-1)$ and $C_p(t)$ respectively. The vertex v_2 is not moved until the next cutting step, when it will renamed as v_0, see figure 3. We update the β matrix and the volume of the involved tetrahedrons to keep the physical validity of the model.

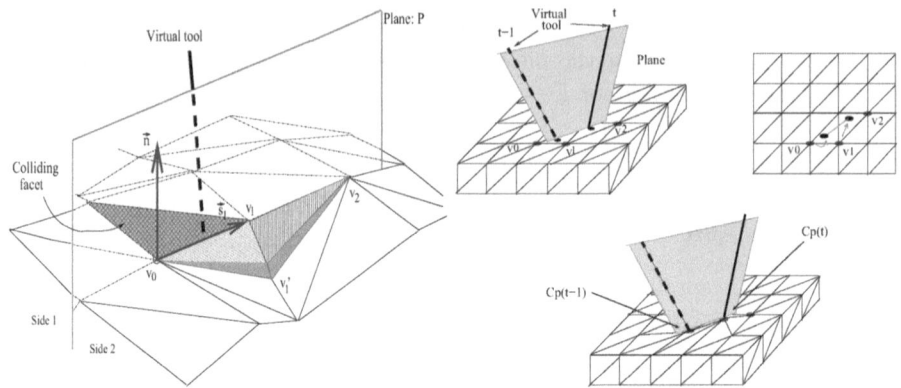

Fig. 3. (left)A set of tetrahedra is put in one side of the plane (Side 1) and another set is put in the other (Side 2). The plane is spanned by vectors s. and n. (right) Repositioning the vertices of the tetrahedrons to fit the cut profile.

3.4 Force Feedback

Haptic interaction was included to increase realism. We solve the different rate frequency problem between the physical (about 20 Hz) and the haptic simulations (about 1 KHz) by separating the haptic and the simulation loop and linking them by a *buffer model* as we have done in [6]. The buffer model computes the haptic forces.

We also model the sensation felt by the user when the material is broken. When a cut is executed in the physical simulation, the haptic rendering *switches* from the buffer model to a model that simulates a haptic cut. We use the curve of the force obtained during fracture proposed by [10]. From it, we propose to compute the haptic force as:

$$F_{haptic}(I) = F_c e^{-\xi I} \tag{13}$$

where F_c is the force of the buffer model at the moment of switching, I is an iteration counter in the haptic loop and ξ is a parameter indicating the lost of energy during the cut. The value of ξ is calculated empirically. The value of I is incremented at each haptic interaction and set it to zero when a cut task has finished.

4   Results

In figure 4 we show a physical simulation of an object representing a human knee graft ligament. It is composed of 100 tetrahedrons simulated using explicit finite elements and non-linear Green formalism ($\lambda = 140000$, $\mu = 11000$, $\psi = 10$, $\phi = 80$, sharpness $\kappa = 0.0004$, damage $\zeta = 1$). A PHANToM device is coupled to the simulation to render the sensation of touching and cutting the object. We use a 800 MHz. processor and we obtain 30 Hz. for the physical and

Fig. 4. Cutting a human ligament using a force feedback device.

graphical simulation. The haptic rendering reaches the 1000 Hz and present a stable behavior. Large deformations presented self-collisions.

The haptic sensation of the cut has been largely influenced by the visual rendering.

5   Conclusion

We have proposed a 3D cutting algorithm for biological tissue using explicit finite elements and a non-linear strain tensor (Green). The algorithm uses a physical criteria based in a stress approach and the interpretation of the user movements to break the virtual object. The data structure allows efficient detection of zero area links between the tetrahedrons. We have integrated a stable haptic approach to give the user a touching sensation of the cut.

The implementation of our approach is a proof-of-concept of this technique. This work has opened a new road of research: object self-collisions caused during a cut.

Acknowledgements

This work has been supported by the National Council of Science and Technology of Mexico (CONACYT) and by the National Research Institute in Computer Science and Control (INRIA), France.

References

1. S. Cotin, H. Delingette and N. Ayache, A Hybrid Elastic Model allowing Real-Time Cutting, Deformations and Force-Feedback for Surgery Training and Simulation, The Visual Computer 2000, vol 16, no 8, pages 437–452.

2. Bielser, D., Maiwald, V. A. and Gross, M. H., Interactive Cuts through 3-Dimentional Soft Tissue, EUROGRAPHICS'99, vol 18, pages 31-38, 1999.
3. Mor, A. and Kanade, T., Modifying Soft Tissue Models : Progressive Cutting with Minimal New Element Creation, MICCAI, Medical Image Computing and Computer Assisted Intervention, oct. USA, 2000.
4. Nienhuys, H. and Vanderstappen, F., Combining finite element deformation with cutting for surgery simulations, EUROGRAPHICS, 2000.
5. James O'Brien and Jessica Hodgins , Graphical Models and Animation of Brittle Fracture , SIGGRAPH Conference Proc. 1999.
6. Mendoza, C. and Laugier, C., Realistic Haptic Rendering for Highly Deformable Virtual Objects, IEEE Virtual Reality (VR), Yokohama Japan, march 2001.
7. Mendoza, C., Laugier, C. and Boux-de-Casson, F., Virtual Reality Cutting Phenomena using Force Feedback for Surgery Simulations, Interactive Medical Image Visualization and Analisys, MICCAI, Holland,2001.
8. Debunne, G., Desbrun, M. and Cani, M. and Barr, A. , Dynamic Real-Time Deformations using Space and Time Adaptive Sampling, Computer Graphics, SIGGRAPH, USA ,2001.
9. Fung, Y.C., A First Course in Continuum Mechanics., Prentice-Hall, Englewood Cliffs, N.J., 1969
10. Mahvash, M. and Hayward, V. , Haptic Rendering of Cutting: A Fracture Mechanics Approach, Haptics-e, http://www.haptics-e.org, vol. 2, no. 3, nov 2001.

Mammography Registered Tactile Imaging

Anna M. Galea and Robert D. Howe

Division of Engineering and Applied Sciences
Harvard University
{galea,howe}@deas.harvard.edu

Abstract. Breast cancer and other breast pathologies often manifest as an area of increased tissue stiffness. The gold standard for breast cancer screening is mammography, which records the radioopacity of tissues in the breast, and therefore depends on factors other than stiffness. Tactile imaging uses an array of pressure sensors to noninvasively record the palpable extent of breast tissue stiffness. Tactile imaging quantifies palpation, and holds promise for increasing the positive predictive value of screening mammography by highlighting areas of abnormal stiffness. We propose a method for registering tactile images obtained in the same plane and immediately after the corresponding mammogram. A finite element model-based approach is presented which is used to account for the spreading of the breast tissue induced by the mammographic compression that is not present in obtaining the tactile image. We devise an algorithm that can register the modeled tactile images to within 6% of the modeled mammograms for breasts of varying size and stiffness. Clinical mammograms and tactile images were collected on 11 subjects, and the registration algorithm applied to the images. The registered tactile images and mammograms correlate well over stiff, radioopaque areas such as glandular and fibrous tissue, and highlight areas of increased stiffness not indicated by the mammogram alone.

1 Introduction

Mammography is the gold standard in screening for breast pathologies. It records the radioopacity of breast tissue, and results in a complicated image which radiologists learn to interpret mainly by experience [1] to detect suspicious areas. Manual palpation, in the form of the Clinical Breast Exam (CBE), is an established adjunct to the screening process [2], since many diseases of the breast, including cancer, manifest as a change in tissue stiffness [3,4]. However, since CBEs rely on a qualitative assessment of the palpable extent of breast tissue and are performed by a clinician different than the radiologist, the information from the CBE is often not used in conjunction with the mammogram.

Tactile Imaging is a new imaging modality that quantifies the palpable extent of soft tissue [5,6,7]. The tactile imager [figure 1] is a passive, pressure-sensitive device that is gently stroked over the area of interest. It records the surface pressures that result from the stress field within the tissues, and thus records a higher pressure in the vicinity of a stiff lesion. The forces required to obtain a tactile image are lower and more easily tolerated than those involved in compression of the breast for a mammogram.

We hypothesize that the information from tactile imaging can add value to breast cancer screening, increasing the positive predictive value of mammography by high-

N. Ayache and H. Delingette (Eds.): IS4TM 2003, LNCS 2673, pp. 183–193, 2003.
© Springer-Verlag Berlin Heidelberg 2003

lighting areas of abnormal stiffness [8]. A mammogram would be viewed alongside a tactile image taken in the same plane at the same time so that the tactile image can provide direct tactile information to the radiologist. In order to obtain images registered in space and time, tactile images to be registered to mammograms should be obtained immediately after the cranio-caudal mammogram, and before the breast was removed from the bottom plate of the mammography machine [Figure 2].

An important first step in establishing and realizing this utility is registering the tactile images to mammograms. Registration of the tactile image and the mammogram requires an understanding of the relative motion of the breast incurred in each imaging modality. The mammogram is obtained with the breast compressed between two parallel plates for the greatest compression and spreading the patient can tolerate. In contrast, the tactile image is obtained by indenting the breast with a small scanhead stroked over the surface of the breast. The breast is stretched as the tactile image is obtained, but due to the lower compression and the small area of contact the total deformation under tactile imaging is different from and less than that under mammography. The difference in the deformation of the breast tissue as the tissue is imaged under each modality must be accounted for in order for registration to occur.

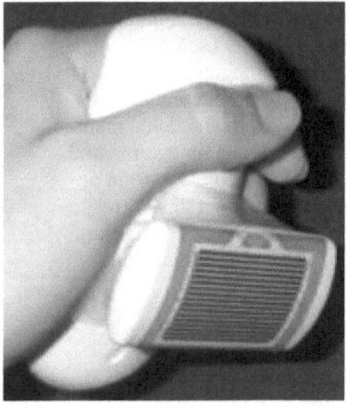
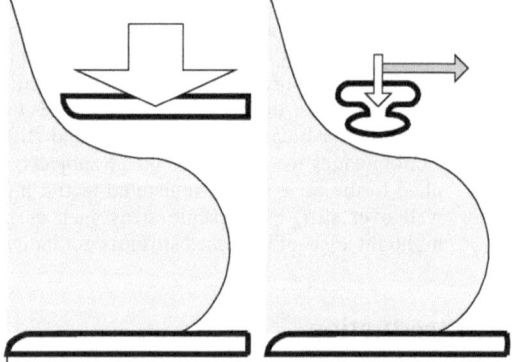

Fig. 1. Tactile Imaging scanhead used in this study. The working portion is an array of passive pressure sensors. When pressed gently and stroked over the tissue of interest it records higher pressures near stiffer regions, thus recording the palpable characteristics of the tissue

Fig. 2. Side view of breast on mammography plate. Left: Mammograms are obtained with the breast laying on a rigid plate and compressed by a top plate. Right: In order to obtain tactile images to register with the mammograms, the tactile images should be obtained with the breast in the same position. The tactile scanhead is gently pressed on the breast and the tactile image recorded as the scanhead is stroked away from the chest wall

In this study we develop an algorithm to account for the difference in deformation inherent between mammography and tactile imaging. We first discuss mechanical modeling of the deformation fields of the two imaging modalities, and the transformation algorithm created to register the modeled tactile imaging deformation to the mammographic deformation. Results of applying this transformation to the model data are presented. Finally, we present a preliminary evaluation of the registration applied to clinically collected mammograms and tactile images.

2 Registration Algorithm Development

2.1 Model Development

To characterize realistic shapes for breast models, breast contours of healthy female subjects were measured with each breast resting gently on a horizontal plate. Curves of the front and side profiles of the left and right breast were generated. Three representative subjects were analyzed that spanned the range of breast size and shape: Small (US Brassiere size A), Medium (C) and Large (DD). Figure 3 shows the three models created by fitting splines to the contour information from the three subjects. The average contours of the left and right breast were used, generating models symmetric about the vertical center plane. The chest wall is assumed to be 1cm beyond the skin surface of the chest.

The breasts were modeled as fat-replaced postmenopausal breasts since most cases of breast cancer occur post menopause. The fat was modeled as a linear elastic isotropic material of 15kPa stiffness, congruent with tissue property measurements found in literature [5,9]. Finite element models were created and meshed in Femap v8.0 (EDS Inc.) and solved using Abaqus Standard 6.2-5 (Abaqus Inc.) with nonlinear geometry formulation.

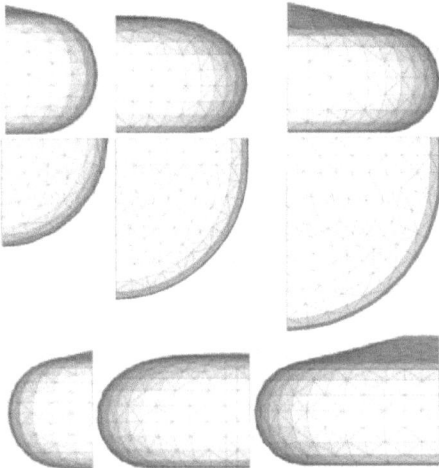

Fig. 3. Symmetric Finite element models of half of the small, medium and large breasts created for this study. From the top down are shown the front, top, and side views

Mammographic and tactile image compressions were applied to the models, and the deformations recorded. Deformations of up to 50% strain were achieved on the postmenopausal breast models under mammographic compression, while the tactile imaging compression was limited to the tactile imaging pressure maximum of 80 Pa which resulted in compressions less than 30%. The tactile imager was modeled as a long cylinder spanning the width of the breast, and was indented into the breast at five positions away from the chest wall. The full compression field from the tactile imager was interpolated between the results for each position. The full mammographic compression and one sample compression of the tactile imager on the medium sized breast are shown in figure 4.

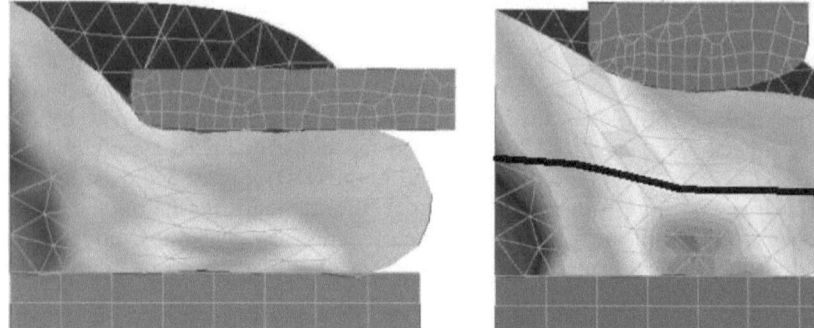

Fig. 4. Medium-sized, post menopausal breast model under mammographic and tactile compression, showing areas of increased von Mises stress at center of the base plate. The contact boundaries were non-slip and the breast tissue was fixed to the chest wall. The tactile imager scanhead was modeled at five locations and the total tactile compression was interpolated between the results. The midline of the model is shown at right as a dark line

2.2 Generating the Registration Algorithm

The tactile and mammographic images are two-dimensional projections of three-dimensional information. The mammogram is a simple projection through the tissue by near-parallel beams of x-rays. The tactile image is a projection of three-dimensional pressure information onto a flat plane that best fits the approximately horizontal surface of the top of the breast as it is compressed by the tactile imager. For each case, we need to collapse the three-dimensional deformation field information of the models into a two-dimensional field for analysis.

The tactile imager records the palpable extent of tissues, and so points below the midline form a small contribution to the tactile image [5]. Figure 6a shows the y displacements (away from the chest wall, per figure 5) of the tactile image compression for points at the finite element nodes in the vertical plane of symmetry of the medium breast model. The results are similar for the other breast models. The displacement field along the midline of a tactile image compression is very similar to the displacement field for points above the midline, which have the greatest contribution to the tactile image. Similarly, from figure 6b we see that the mammographic midline displacements are an average of the displacement of all points in the mammogram, which contribute to the resulting mammogram. Thus the deformations of the midline [figure 4] are representative of the deformations of the tissue imaged in both the tactile image and the mammogram. We focus our study on the deformation field of the points on the horizontal midplane of the breast in order to determine an algorithm to register the tactile image points to their corresponding points on the mammogram.

Figure 7 shows the displacement field for the horizontal midplane of the tactile image and mammogram of the medium breast. We can collapse the displacement in y (away from the chest wall) onto one axis [figure 8a] and note a good fit to a linear trend with zero difference at the edge of the mammography plate ($y = 0$). As such we expect the registration algorithm required to take the point at y in the tactile image to y' in the mammogram to have the form

$$y' = y\,(\,1 + K_{yy}y\,).$$

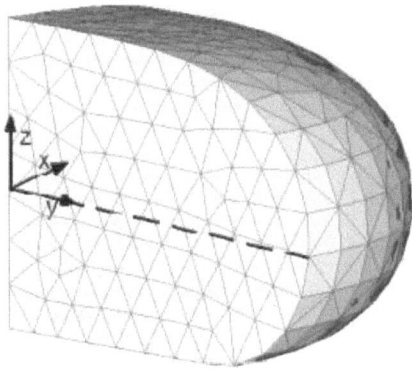

Fig. 5. Cartesion coordinates used for model and data analysis. The flat back of each model, corresponding to the rigid chest wall, is at $y = 1$ cm, so that the back edge of the mammography plate used clinically is at $y = 0$. The dashed line connects the chest wall to the nipple along the y axis

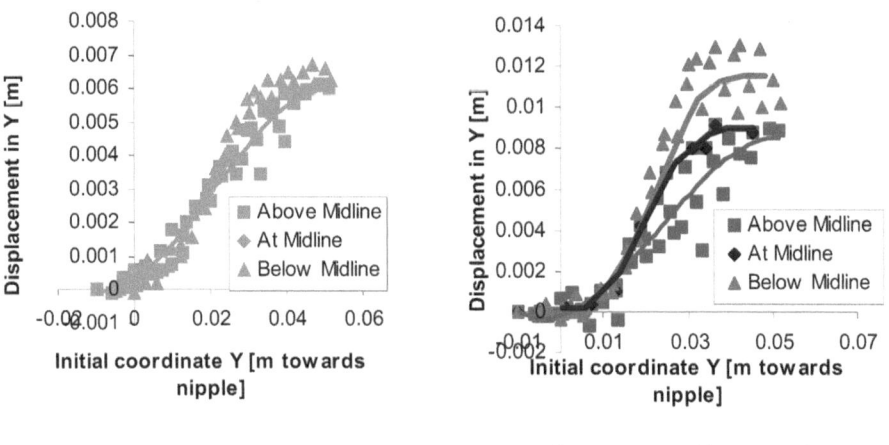

a) Tactile Image Translation

b) Mammographic Translation

Fig. 6. Forward translation of points in the vertical midplane of the medium model. Points at the finite element nodes are shown. The midline is representative of the points which generate each image

Performing a similar analysis for the difference in the x-displacements (along the chest wall, away from the center) yields figure 8b, showing a much larger spread of the data. Separating the data into bins along the y-axis shows a clear dependence on the y-position of the point in the tactile image. As such, we expect the registration algorithm in the x direction (from x in the tactile image to x' in the mammogram) to take the form

$$x' = x\,(\,1 + K_{xx}x\,)\,(\,1 + K_{xy}y\,).$$

The parameters K_{yy}, K_{xx}, and K_{xy} must depend on the size of the images and the total required deformation. We denote the maximum excursion of the mammogram and the tactile image in the x direction as X_{Mm} and X_{Mt} respectively and in the y-direction as Y_{Mm} and Y_{Mt} and use these values in determining the constants K_{yy}, K_{xx}, and K_{xy}.

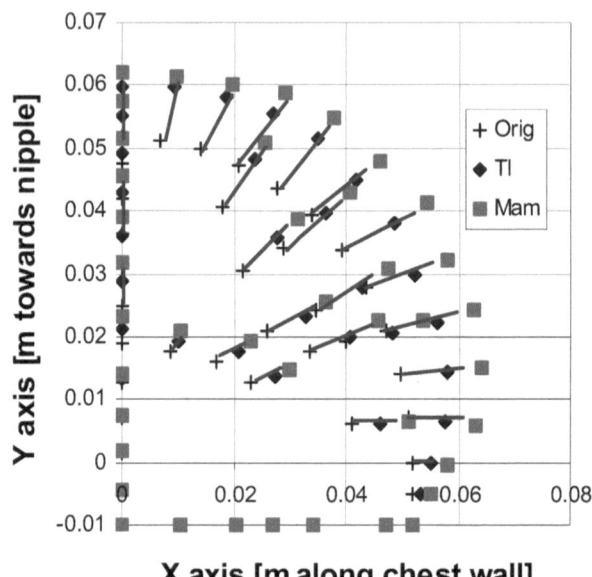

Fig. 7. Mammographic and tactile imaging displacement fields for the modeled medium breast. The solid lines indicate the average translation of the original point

The parameter K_{yy} needed to stretch the tactile image in the y-direction is dependent on the difference between the peak values for the mammogram and the tactile image. We therefore propose the following transformation in the y-direction:

$$y' = y\left(1 + \frac{y}{Y_{Mt}} \frac{Y_{Mm} - Y_{Mt}}{Y_{Mt}}\right). \tag{1}$$

Similarly, in the x-direction we expect K_{xx} to depend on the difference $X_{Mm} - X_{Mt}$. The parameter K_{xy} should, much like K_{yy}, depend on Y_{Mm}-Y_{Mt}, but examination of the various sized models shows a dependence on X_{Mm}-X_{Mt} as well. The final transformation in the x-direction was found to be

$$x' = x\left(1 + \frac{x}{X_{Mt}} \frac{X_{Mm} - X_{Mt}}{X_{Mt}}\right)\left(1 + \frac{y}{Y_{Mm}}\left(\frac{Y_{Mm} - Y_{Mt}}{2[2X_{Mm} - 2X_{Mt})]}\right)\right)^3. \tag{2}$$

The third power is an empirical constant that accounts for the cubic relationship between the volume of tissue to be compressed and the linear difference between the amount of deformation in the two modalities.

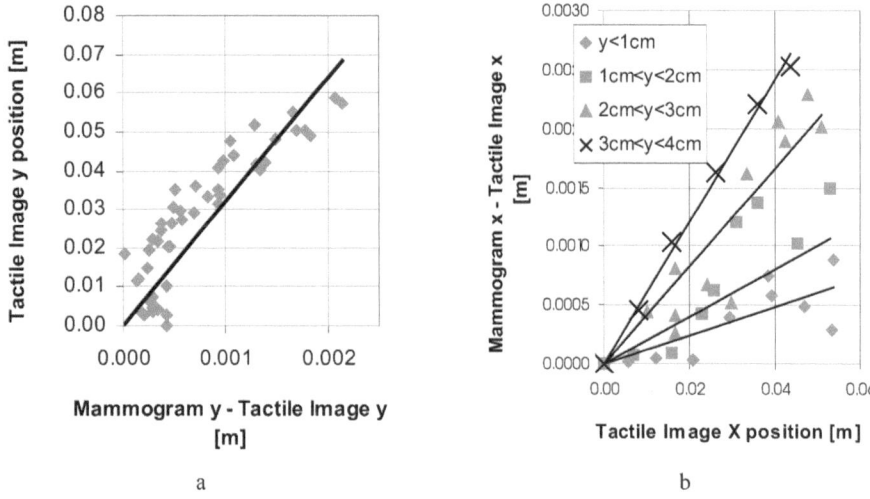

Fig. 8. Difference in translation in y (a) and x (b) between the tactile imaging and the mammographic compression. The difference in y shows a linear growth with the tactile imaging y-position, while the difference in x shows a linear relationship to both x and y

2.3   Model Results

The registration algorithm in equations 1 and 2 was applied to the deformation fields in the modeled tactile images. The resulting deformation matched the mammographic deformation well (table 1). We also applied the registration algorithm to deformations from a fourth model, of a medium-sized pre-menopausal breast which incorporated a cone of glandular tissue with its base at the chest wall and its tip at the nipple (figure 9). The glandular tissue was modeled as linear elastic with a stiffness three times greater than the surrounding fat [5]. The stiffer glandular tissue did not affect the quality of the registration. As seen from table 1 the mean absolute error of the registration is better than that for the other models, most likely because the absolute deformations are smaller for this stiffer model. The results from this premenopausal model are shown in figure 10.

Table 1. Error between the registered tactile image (equations 1 and 2) and the corresponding points on the mammogram for the midplane of the finite element models. Error is presented as percentage of final position

Model	Mean Absolute Registration Error
Small postmenopausal	5.4%
Medium postmenopausal	2.7%
Large postmenopausal	4.2%
Medium premenopausal	1.6%

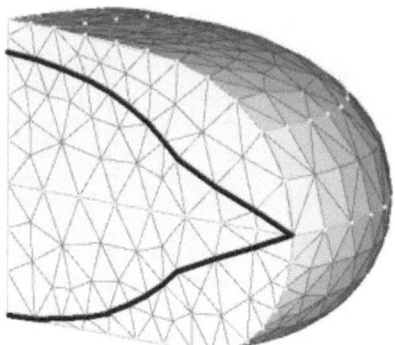

Fig. 9. Medium pre-menopausal breast model showing outline of glandular tissue cone with its base at the chest wall and apex at the nipple

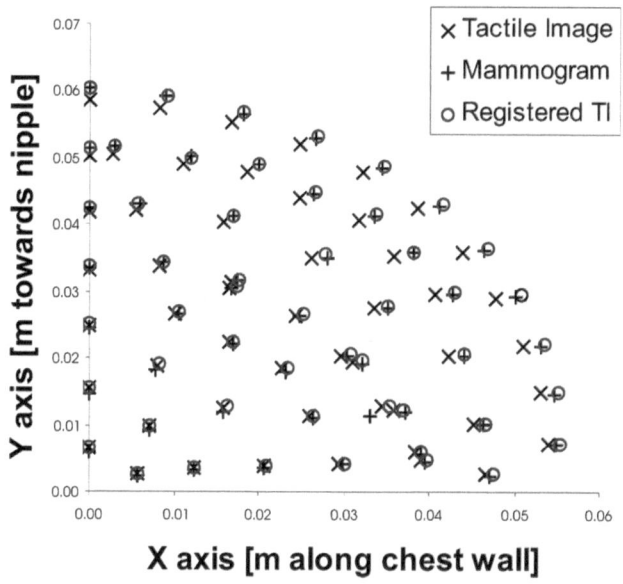

Fig. 10. Application of registration algorithm to the horizontal midplane tactile imaging deformations of the pre-menopausal medium breast model shown in figure 11. Perfect registration occurs when the open circles completely enclose the vertical crosses

3   Preliminary Application to Clinical Data

The algorithm generated in section 2 registers tactile images to mammograms for breast cancer screening. A full evaluation of the registration algorithm will require many subjects, most effectively by imaging subjects with known abnormalities that image well in both mammography and tactile imaging. As a preliminary study, we tested the registration on mammograms and tactile images from subjects undergoing breast cancer screening in order to establish a protocol for obtaining and registering tactile images in a screening situation.

3.1 Clinical Data Acquisition

Subjects were recruited who were scheduled to undergo a routine mammogram. The subjects varied in age from premenopausal to postmenopausal, in size from small to large breasts, and included women with previous positive and negative biopsies. Tactile images were recorded on each breast immediately after the cranio-caudal mammogram, before the subject moved relative to the bottom plate of the mammogram.

The tactile images obtained had a 30 x 30 cm field of view, which was sufficient to record the entire area of the largest mammographic plate. The tactile image resolution was 0.5mm/pixel [7]. The mammograms were digitized to the same resolution, and the top left corner of the film assigned to $(x,y) = (1,1)$ cm, since the film was 1 cm smaller than the mammography plate on each side.

The tactile imager used in this study uses a magnetic tracker for positioning. Due to the magnetic interference of the mammography machine, the raw tactile images obtained required a preliminary calibration step in order to align the x and y axes with those of the mammogram. Position calibration information was obtained from the mammography plate by obtaining a tactile image of soft foam over the edges of the mammography plate and the mammography markers used in each mammogram. A third-order transformation was applied to the raw tactile images based on the calibration information acquired from the plate edges and the fiducials. This transformation was a simple polynomial fit that satisfies the laws of magnetism [10] and is based on the finding of incremental magnetic positioning error [11].

3.2 Clinical Data Results

Tactile images exhibit increased signal intensity near stiff areas, whereas mammograms record tissue radioopacity. Glandular and fibrous tissue, for example, are both stiffer than fat and are also more radioopaque. Therefore some degree of correlation is expected between the two signals. We expect the tactile image to pick up instances of stiffness that are not radioopaque [12], such as some older scars [13].

The registration algorithm in equations (1) and (2) was applied to register the calibrated tactile images to the digitized mammograms and the resulting images compared. Figure 11 shows the registered images for a subject (one breast). The registered tactile image shows increased palpation intensity in the regions the mammogram shows higher glandularity, and a correlation verifies the images are well-registered (correlation peak within 3mm of the center). Figure 12 shows another mammogram-tactile image pair. The correlation between these two images is far poorer. The tactile image clearly exhibits more extensive stiff regions than the mammogram implies. Note, however, that near the base of the mammogram there are a few small very bright dots. These are clips left from previous biopsies and indicate areas of scarring. Scar is stiffer than the surrounding tissue, but does not image well in mammography a few years post-surgery [13].

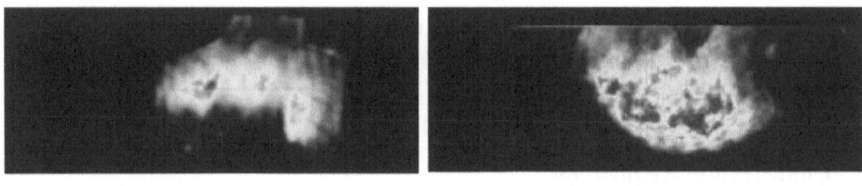

Registered Tactile Image   Mammogram

Fig. 11. Left: Tactile image after unwarping and registration shown for a 10cm by 20cm area. Right: the corresponding mammogram. The x direction increases to the right, and the y direction (towards the nipple) increases downwards from the chest wall. All images are represented by a single intensity value at each pixel, where the dark background has the lowest value, the value increases through the pale areas, and dark areas surrounded by light are the highest. The mammogram starts at $y=1cm$ since the edge of the plate is approximately 1cm behind the edge of the mammographic film

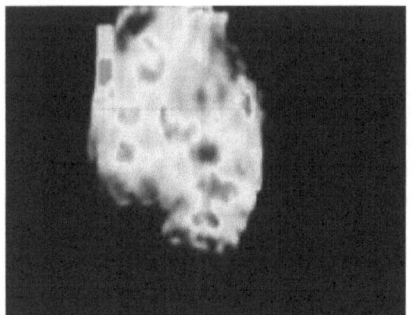
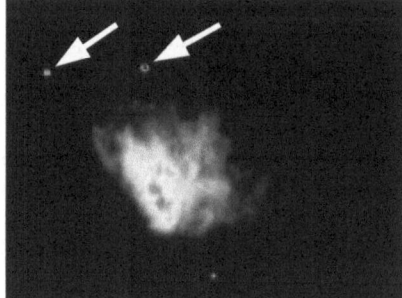

Registered Tactile Image   Mammogram

Fig. 12. A registered tactile image and mammogram pair shown for a 15cm by 15cm area. The x direction increases to the right, and the y direction (towards the nipple) increases downwards from the chest wall. The tactile image shows a lot more bright regions than the mammogram, since this subject has had several biopsies which have resulted in scarring. In the mammogram the only evidence of the scarring are the two small bright dots indicated by the arrows, which are left over from the earlier biopsies

4   Discussion

The registration of tactile images to mammograms is inherently dependent on the amount of mammographic compression, the uniformity of that compression, and the breast size, shape, and stiffness characteristics. The registration algorithm generated in this study fits the four finite element models' midline deformations well, though it is not possible to create an algorithm with constants derived from the simple parameters used in this study that would work perfectly for every breast. The parameters used in this study, namely the extent from and along the chest wall of each image, are simple edge parameters readily available in a clinical setting, The algorithm used in this study uses minimal information from the mammogram with no registration bias in order to generate results that are accurate to within a few percent.

The results of this study show promise in increasing the efficacy of breast cancer screening by adding a new, inexpensive imaging modality registered to the current standard. The next stage in this study will involve using the protocol developed to obtain tactile images and mammograms on more subjects beyond the preliminary study presented here. The registered images will be presented to radiologists, who routinely study mammograms, to highlight sections of interest in each mammogram with and without the tactile image. It is from this that we will realize an increase in the positive predictive value of screening mammography.

Acknowledgements

The clinical data collected was supported by Assurance Medical Inc. and Dr. E. Dalton. The authors would like to thank Camilla Lau for assistance with the finite element modeling of this work.

References

1. Jatoi I.: Breast Cancer Screening. Medical Intelligence unit, Chapman & Hall, New York (1997)
2. Kopans D.B.: Clinical Breast Examination for Detecting Breast Cancer. JAMA. 283 (2000) 1688
3. Lester S., Cotran R.,: The Breast. Chapter 25 in Robbins Pathologic Basis of Disease, 6th Ed., WB Saunders Co., Philadelphia (1999)
4. Ronnov-Jessen L., Petersen O., Bissel M.: Cellular Changes Involved in Conversion of Normal to Malignant Breast: Importance of the Stromal Reaction. Physiological Reviews 76 (1996) 69-125
5. Wellman, P S.: Tactile Imaging. Doctoral Thesis, Division of Engineering and Applied Sciences, Harvard University: (1999)
6. Wang, Y., Nguyen, C., Srikanchana, R., Geng, Z., Freedman, M.T.: Tactile Mapping of Palpable Abnormalities for Breast Cancer Diagnosis. Proc. IEEE Intl. Conf. Robotics Automation. (1999) 1305-9
7. Wellman P.S., Howe R.D., Dewagan N., Cundari M.A., Dalton E., Kern K.A.: Tactile Imaging: A Method for Documenting Breast Masses, Proc. First Joint Biomed Eng Soc/IEEE Eng Med Bio Soc Conf, (1999) 1131-2.
8. US Dept of Health and Human Services: Clinical Practice Guideline. High-Quality Mammography: Information for referring providers. Agency for Health Care Policy and Research, Rockville, MD (1994)
9. Krouskop, T A, Price, R E, Wheeler, T, Younes, P S: Modulus Variations in Breast Tissues. Proc of the 1st Int. Conf. On the Ultasonic Measurement and Imaging of Tissue Elasticity. Niagara Falls, 2002
10. DiBartolo, B.: Classical Theory of Electromagnetism. Prentice Hall, USA (1991)
11. Milne, A.D., Chess, D.G., Johnson, J.A., King, G.J.W.: Accuracy of an Electromagnetic Tracking Device: A Study of the Optimal Operating Range and Metal Interference. J. Biomechanics, 29 (1996) 791-7936
12. Wellman, P S, Dalton E P, et al.: Tactile Imaging of Masses: First Clinical Report. Archives of Surgery, 136 (2001) 204-08
13. Krag, D. Personal Communication. February 2003.

Realistic Haptic Interaction in Volume Sculpting for Surgery Simulation

Andreas Petersik[1], Bernhard Pflesser[1], Ulf Tiede[1],
Karl-Heinz Höhne[1], and Rudolf Leuwer[2]

[1] Institute of Mathematics and Computer Science in Medicine (IMDM),
University Hospital Hamburg-Eppendorf, Germany
{petersik,pflesser,tiede,hoehne}@uke.uni-hamburg.de
[2] ENT-Clinic,
University Hospital Hamburg-Eppendorf, Germany
leuwer@uke.uni-hamburg.de

Abstract. Realistic haptic interaction in volume sculpting is a decisive prerequisite for successful simulation of bone surgery. We present a haptic rendering algorithm, based on a multi-point collision detection approach which provides realistic tool interactions. Both haptics and graphics are rendered at sub-voxel resolution, which leads to a high level of detail and enables the exploration of the models at any scale. With a simulated drill bony structures can be removed interactively. The characteristics of the real drilling procedure like material distribution around the drill are considered to enable a realistic sensation. All forces are calculated at an extra high update rate of 6000 Hz which enables rendering of drilling vibrations and stiff surfaces. As a main application, a simulator for petrous bone surgery was developed. With the simulated drill, access paths to the middle ear can be studied. This allows a realistic training without the need for cadaveric material.

1 Introduction

In surgery simulation it is often not sufficient to deal only with a graphical representation of the anatomy. For most applications the sense of touch is necessary for realistic interaction. In contrast to our other senses it allows us to simultaneously explore and interact with our environment. Most applications of surgery simulation concentrate on the simulation of elastic deformations of soft tissue. In contrast to that, the simulation of material removal (sawing, milling, drilling) as performed in bone surgery is a less developed field and existing systems do not provide the 'look and feel' close to the real procedure.

Realistic look and feel means that on one hand the irregular shaped cut surfaces must have realistic shape and texture in the visual representation, on the other hand the algorithm for haptic rendering must be able to provide detailed sensations of all structures with no delay. Both requirements are fulfilled best by using a volume model. As we have shown earlier, it allows very realistic display of cut surfaces [8]. For realistic haptic rendering it has the decisive advantage over a polygonal representation that computation time for collision detection is independent of the complexity of the scene.

N. Ayache and H. Delingette (Eds.): IS4TM 2003, LNCS 2673, pp. 194–202, 2003.

Existing algorithms for haptic rendering are mostly point-based, i.e. only one point of the virtual tool is used to calculate collisions and forces. While this might be sufficient for simulation of deformations by the tip of a tool, it leads to problems in the case of material removal:

- In complex scenes (e.g. with sharp edges) the shape of the tool plays a decisive role for the haptic sensation and cannot be simulated by one single point.
- The virtual tool can reach points which cannot be reached by the simulated real world tool. (A large drill could enter a small hole.)

The purpose of the work presented in this paper is to develop algorithms which can calculate forces for haptic feedback with high resolution and realism which are suitable for material removal from rigid objects. Realistic feeling should be provided while touching the model as well as while actively modifying the model with virtual tools. The algorithms should be able to handle models of arbitrary complexity. The feasibility of the algorithms is to be demonstrated with a petrous bone surgery simulator, where different types of virtual drills can be used both to touch and to surgically modify the anatomic model.

2   Related Work

An early approach of haptic interaction with volume visualization was presented in [1]. While this approach allows exploration and modification of the volumetric data, the forces generated are not intended to be a realistic simulation of interacting with materials. Rather, the intent is to convey additional information to the user about the data being explored. Furthermore this approach uses a single-point contact model which does not fulfill our needs for a realistic interaction.

As stated earlier, single-point interaction is not realistic in that it does not prohibit unrealistic situations like entering a small hole with a large drill. Therefore multi-point collision detection approaches were developed [2,3]. With these approaches, more realistic simulations of tool-object interactions can be achieved.

The voxel-based approach to haptic rendering presented in [2] enables 6-DOF interaction of rigid tool within an arbitrary complex environment of static objects. This approach provides a realistic haptic feedback in regard to the exact shape of the tool. One drawback of this method is that the haptic rendering is at the resolution of the original voxels only which is not sufficient for medical applications. Also modification of the static environment is not included.

In [4] and [5] simulators for petrous bone surgery are presented, but they lack of high resolution rendering. Most structures which are important for a successful petrous bone surgery are not visible.

3   Methods

In order to achieve a realistic haptic surface rendering, collisions between the tool and the static scene must be computed and a collision-free position must

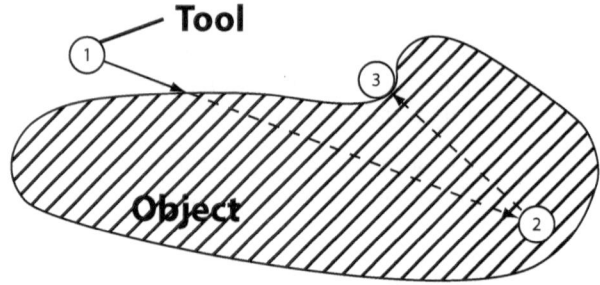

Fig. 1. To get a realistic haptic rendering of surfaces, position 3 must be calculated based on the last device position 1 and the new device position 2.

be determined. Figure 1 shows a situation where the user moved the virtual tool from position 1 to position 2. Since the tool could not reach position 2 in reality, the rendering algorithm must calculate position 3 which would have been reached in reality. Then a force which pushes the haptic device to position 3 must be applied.

In order to develop a simulator for petrous bone surgery which allows realistic drilling into the mastoid bone the following points concerning haptics had to be considered:

- Haptic rendering should be based on a multi-point collision detection approach to allow realistic tool-object interactions for passive interaction.
- For material removal an algorithm is needed which works with sub-voxel resolution to be able to simulate the effect of small tools as they are used in petrous bone surgery.
- To enable realistic haptic interactions while modifying the models, an algorithm is needed which calculates realistic drilling forces based on parameters like: amount of removed material, distribution of material around the drill etc.

3.1   Data Representation

The model of the petrous bone was created from CT-data and is represented by a volume of attributed voxels (volume elements) which have a size of $0.33mm^3$. The attributes are density values and membership to an organ. The membership to an organ is determined during the semi-automatic, threshold based segmentation process.

Since our voxel-based representation does not contain an explicit representation of the object surfaces, the surfaces must be calculated based on the segmentation data. This is done by a ray-casting algorithm [6] which renders iso-surfaces at sub-voxel resolution based on the partial volume effect and density value of the voxels. The sub-voxel approach leads to very detailed surfaces which can be explored at any scale.

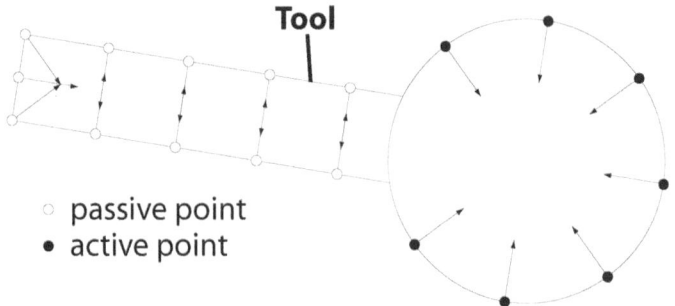

Fig. 2. Tool representation by surface points and inward pointing normal vectors.

3.2   Tool Representation

For multi-point collision detection the tool is represented by a number of sample points P_i which are distributed at preferably equal distances over the tool surface. Each of these points is checked whether it collides with the objects or not. Additionally every point has an associated normal vector n_i which is pointing to the inside of the tool.

The inward pointing vectors n_i describe the tool's curvature and can be used by the collision detection algorithm to find the static object's surface.

In our implementation of the petrous bone surgery simulator we are using a sphere-shaped tool, which simulates a drill (Fig. 2). To get an adequate representation of the tool's shape while reducing computation to a minimum, we are using 56 sample points on the tool's surface.

3.3   Haptic Surface Rendering

Our multi-point collision detection algorithm was inspired by the work described in [2]. However our approach differs from this work in several points. While our model is also using a voxel representation for the static objects, the exact location of the surfaces is calculated by a ray-casting algorithm at sub-voxel resolution (see 3.1). This leads to a more precise calculation of force direction and surface location. The algorithm presented in [2] cannot provide the precision which is needed in our cases, since the static objects are voxelized in a binary manner. Since our sub-voxel rendering approach is generating iso-surfaces the problems with discontinuities at voxel boundaries is eliminated. Another improvement we have made is the representation of the dynamic object. While the dynamic object in [2] is voxelized and the center points of the voxels are used for the collision detection, we are using sample points which are located exactly on the surface of the dynamic object, which improves the resolution further.

Collision Detection Algorithm. To calculate the collision force direction, all surface points of the tool are checked, whether they are inside or outside the object. All surface points P_i which are inside the volume (Fig. 3, filled dots) are

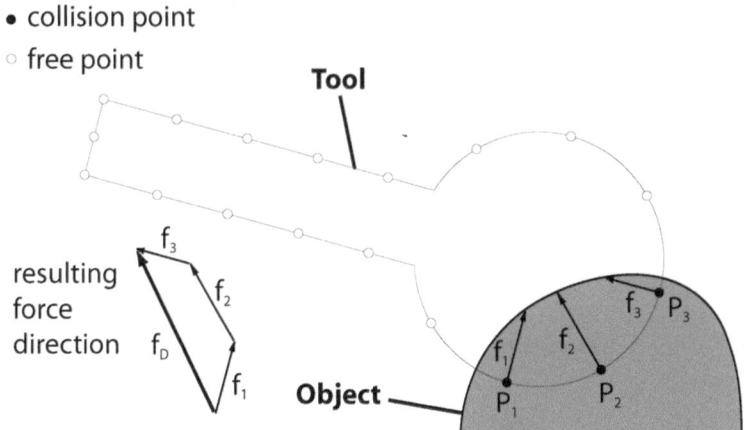

Fig. 3. Calculation of forces during passive interaction. The resulting force direction f_D is calculated by adding the three vectors f_1, f_2 and f_3.

traced along the inward pointing normal until the surface is found. All found vectors f_i are added and the direction of the sum vector f_D is the direction of the force vector which must be applied to the haptic device. As stated in [2], the summation of the found vectors would lead to force instabilities. To avoid this we decided not to use force summing for the calculation of the magnitude of the force vector. Instead we are searching for the longest projection of the vector f_i on the vector f_D.

Proxy Object Algorithm. The previously described algorithm works only for small tool object penetrations, when the surface can be found for all collision points. To overcome that limitation a modified proxy object algorithm was implemented. The idea behind a proxy object is to represent the device position by a virtual object which never penetrates objects [7].

Searching for the local minimum to update the proxy position as described in [7] would be computationally too expensive in our model. Thus a simplified algorithm was implemented. Whenever more than a certain number of surface sample points of the dynamic object are in contact with an object or one inward pointing vector is completely immersed, the way between proxy and current position is traced until the object surface is found. Starting from this position the way back to the proxy is traced until the number of contacts is below the limit so that the force vector can be calculated as described in chapter 3.3.

3.4   Volume Interaction

The sense of touch allows not only the exploration of the anatomy but also the interactive modification of the objects we are touching. In petrous bone surgery the drill allows the surgeon to feel and to drill the bone.

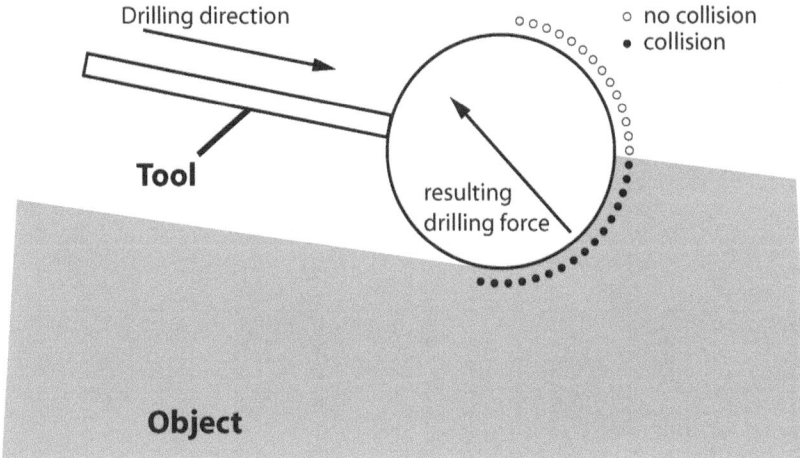

Fig. 4. Calculation of forces during drilling. The resulting force is dependent on collisions with sample points around the drill.

Our freeform volume modification algorithm which is described in detail in [8] is working with sub-voxel resolution. This produces realistic structures even when using very small tools.

Calculation of Drilling Forces. During volume modification, the modified volume must be modelled and the visual renderer must redraw the modified region. Both steps consume too much time to be able to perform the collision detection based on the modified volume. Thus drilling forces must be calculated by a simplified algorithm, which can run asynchronously to the modification process.

To get a realistic force during drilling, we apply as a first approximation a force which is opposite to the drilling direction and the speed of the drill movement. The vector of this force is pointing from the current position of the tool P_C to the last stored position P_L.

Additionally we consider the material distribution around the drill and the amount of removed material. The material distribution and the amount of removed material is calculated by looking for collisions at sample points in front of the drill (Fig. 4).

The balance point b of the mass removal in tool coordinates can be calculated by adding all colliding active points of the tool A_i and dividing the sum by the number c of colliding vectors found:

$$b = \frac{\sum_{i=1}^{c} A_i}{c} \tag{1}$$

The direction of the drilling force vector f can now be calculated as follows:

$$f = P_C - b \cdot \frac{\|P_C - P_L\|}{\|b\|} \tag{2}$$

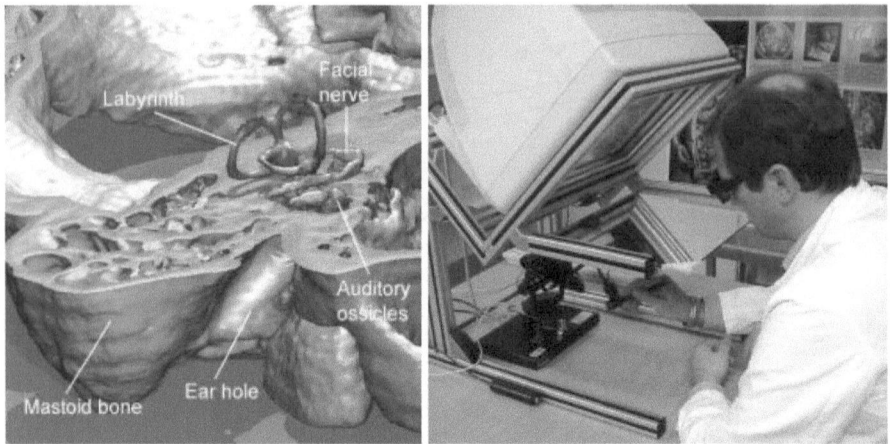

Fig. 5. The petrous bone model was built from CT data (left). The setup of the simulator together with stereoscopic viewing allows a realistic handling close to the real procedure (right).

In figure 4 you can see one example for such a calculation: Material will be removed and the vector of the drilling force will be rotated about 45 degrees because only one quarter of the spherical drill is in contact with the material.

4   Implementation

The petrous bone surgery simulator was integrated in the VOXEL-MAN [9] system, which provides the anatomic model, high-quality visual rendering and also free-form volume modification. The system was implemented on a dual processor AMD PC with two AthlonMP 1900+ processors. The PC is equipped with 1GB DDRAM. As haptic device we are using a 3-DOF Phantom Premium 1.0A (Sensable Technologies Inc.). Our system is running SuSE Linux 8.0. For connecting the device directly to the system, we are using the open-source Phantom Linux-driver (http://decibel.fi.muni.cz/phantom), which enables haptic update rates between 1 and 10kHz. To enable stereoscopic viewing we are using an Nvidia Quadro2MXR graphics board in combination with ELSA Revelator shutter glasses.

5   Results and Conclusions

With the described algorithms for haptic material removal interactions, haptic realism can be achieved through:

- congruent sub-voxel resolution both for visualization and haptic rendering
- multi-point collision detection

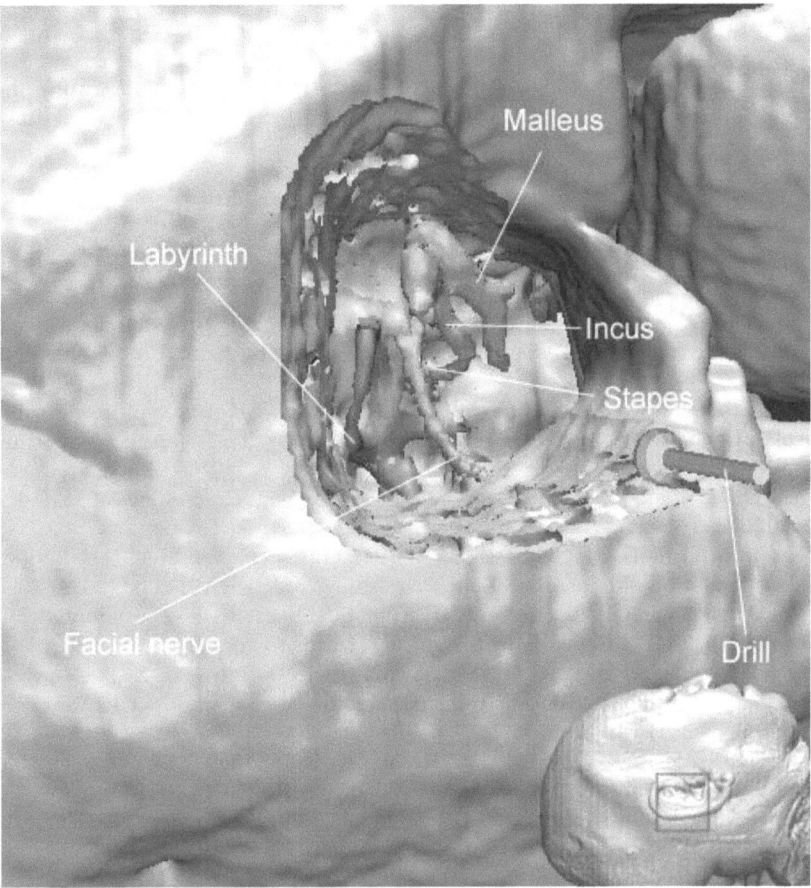

Fig. 6. Simulation of petrous bone surgery. The haptic sensation is congruent to the detailed visualization of the irregular structures.

- high haptic update rate of 6000Hz
- consideration of material distribution around the drill
- simulation of drilling vibrations

The implemented simulator for petrous bone surgery has shown that the developed algorithms allow a realistic simulation of material removal procedures which could be used in other medical areas like craniofacial surgery or dentistry.

Of course a quantitative validation of the algorithms is difficult. So far we rely on the judgement of our ENT-surgeons that the haptic feeling of the simulation is nearly indistinguishable from that of the real procedure.

The next step of our research will be the extension of the algorithms to be able to deal with more complex shapes and functions of the tool. An example for this is a saw blade with an active face and a passive backside.

Acknowledgements

We are grateful for the help of Andreas Pommert, Martin Riemer, Rainer Schubert and all members of the VOXEL-MAN team.

References

1. Avila, R.S., Sobierajski, L.M.: A haptic interaction method for volume visualization. Proceedings of Visualization 96 (1996) 197–204
2. McNeely, W.A., Puterbaugh, K.D., Troy, J.J.: Six degree-of-freedom haptic rendering using voxel sampling. Computer Graphics (SIGGRAPH99 Proceedings) (1999) 401–408
3. Kim, Y.J., Otaduy, M.A., Lin, M.C., Dinesh, M.: Six-degree-of-freedom haptic display using localized contact computations. Proceedings of 10th Symposium on Haptic Interfaces (2002) 209–216
4. Wiet, G.J., Bryan, J.: Virtual temporal bone dissection. Proceedings of MMVR2000 (2000) 378–384
5. Agus, M., Giachetti, A., Gobbetti, E., Zanetti, G., Zorcolo, A.: Real-time haptic and visual simulation of bone dissection. Proceedings of IEEE VR 2002 (2002) 209–216
6. Tiede, U., Schiemann, T., Höhne, K.H.: High quality rendering of attributed volume data. In Ebert, D., Hagen, H., Rushmeier, H., eds.: Proc. IEEE Visualization '98, Research Triangle Park, NC (1998) 255–262 (ISBN 0-8186-9176-X).
7. Zilles, C., Salisbury, K.: A constraint-based god object method for haptics display. Proceedings of IEEE/RSJ (1995)
8. Pflesser, B., Petersik, A., Tiede, U., Höhne, K.H., Leuwer, R.: Volume cutting for virtual petrous bone surgery. Comput. Aided Surg. **7** (2002) 74–83
9. Höhne, K.H., Pflesser, B., Pommert, A., Riemer, M., Schiemann, T., Schubert, R., Tiede, U.: A new representation of knowledge concerning human anatomy and function. Nat. Med. **1** (1995) 506–511

Capturing Brain Deformation

Simon K. Warfield, Florin Talos, Corey Kemper, Lauren O'Donnell, Carl-Fredrik
Westin, William M. Wells, Peter McL. Black, Ferenc A. Jolesz, and Ron Kikinis

Computational Radiology Laboratory, Surgical Planning Laboratory, Department of Radiology,
Department of Surgery, Harvard Medical School and Brigham and Women's Hospital, 75
Francis St., Boston, MA 02115 USA, Massachusetts Institute of Technology
warfield@bwh.harvard.edu

Abstract. A critical challenge for the neurosurgeon during surgery is to be able
to preserve healthy tissue and minimize the disruption of critical anatomical structures while at the same time removing as much tumor tissue as possible. Over the
past several years we have developed intraoperative image processing algorithms
with the goal of augmenting the surgeon's capacity to achieve maximal tumor resection while minimizing the disruption to normal tissue. The brain of the patient
often changes shape in a nonrigid fashion over the course of a surgery, due to loss
of cerebrospinal fluid, concomitant pressure changes, the impact of anaesthetics
and the surgical resection itself. This further increases the challenge of visualizing and navigating critical brain structures. The primary concept of our approach
is to exploit intraoperative image acquisition to directly visualize the morphology
of brain as it changes over the course of the surgery, and to enhance the surgeon's
capacity to visualize critical structures by projecting extensive preoperative data
into the intraoperative configuration of the patient's brain.
Our approach to tracking brain changes during neurosurgery has been previously
described. We identify key structures in volumetric preoperative and intraoperative scans, and use the constraints provided by the matching of these key surfaces
to compute a biomechanical simulation of the volumetric brain deformation. The
recovered volumetric deformation field can then be applied to preoperative data
sets, such as functional MRI (fMRI) or diffusion tensor MRI (DT-MRI) in order to warp this data into the new configuration of the patient's brain. In recent
work we have constructed visualizations of preoperative fMRI and DT-MRI, and
intraoperative MRI showing a close correspondence between the matched data.
A further challenge of intraoperative image processing is that augmented visualizations must be presented to the neurosurgeon at a rate compatible with surgical
decision making. We have previously demonstrated our biomechanical simulation of brain deformation can be executed entirely during neurosurgery. We used
a generic atlas to provide surrogate information regarding the expected location of
critical anatomical structures, and were able to project this data to match the patient and to display the matched data to the neurosurgeon during the surgical procedure. The use of patient-specific DTI and fMRI preoperative data significantly
improves the localization of critical structures. The augmented visualization of
intraoperative data with relevant preoperative data can significantly enhance the
information available to the neurosurgeon.

N. Ayache and H. Delingette (Eds.): IS4TM 2003, LNCS 2673, pp. 203–217, 2003.
© Springer-Verlag Berlin Heidelberg 2003

1   Introduction

A critical challenge for the neurosurgeon during surgery is to be able to accurately identify, navigate to and remove intracranial lesions while minimizing damage to healthy
brain tissue. An overriding goal is the preservation of neurological function. This requires precise delineation of the morphology and functional anatomy of the patient's
brain and the margins of the lesion. Unfortunately, the similarity of the visual appearance of healthy and diseased brain tissue (such as with infiltrating tumors) and the
inability of the surgeon to see critical structures obscured by the brain surface poses
difficulties during the operation. Particularly important regions, such as functionally
eloquent regions and white matterfiber tracts, may not be visually recognizable at all.
While complete resection of diseased tissue is regarded as the best surgical outcome,
and is believed to be correlated with improved patient outcome, it is made extremely
difficult by the restricted capacity to perceive lesion (e.g. tumor) boundaries [1].

Image-guided neurosurgery (IGNS) techniques have improved rapidly over the past
decade and this has lead to the development of sophisticated minimally-invasive procedures. These procedures are executed in operating rooms specially equipped with
imaging and navigation systems. These systems are used to provide real-time imaging
on demand to the surgeon, and offer the possibility of augmented three-dimensional
navigation and quantitative monitoring of the progress of the surgery. The improved visualization of critical structures, and the availability of imaging modalities which show
improved contrast between lesion and healthy tissue has allowed the planning and execution of surgical procedures with greater precision.

Considerable previous work in support of IGNS has been concerned with improving image acquisition, registration and display [2]. Novel algorithm development has
focused on improving imaging speed and quality. Due to the constraints of an operating
room, intraoperative imaging typically results in images with lower signal to noise ratio
and less fiexibility in the choice of imaging modality than conventional imaging done
outside the operating room. In order to improve the richness of the visual information
presented to the surgeon the incorporation of several forms of preoperative imaging
information, including conventional morphologic MRI and other modalities has been
explored. Several research groups have explored models for using imaging data to aid
in the estimation or prediction of the extent of brain deformation [3–13].

We have previously developed and described an algorithm which enables the projection of preoperative images onto intraoperative images [14], allowing fusion of images from multiple imaging modalities and with multiple contrast types. The algorithm
tracks surfaces of the brain and ventricles in intraoperatively acquired images, allowing
the projection of preoperative images into the configuration of the patient's brain during
the neurosurgical procedure. A volumetric deformationfield is then inferred from the
surface changes. This field captures nonrigid deformations of the shape of the brain due
to factors such as brain swelling, cerebrospinal fluid loss, anaesthetic agents and the
actions of the neurosurgeon. We have also investigated an extension of this technique
to deal with surgical resection [15]. In order for this approach to have practical appli-
cability during neurosurgery, it must meet the real-time constraints of neurosurgery. If
the updated images arrive too late to provide meaningful guidance to the neurosurgeon
then their value is lost. We have recently reported the successful projection of anatom-

ical information from a generic digital atlas onto patient specific imaging data during neurosurgical procedures [16], at a rate compatible with the requirements of neurosurgical decision making. We have also carried out successful alignment of patient-specific DTI and fMRI preoperative data into the intraoperative configuration of the patient's brain. This can significantly enhance the information available to the neurosurgeon.

2   Method

This section provides an overview and summary of our previously reported image analysis procedures. Image analysis takes place both before and during the IGNS procedure. The goal of enhanced data interpretation is achieved by algorithms for image segmentation and registration. In order to rapidly and accurately segment anatomical structures from the volumetric imaging data acquired during IGNS we exploit preoperative data acquisitions. Preoperative segmentation is also used for preoperative planning of the surgical approach. Intraoperative segmentation can be used for quantitative monitoring of surgical progression, which is especially useful when dealing with cryoablation or laser ablation. Image registration is used to project patient-specific preoperative data onto the intraoperative data in two steps. The first step is to estimate a rigid transformation (translation and rotation differences) and the second is to carry out a biomechanical simulation of the volumetric brain deformation. The volumetric deformation is inferred through a biomechanical simulation with boundary conditions established via surface matching.

2.1   Preoperative Data Acquisition, Image Segmentation and Fusion

There is usually considerably more time available for preoperative data acquisition and analysis than for intraoperative data acquisition and analysis. Therefore preoperative data acquisition can be far more comprehensive than intraoperative data acquisition, and more extensive segmentation may be carried out.

At our institution we use a range of segmentation strategies, including interactive [17], semi-automated [18, 19] and automated [20–22] approaches. We select the most accurate, robust approach available for the type of preoperative data and the particular critical structures of interest in a particular surgical case.

Diffusion tensor MRI is processed to enable the display of white matter fiber tracts [23, 24], and functional MRI is analysed to indicate regions of activation [25, 26]. An illustration of the visualization of preoperative MRI, together with segmentation of the tumor (shown in green) and ventricles (blue), critical white matter fiber tracts (shown in yellow) and functionally eloquent regions detected by fMRI (shown in red) is seen in Figure 1.

2.2   Intraoperative Image Processing

Intraoperative image processing consists of acquiring a new intraoperative acquisition of one or more volumetric data sets, constructing a segmentation of the intraoperative acquisition, computing a rigid registration of the patient-specific preoperative data to

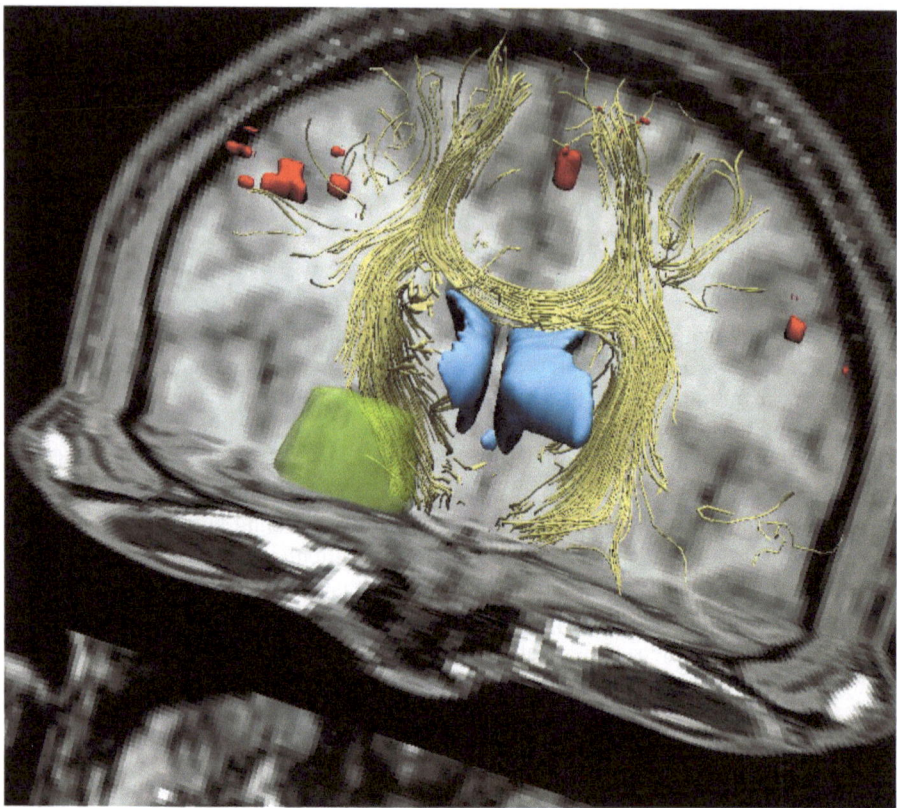

Fig. 1. Preoperative data acquisition and analysis is carried out to enable enhanced preoperative surgical planning and to provide patient-specific assessment of critical functionally eloquent regions and fiber tracts. Preoperative data can be projected into the configuration of the patient as the brain changes shape over the course of the surgery. In this image we see the preoperative MRI together with segmentation of the tumor (shown in green) and ventricles (blue), critical white matter fiber tracts (shown in yellow) and functionally eloquent regions detected by fMRI (shown in red).

the new acquisition, identifying the correspondences between key surfaces of the preoperative and intraoperative data, solving a biomechanical model to infer a volumetric deformation field, applying the deformation to the preoperative data and constructing a new visualization merging critical structures and information from the preoperative data with the intraoperative data.

Segmentation of Intraoperative Volumetric Images. In obtaining the results presented below, a rapid segmentation of the brain was obtained through a binary curvature driven evolution algorithm [19]. The region identified as brain was then interactively corrected to remove any portion of misclassified skin and muscle using the software described by Gering et al [27]. This was then repeated to obtain a segmentation of the lateral ventricles of the subject. This approach allows the neurosurgeon to inspect the segmentations as they are constructed during the surgery and enhances the surgeon's confidence in the quality and availability of the segmentations.

We have also experimented with rapid and automated intraoperative segmentation [28, 20] utilizing tissue classification in a multi-channel feature space with a model of expected anatomy. It is our expectation that this automated approach will be preferable once sufficient validation experiments have been carried out.

Validation experiments can aid in demonstrating that a segmentation approach is identifying the intended target and hence that preoperative planning and intraoperative navigation and monitoring of therapy can proceed on the basis of sufficiently accurate data. Currently much of the surgical planning, navigation and quantitative monitoring that we carry out is based upon interactive or semi-automatic segmentation by domain experts. We have recently developed a novel algorithm, STAPLE (Simultaneous Truth and Performance Level Estimation) [29] which analyses segmentations obtained by multiple experts or repeated segmentations by a single expert and estimates the hidden "true" segmentation that the experts are attempting to generate as well as the sensitivity and specificity performance parameters that describe the quality of each segmentation in achieving the hidden "true" segmentation.

2.3   Unstructured Mesh Creation and Surface Representation

From the preoperative segmentation, we extract an explicit representation of the surface of the brain and ventricles. We also create a volumetric unstructured tetrahedral mesh representing the domain over which our biomechanical simulation of deformation will ultimately be executed. We generate a tetrahedral mesh of the brain and ventricles from segmentations of each volumetric intraoperative acquisition carried out during the surgery.

We have implemented a tetrahedral mesh generator specifically suited for labeled 3D medical images, building upon previously described techniques [30, 31]. The mesh generator can be seen as the volumetric counterpart of a marching tetrahedra surface generation algorithm. A detailed explanation of the algorithm has been previously reported [10, 32, 14]. We can readily associate different biomechanical model parameters with different anatomical structures since the preoperative segmentation and tetrahedral mesh nodes have the same coordinate system.

Rigid Registration of Preoperative Volumetric Images to Intraoperative Volumetric Images.   For rigid registration we use a fast and parallel implementation of an extremely robust algorithm which is based upon aligning segmented image data [33]. We register the brain segmentations of the preoperative and intraoperative scans. This allows us to capture rigid transformation changes (rotation and translation) between the preoperative data and intraoperative acquisitions.

Volumetric Biomechanical Simulation of Brain Deformation.   As the surgical procedure progresses the brain shape changes under the influence of the surgical intervention (craniotomy, tissue resection, cerebrospinal fluid drainage are some of the factors).

Over the course of a neurosurgical procedure, the surgeon acquires a new volumetric MRI when it is thought to be necessary to review the current configuration of the entire brain. A volumetric deformation field to enable the projection of previous acquisitions to this new scan is computed by first matching surfaces from a previous scan to the new acquisition, and then inferring the volumetric deformation based upon these surface correspondences. The primary concept is to apply forces to the volumetric model that will produce the same displacement field at the surface as was obtained by the surface matching. The biomechanical model then allows the computation of the deformation throughout the volume.

Measuring Surface Correspondences. An active surface algorithm iteratively deforms surfaces of both the brain and the lateral ventricles from the earlier acquisition to match that of the current acquisition. This is done iteratively by applying forces derived from the volumetric data to an elastic membrane model of the surface. The derived forces are a decreasing function of the image intensity gradients, so as to be minimized at the edges of objects in the volume. To increase robustness and the convergence rate of the process, we have included prior knowledge about the expected gray level and gradients of the objects being matched. This algorithm has been fully described previously [34, 15].

Biomechanical Simulation of Volumetric Brain Deformation. We treat the brain as an homogeneous linear elastic material. The deformation energy of an elastic body, under no initial stresses or strains, subject to externally applied forces can be described by the following model [35]:

$$\mathbf{E} = \frac{1}{2} \int_\Omega \sigma^T \varepsilon \, d\Omega + \int_\Omega \mathbf{F}^T \mathbf{u} \, d\Omega \qquad (1)$$

where $\mathbf{F} = \mathbf{F}(x,y,z)$ is the vector representing the forces applied to the elastic body (forces per unit volume, surface forces or forces concentrated at the nodes of the mesh), $\mathbf{u} = \mathbf{u}(x,y,z)$ the displacement vector field we wish to compute, and Ω the body on which one is working described by a mesh of tetrahedral elements. ε is the strain vector and σ the stress vector, linked to the strain vector by the constitutive equations of the material. For the above model, this relation is described as

$$\sigma = \left(\sigma_x, \sigma_y, \sigma_z, \tau_{xy}, \tau_{yz}, \tau_{xz} \right)^T = \mathbf{D}\varepsilon,$$

where $\mathbf{D}$ is the elasticity matrix characterizing the material's properties [35]. Strain is related to displacement by the assumption $\varepsilon = \mathbf{L}^T \mathbf{u}$ where $\mathbf{L}$ is a linear operator.
   A finite element discretization is used to obtain a mesh over the volumetric image domain. The mesh elements we use to represent volumetric data are tetrahedral and hence each element is defined by four mesh nodes. The continuous displacement field $\mathbf{u}$ everywhere within element e of the mesh is defined as a function of the displacement at the element's nodes $\mathbf{u}_i^e$ weighted by the element's interpolating functions $N_i^e(\mathbf{x})$,

$$\mathbf{u}(\mathbf{x}) = \sum_{i=1}^{N_{nodes}} N_i^e(\mathbf{x})\mathbf{u}_i^e \quad . \qquad (2)$$

We use linear interpolating functions to define the displacement field inside each element. Hence, the interpolating function of node i of tetrahedral element e is defined as

$$N_i^e(\mathbf{x}) = \frac{1}{6V^e}\left(a_i^e + b_i^e x + c_i^e y + d_i^e z\right) \tag{3}$$

The computation of the volume of the element V^e and the interpolation coefficients are detailed in [35, pages 91–92].

The volumetric deformation of the brain is found by solving for the displacement field that minimizes the energy described by Equation 1. Defining matrix $\mathbf{B}_i = ^e\mathbf{L}N_i^e$ for every node i of each element e, solving for

$$\frac{\partial E\left(\mathbf{u}_1^e,\ldots,\mathbf{u}_{N_{nodes}}^e\right)}{\partial \mathbf{u}_i^e} = 0 \quad ; \quad i = 1,\ldots,N_{nodes} \tag{4}$$

yields the following equation:

$$\int_\Omega \sum_{j=1}^{N_{nodes}} \mathbf{B}_i^{eT}\mathbf{DB}_j^e \mathbf{u}_j^e \, d\Omega = -\int_\Omega \mathbf{F}N_i^e \, d\Omega \quad ; \quad i = 1,\ldots,N_{nodes}. \tag{5}$$

This can be written as a linear equation system, which can be solved for the displacements resulting from the forces applied to the body:

$$\mathbf{Ku} - -\mathbf{F}. \tag{6}$$

The displacements at the boundary surface nodes are fixed to match those generated by the active surface model. Let $\tilde{\mathbf{u}}$ be the vector representing the displacement to be imposed at the boundary nodes. The elements of the rows of the stiffness matrix $\mathbf{K}$ corresponding to the nodes for which a displacement is to be imposed are set to zero and the diagonal elements of these rows to one. The force vector $\mathbf{F}$ is set to equal the displacement vector for the boundary nodes: $\mathbf{F} = \tilde{\mathbf{u}}$ [35]. In this way solving Equation 6 for the unknown displacements will produce a deformation field over the entire mesh that matches the prescribed displacements at the boundary surfaces.

2.4   Computational Considerations

The volumetric deformation of the brain is found by solving for the displacement field that minimizes the energy described by Equation 1, after fixing the displacements at the surface to match those generated by the active surface model.

At each node of the tetrahedral mesh, three variables representing the displacements along each coordinate axis are to be determined. Each variable gives rise to one row and one column in the global K matrix. The rows of the matrix are divided equally amongst the CPUs available for computation and the global matrix is assembled in parallel. Each CPU assembles the local $\mathbf{K}^e$ matrix for each element in its subdomain. Each CPU has an equal number of rows to process but because the connectivity of the mesh is irregular, some CPUs may do more work than other CPUs.

Following the assembly of the matrix, the boundary conditions determined by the surface matching are applied. The global K matrix is adjusted such that rows associated with variables that are determined consist of a single non-zero entry of unit magnitude on the diagonal.

We solve the volumetric biomechanical brain model system of equations (and the active surface membrane model equations) with the Portable, Extensible Toolkit for Scientific Computation (PETSc) package [36, 37] using the Generalized Minimal Residual (GMRES) solver with block Jacobi preconditioning. During neurosurgery, the system of equations was solved on a Sun Microsystems SunFire 6800 symmetric multi-processor machine with 12 750MHz UltraSPARC-III (8MB Ecache) CPUs and 12 GB of RAM. This hardware platform gives us sufficient compute capacity to execute the intraoperative image processing during a neurosurgery.

2.5   Augmented Intraoperative Visualization

Once the volumetric deformation field has been obtained, we can apply this to earlier data to warp it into the current configuration of the patient's brain. We can project both imaging data, such as preoperative MRI or nuclear medicine scans, and segmentations. We display the imaging data by texture mapping onto flat planes. This allows ready comparison of the current intraoperative scan and earlier scans. We construct new triangle models to enable surface rendering of segmented structures in 3D. This allows ready appreciation of the 3D anatomy of these segmented structures together with the imaging data in the form of planes passing through or over the 3D triangle models [27]. We can also apply the deformation field to the vertices of previously obtained triangle models of preoperative data, such as white matter fiber tracts, or regions of functional activation after applying an appropriate rigid transformation to those models. This allows us to warp the triangle models into the current configuration of the patient brain.

These visualizations are rendered on a Sun Microsystems workstation with hardware accelerated triangle rendering and texture mapping and displayed on LCD screens attached to the open magnet MRI scanner. The surgeon can then visualize directly the surface of the patient's brain, the volumetric representation of the internal structure of the brain from IMRI augmented with earlier image acquisitions, and 3D surface models of selected structures in the correct alignment [16]. This augments the surgeon's ability to see the relationship between the patient brain morphology, white matter structure and functionally active regions. This allows the neurosurgeon to better appreciate the spatial relationship between the tumor, the region to be resected, and the critical structures.

3   Results

In this section we provide examples of preoperative data aligned with the intraoperative configuration of the patient's brain. The accuracy of the interpretation of these visualizations depends critically upon the accuracy of the registration process which has been previously described and validated [14, 15] and also upon the accuracy of the segmentation procedure. We provide illustrative results to indicate the procedure for carrying out validation of segmentations of intraoperative therapy delivery.

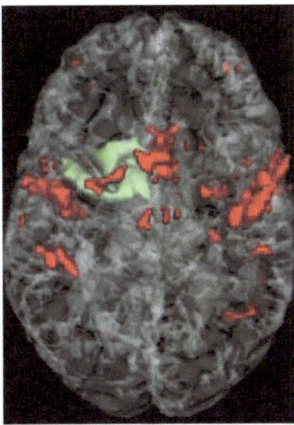

(a) Preoperative fMRI (red) illustrating activation near motor regions shown in the preoperative configuration of the patient's brain.

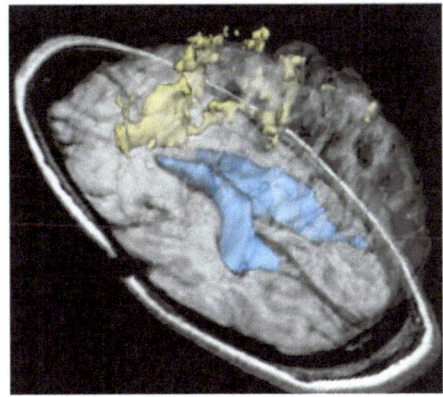

(b) fMRI (yellow) projected into the intraoperative brain configuration. Lateral ventricles are visible (blue).

Fig. 2. Illustration of projection of fMRI activation from the preoperative configuration into the intraoperative configuration of the patient's brain. Considerable brain shift is visible in the frontal region.

3.1   Visualization of Preoperative Scans Projected into Intraoperative Configuration

Figure 2 illustrates the projection of preoperative fMRI activation from the initial configuration of the brain into the intraoperative configuration of the brain. The magnitude of the brain deformation in the frontal region can be clearly seen on the cross-sectional MRI.

Figure 3 shows two-dimensional cross-sectional images which illustrate the projection of preoperative data into the intraoperative configuration following resection of tumor. The brain shift is largest in the frontal region, and relatively small in other regions. Preoperative DTI and fMRI data were projected to match the intraoperative configuration, and are illustrated in Figure 4.

3.2   Validation of Segmentation of Intraoperative Therapy Delivery

Figure 5 illustrates the procedure used to validate segmentations of intraoperative imaging data. In this example we examined repeated segmentations of a region of cryoablation of a tumor from MRI. Five repeated segmentations were carried out by a single rater, first with limited training in the interpretation of the images, and then after a training session with an experienced radiologist. Both before and after training, the STAPLE [29] algorithm was used to create a probabilistic estimate of the underlying true cryoablation region. Using this technique, we were able to observe that before training the rater was much more variable than after training. The STAPLE algorithm can be used to compare interactive segmentations and semi-automatic or automated algorithms, and

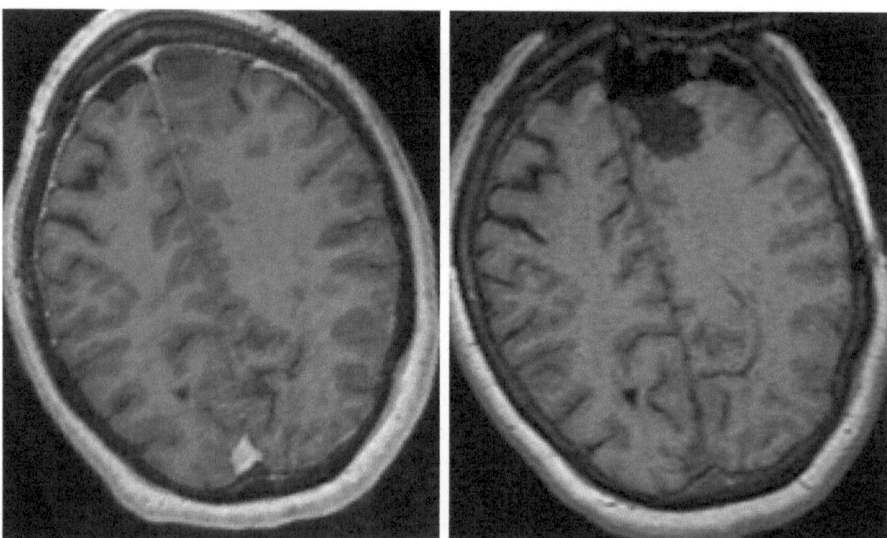

(a) Preoperative SPGR of the brain of a patient. (b) Intraoperative SPGR of the brain of a patient at the end of surgery following resection.

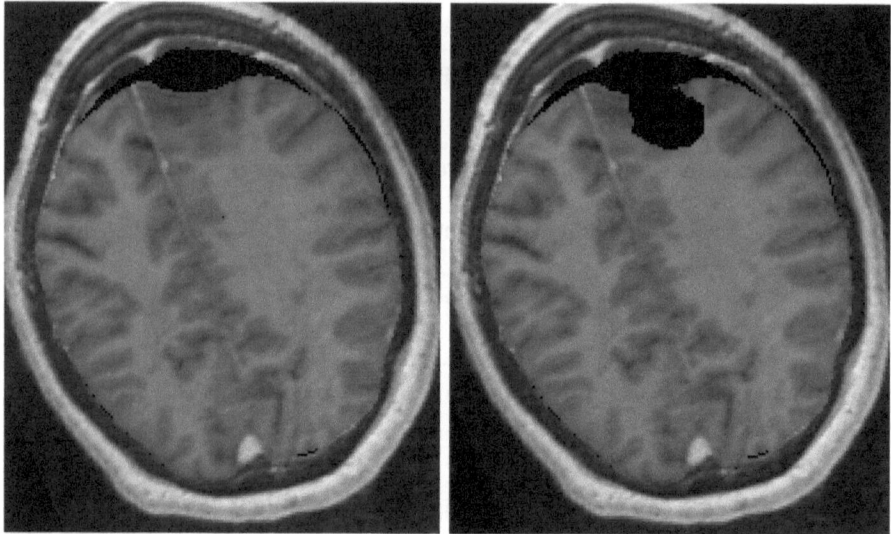

(c) Preoperative SPGR warped to match the intraoperative scan. (d) Warped preoperative patient scan clipped in the region of the resection cavity. A close match to image (b) can be seen.

Fig. 3. Illustration of projection of preoperative scan onto the intraoperative scan and clipping of the projected data to account for resection. The warped clipped preoperative data closely matches the intraoperative conı guration of the patient brain.

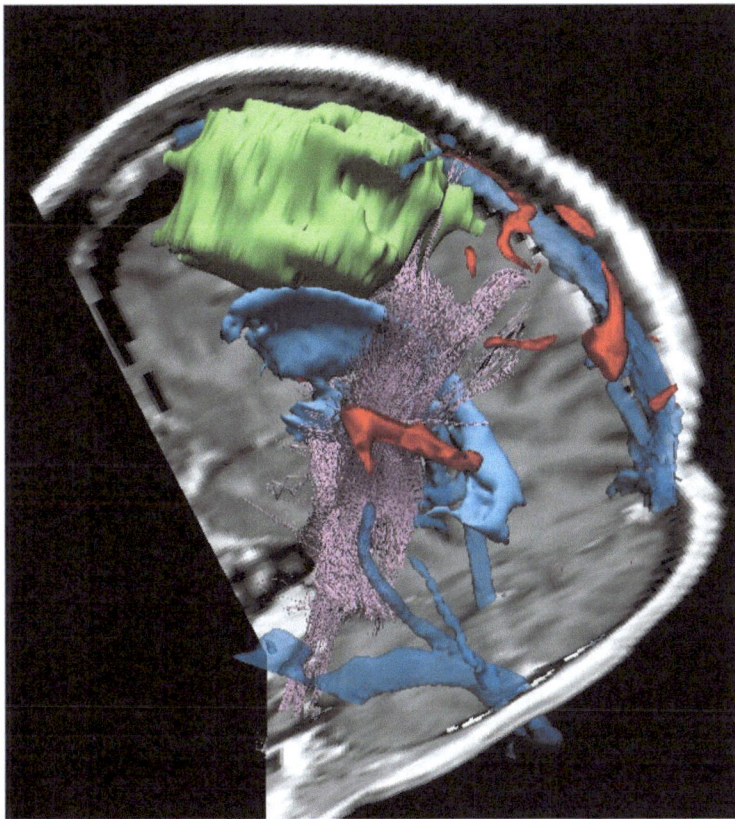

Fig. 4. Illustration of alignment of preoperative DTI and fMRI to match the intraoperative configuration of the brain of the patient following resection. The preoperative extent of the tumor is rendered in green, the warped preoperative MRI is shown on the cross-sectional planes, lateral ventricles and major vessels are rendered in blue, functionally activated regions detected by fMRI are shown in red and major fiber tracts are shown in pink.

to validate segmentations used for preoperative planning, intraoperative navigation and monitoring of therapy delivery.

4   Discussion and Conclusion

Brain deformations occur during image guided neurosurgery and can limit the effectiveness of preoperative imaging for intra-treatment planning, monitoring and surgical navigation. Intraoperative nonrigid registration can re-establish the spatial correspondence of preoperative and intraoperative imaging data. This can allow quantitative monitoring of therapy application, including the ability to make quantitative comparisons with a preoperatively-defined treatment plan. This also enables augmented visualizations that present to the surgeon both intraoperative imaging data and preoperative imaging data

214   Simon K. Warfield et al.

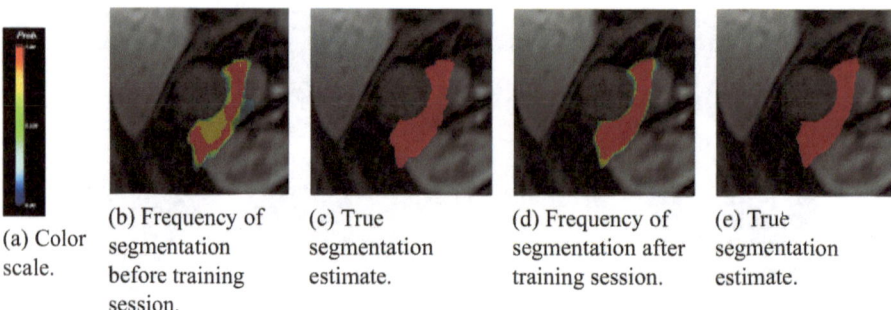

(a) Color scale.

(b) Frequency of segmentation before training session.

(c) True segmentation estimate.

(d) Frequency of segmentation after training session.

(e) True segmentation estimate.

Fig. 5. Illustration of validation of segmentation of intraoperative images. The color scale is used to display the probability of each of the segmented voxels being the target region, where a probability of 0.0 is represented by dark blue and a probability of 1.0 by deep red. The same MR image is shown four times with different overlays. The region of therapy delivery was segmented five times by a single operator both before and after a training session with an expert radiologist. The overlays show the frequency of segmentation of each voxel before training (b) and after training (d), and the estimate of the true region of the therapy delivery found using the STAPLE algorithm on the basis of these segmentations before (c) and after training (e).

projected into alignment with the current configuration of the brain of the patient. This can significantly enhance the information available to the neurosurgeon. Ultimately, such approaches may provide a truly integrated, multimodality environment for quantitative monitoring of intraoperative therapy delivery and enable improved neurosurgical decision making and hence improved patient outcomes.

Acknowledgements

This investigation was supported by NIH grants P41 RR13218,
P01 CA67165, R01 CA86879, and by a research grant from the Whitaker Foundation.

References

1. F. Jolesz, "Image-guided Procedures and the Operating Room of the Future," *Radiology*, vol. 204, pp. 601–612, May 1997.
2. A. Nabavi, P. M. Black, D. T. Gering, C. F. Westin, V. Mehta, R. S. Pergolizzi, M. Ferrant, S. K. Warfield, N. Hata, R. B. Schwartz, W. M. Wells III, R. Kikinis, and F. A. Jolesz, "Serial Intraoperative MR Imaging of Brain Shift," *Neurosurgery*, vol. 48, pp. 787–798, Apr 2001.
3. A. Hagemann, K. Rohr, H. S. Stiel, U. Spetzger, and J. M. Gilsbach, "Biomechanical modeling of the human head for physically based, non-rigid image registration," *IEEE Transactions On Medical Imaging*, vol. 18, no. 10, pp. 875–884, 1999.
4. O. Skrinjar and J. S. Duncan, "Real time 3D brain shift compensation," in *IPMI'99*, pp. 641–649, 1999.
5. M. Miga, K. Paulsen, J. Lemery, A. Hartov, and D. Roberts, "In vivo quantification of a homogeneous brain deformation model for updating preoperative images during surgery," *IEEE Transactions On Medical Imaging*, vol. 47, pp. 266–273, February 1999.

6. O. Skrinjar, C. Studholme, A. Nabavi, and J. Duncan, "Steps Toward a Stereo-Camera-Guided Biomechanical Model for Brain Shift Compensation," in *Proceedings of International Conference of Information Processing in Medical Imaging*, pp. 183–189, 2001.

7. M. Ferrant, S. K. Warfield, A. Nabavi, B. Macq, and R. Kikinis, "Registration of 3D Intraoperative MR Images of the Brain Using a Finite Element Biomechanical Model," in *MICCAI 2000: Third International Conference on Medical Robotics, Imaging And Computer Assisted Surgery; 2000 Oct 11–14; Pittsburgh, USA* (A. M. DiGioia and S. Delp, eds.), (Heidelberg, Germany), pp. 19–28, Springer-Verlag, 2000.

8. D. Hill, C. Maurer, R. Maciunas, J. Barwise, J. Fitzpatrick, and M. Wang, "Measurement of intraoperative brain surface deformation under a craniotomy," *Neurosurgery*, vol. 43, pp. 514–526, 1998.

9. N. Hata, *Rigid and deformable medical image registration for image-guided surgery*. PhD thesis, University of Tokyo, 1998.

10. M. Ferrant, S. K. Warfield, C. R. G. Guttmann, R. V. Mulkern, F. A. Jolesz, and R. Kikinis, "3D Image Matching Using a Finite Element Based Elastic Deformation Model," in *MICCAI 99: Second International Conference on Medical Image Computing and Computer-Assisted Intervention; 1999 Sep 19–22; Cambridge, England* (C. Taylor and A. Colchester, eds.), (Heidelberg, Germany), pp. 202–209, Springer-Verlag, 1999.

11. N. Hata, A. Nabavi, W. M. Wells, S. K. Warfield, R. Kikinis, P. M. Black, and F. A. Jolesz, "Three-Dimensional Optical Flow Method for Measurement of Volumetric Brain Deformation from Intraoperative MR Images," *J Comput Assist Tomogr*, vol. 24, pp. 531–538, Jul 2000.

12. K. Paulsen, M. Miga, F. Kennedy, P. Hoopes, A. Hartov, and D. Roberts, "A Computational Model for Tracking Subsurface Tissue Deformation During Stereotactic Neurosurgery," *IEEE Transactions On Medical Imaging*, vol. 47, pp. 213–225, February 1999.

13. M. I. Miga, D. W. Roberts, F. E. Kennedy, L. A. Platenik, A. Hartov, K. E. Lunn, and K. D. Paulsen, "Modeling of Retraction and Resection for Intraoperative Updating of Images," *Neurosurgery*, vol. 49, pp. 75–85, July 2001.

14. M. Ferrant, A. Nabavi, B. Macq, F. A. Jolesz, R. Kikinis, and S. K. Warfield, "Registration of 3D Intraoperative MR Images of the Brain Using a Finite Element Biomechanical Model," *IEEE Trans Med Imag*, vol. 20, pp. 1384–1397, Dec 2001.

15. M. Ferrant, A. Nabavi, B. Macq, P. M. Black, F. A. Jolesz, R. Kikinis, and S. K. Warfield, "Serial Registration of Intraoperative MR Images of the Brain," *Med Image Anal*, vol. 6, no. 4, pp. 337–359, 2002.

16. S. K. Warfield, F. Talos, A. Tei, A. Bharatha, A. Nabavi, M. Ferrant, P. M. Black, F. A. Jolesz, and R. Kikinis, "Real-Time Registration of Volumetric Brain MRI by Biomechanical Simulation of Deformation during Image Guided Neurosurgery," *Comput Visual Sci*, vol. 5, pp. 3–11, 2002.

17. D. Gering, A. Nabavi, R. Kikinis, W. Grimson, N. Hata, P. Everett, F. Jolesz, and W. Wells, "An Integrated Visualization System for Surgical Planning and Guidance using Image Fusion and Interventional Imaging," in *MICCAI 99: Proceedings of the Second International Conference on Medical Image Computing and Computer Assisted Intervention*, pp. 809–819, Springer Verlag, 1999.

18. R. Kikinis, M. E. Shenton, G. Gerig, J. Martin, M. Anderson, D. Metcalf, C. R. G. Guttmann, R. W. McCarley, W. E. Lorenson, H. Cline, and F. Jolesz, "Routine Quantitative Analysis of Brain and Cerebrospinal Fluid Spaces with MR Imaging," *Journal of Magnetic Resonance Imaging*, vol. 2, pp. 619–629, 1992.

19. A. Yezzi, A. Tsai, and A. Willsky, "Medical image segmentation via coupled curve evolution equations with global constraints," in *Mathematical Methods in Biomedical Image Analysis*, (New York), pp. 12–19, IEEE, 2000.

20. S. K. Warfield, M. Kaus, F. A. Jolesz, and R. Kikinis, "Adaptive, Template Moderated, Spatially Varying Statistical Classification," *Med Image Anal*, vol. 4, pp. 43–55, Mar 2000.
21. M. R. Kaus, S. K. Warfield, A. Nabavi, E. Chatzidakis, P. M. Black, F. A. Jolesz, and R. Kikinis, "Segmentation of MRI of meningiomas and low grade gliomas," in *MICCAI 99: Second International Conference on Medical Image Computing and Computer-Assisted Intervention; 1999 Sep 19–22; Cambridge, England* (C. Taylor and A. Colchester, eds.), (Heidelberg, Germany), pp. 1–10, Springer-Verlag, 1999.
22. S. K. Warfield, R. V. Mulkern, C. S. Winalski, F. A. Jolesz, and R. Kikinis, "An Image Processing Strategy for the Quantification and Visualization of Exercise Induced Muscle MRI Signal Enhancement," *J Magn Reson Imaging*, vol. 11, pp. 525–531, May 2000.
23. C. F. Westin, S. E. Maier, H. Mamata, A. Nabavi, F. A. Jolesz, and R. Kikinis, "Processing and visualization for diffusion tensor MRI," *Med Image Anal*, vol. 6, no. 2, pp. 93–108, 2002. 1361-8415 Journal Article.
24. L. O'Donnell, S. Haker, and C.-F. Westin, "New Approaches to Estimation of White Matter Connectivity in Diffusion Tensor MRI: Elliptic PDEs and Geodesics in a Tensor-Warped Space," in *MICCAI 2002: Fifth International Conference on Medical Image Computing and Computer Assisted Intervention*, (Heidelberg, Germany), pp. 459–466, Springer-Verlag, 2002.
25. A. Tsai, J. Fisher, C. Wible, W. M. Wells, J. Kim, and A. S. Willsky, "Analysis of functional mri data using mutual information," in *MICCAI 1999: Second International Conference on Medical Image Computing and Computer Assisted Intervention*, (Heidelberg, Germany), pp. 473–480, Springer-Verlag, 1999.
26. J. Fisher, E. Cosman, C. Wible, and W. Wells, "Adaptive entropy rates for fmri time-series analysis," in *MICCAI 2001: Fourth International Conference on Medical Image Computing and Computer Assisted Intervention*, (Utrecht, the Netherlands), pp. 905–912, Springer-Verlag, 2001.
27. D. Gering, A. Nabavi, R. Kikinis, N. Hata, L. O'Donnell, W. Grimson, F. Jolesz, P. Black, and W. Wells III, "An integrated visualization system for surgical planning and guidance using image fusion and an open MR," *J Magn Reson Imaging*, vol. 13, pp. 967–975, Jun 2001.
28. S. K. Warfield, F. A. Jolesz, and R. Kikinis, "Real-Time Image Segmentation for Image-Guided Surgery," in *SC 1998: High Performance Networking and Computing Conference; 1998 Nov 7–13; Orlando, USA*, no. 1114, (New York), pp. 1–14, IEEE, 1998.
29. S. K. Warfield, K. H. Zou, and W. M. Wells, "Validation of Image Segmentation and Expert Quality with an Expectation-Maximization Algorithm," in *MICCAI 2002: Fifth International Conference on Medical Image Computing and Computer-Assisted Intervention; 2002 Sep 25–28; Tokyo, Japan*, (Heidelberg, Germany), pp. 298–306, Springer-Verlag, 2002.
30. W. Schroeder, K. Martin, and B. Lorensen, *The Visualization Toolkit: An Object-Oriented Approach to 3D Graphics*. Prentice Hall PTR, New Jersey, 1996.
31. B. Geiger, "Three dimensional modeling of human organs and its application to diagnosis and surgical planning," Tech. Rep. 2105, INRIA, 1993.
32. M. Ferrant, A. Nabavi, B. Macq, and S. K. Warfield, "Deformable Modeling for Characterizing Biomedical Shape Changes," in *DGCI2000: Discrete Geometry for Computer Imagery; 2000 Dec 13–15; Uppsala, Sweden* (G. Borgefors, I. Nyström, and G. Sanniti di Baja, eds.), vol. 1953 of *Lecture Notes in Computer Science*, (Heidelberg, Germany), pp. 235–248, Springer, 2000.
33. S. K. Warfield, F. Jolesz, and R. Kikinis, "A High Performance Computing Approach to the Registration of Medical Imaging Data," *Parallel Computing*, vol. 24, pp. 1345–1368, Sep 1998.

34. M. Ferrant, O. Cuisenaire, and B. Macq, "Multi-Object Segmentation of Brain Structures in 3D MRI Using a Computerized Atlas," in *SPIE Medical Imaging '99*, vol. 3661-2, pp. 986–995, 1999.

35. O. C. Zienkiewicz and R. L. Taylor, *The Finite Element Method: Basic Formulation and Linear Problems*. McGraw Hill Book Co., New York, 4th ed., 1994.

36. S. Balay, W. D. Gropp, L. C. McInnes, and B. F. Smith, "Efficient management of parallelism in object oriented numerical software libraries," in *Modern Software Tools in Scientific Computing* (E. Arge, A. M. Bruaset, and H. P. Langtangen, eds.), pp. 163–202, Birkhauser Press, 1997.

37. S. Balay, W. D. Gropp, L. C. McInnes, and B. F. Smith, "PETSc 2.0 users manual," Tech. Rep. ANL-95/11 - Revision 2.0.28, Argonne National Laboratory, 2000.

Left Ventricle Composite Material Model for Stress-Strain Analysis

Zhenhua Hu[1], Dimitris Metaxas[2], and Leon Axel[3]

[1] Department of Computer and Information Science
University of Pennsylvania, Philadelphia, PA 19104, USA
zhhu@seas.upenn.edu
[2] The Center of Computational Biomedicine, Imaging and Modeling
Depts. of Biomedical Engineering and Computer Science
Rutgers University, New Brunswick, NJ 08854-8019, USA
dnm@cs.rutgers.edu
[3] Department of Radiology
New York University, New York City, NY 10016, USA
leon.axel@med.nyu.edu

Abstract. Mechanical properties of the myocardium have been investigated intensively in the last four decades. Due to the nonlinearity and history dependence of the myocardial deformation, many complex strain energy functions have been used to describe the stress-strain relationship of myocardium. These functions are good at fitting in-vitro experimental data from myocardial stretch testing. However it is difficult to model in-vivo myocardium by using the strain energy functions. In a previous paper [24], we have implemented transversely anisotropic material model to estimate in-vivo strain-stress analysis in the myocardium. In this work, the fiber orientation is updated at each time step from the end of diastole to the end of systole, and the stiffness matrix is recalculated using the current fiber orientation. We also extended our model to include residual ventricular stresses and time dependent blood pressure in the left ventricle cavity.

1 Introduction

The heart wall consists primarily of locally parallel muscle cells, a complex vascular network, and a dense plexus of connective tissue [1]. The cardiac muscle cell is the predominant component of myocardium, normally occupying around 70% of heart wall volume. The muscle cells are tied together by a collagenous network and bundled together into fibers. Two major groups of myocardial collagen have been discovered [2]: one group provides myocyte-to-myocyte and myocyte-to-capillary connections, while another group surrounds the muscle fibers. Systematic measurements of muscle fiber orientations of canine heart were carried out by Streeter [3]. His main finding was that fiber directions generally vary in a continuous manner from +60° on the endocardium to -60° on the epicardium. Other studies have shown similar transmural fiber distribution in other mammalian species [1]. More collagen organizes the fibers into sheets, which are loosely tied together. It has been shown that the sheet orientation varies consistently within the wall [4,5].

The structure of myocardium reveals that it is a composite material. The material properties of heart wall were initially estimated based on uniaxial tests, mostly performed with papillary muscles [6,7]. Biaxial tests were subsequently carried out [8,9], showing that the myocardium is anisotropic, and the corresponding constitutive relations have been proposed [10,11], which have included that the myocardium is anisotropic. To model the material properties of heart wall, different complex forms of strain energy function have been used to fit the excised tissue [1]. However these strain energy functions do not offer us a direct quantitative relation between stress and strain and it is difficult to understand the real meaning of the parameters in these functions. In some other research group's work [29], linear elastic energy function has been used together with the active forces. As an alternative, we can use a fiber-reinforced composite material model for the stress estimation of the in-vivo left ventricle. Composite materials have been studied extensively in the past decades, and the modelling of their strain-stress relationship is standardized. By using this composite material model, we can estimate parameters such as Young's modulus and the Poisson ratio, which are more understandable and meaningful than the strain energy function parameters.

In a previous paper [24], we used reconstruction of the displacements of the left ventricular myocardium from MRI-SPAMM tagging [12] and physics-based deformable model method [13] based on previous work of Haber [14,15]. The finite element method [16] was used to calculate the strain and stress. In our model, we studied the heart contraction cycle from the end of diastole to the end of systole. The mechanical properties of the fiber-reinforced composite material are greatly affected by the fiber orientation. Because human fiber data is not available, we have used animal fiber data measured by other researchers [3,17]; there is good agreement between different species that have been studied. The measured fiber orientation data from animal hearts is applied in our heart model at the end of diastole. In our previous paper, the fiber orientation was simplified to be constant over one contract cycle. However, the fiber moves along with the heart wall in which it is embedded, and its orientation changes at every time step during the heart contraction cycle. In this paper, we updated the fiber orientation of all finite elements in each time step. Assuming the fiber's local material coordinates are constant, we can calculate the spatial coordinates of representative points on the fiber at each time step and then compute the current fiber orientation. To simplify the computation, we model the heart muscle stiffness as piece-wise linear; the Young's modulus depends on how much the muscle has deformed. Other researchers have shown that residual stress exists in the left ventricle [25,26,27], which reduces the endocardial stress concentrations during diastole. In this work, we implemented the residual stress at the end of diastole to make our model more realistic. The principal boundary condition in our model is blood pressure in the left ventricle cavity. The change of blood pressure over the contraction cycle is an important aspect of cardiac function. We have included time-variant blood pressure in our extended model.

2 Composite Material Model

2.1 Definition

A composite is a structural material which consists of two or more constituents, which are combined at a macroscopic level and are not soluble in each other [18]. One constituent is called the reinforcing phase, which is embedded in another constituent called the matrix. In our model, the reinforcing phase is muscle fiber and the matrix is collagen.

2.2 Elastic Moduli Evaluation

Consider a representative volume element from a unidirectional lamina which consists of the fiber surrounded by the matrix (Fig. 1). Assume the fiber and matrix volume fractions are V_f and V_m, respectively, the Young's moduli of the fiber and matrix are E_f and E_m, respectively, the Poisson ratios are v_f and v_m, respectively, and the in-plane shear moduli are G_f and G_m, respectively. For the composite, the Young's moduli along the fiber orientation and across the fiber are, respectively [18]:

$$E_1 = E_f V_f + E_m V_m \qquad E_2 = \left(\frac{V_f}{E_f} + \frac{V_m}{E_m} \right)^{-1}. \tag{1}$$

The Poisson ratio and the in-plane shear modulus are, respectively [18]:

$$v_{12} = v_f V_f + v_m V_m \qquad G_{12} = \left(\frac{V_f}{G_f} + \frac{V_m}{G_m} \right)^{-1}. \tag{2}$$

2.3 Strain-Stress Relation

The stress-strain relation is given by [19]:

$$\underline{\varepsilon} = \begin{bmatrix} 1/E_1 & -v_{12}/E_1 & -v_{12}/E_1 & 0 & 0 & 0 \\ -v_{12}/E_1 & 1/E_2 & -v_{23}/E_2 & 0 & 0 & 0 \\ -v_{12}/E_1 & -v_{23}/E_2 & 1/E_2 & 0 & 0 & 0 \\ 0 & 0 & 0 & 2(1+v_{23})/E_2 & 0 & 0 \\ 0 & 0 & 0 & 0 & 1/G_{12} & 0 \\ 0 & 0 & 0 & 0 & 0 & 1/G_{12} \end{bmatrix} \underline{\sigma} = C\underline{\sigma} . \tag{3}$$

where E_1 is the Young's modulus along fiber direction, E_2 is the Young's modulus along cross-fiber direction, v_{12} and v_{23} are the corresponding Poisson ratios, and G_{12} is the shear modulus. Both Young's moduli are assumed piece-wise linear as shown qualitatively below (Fig. 2). Poisson ratios are assumed to be 0.4 since myocardium is approximately incompressible [20], the lack of full incompressibility can be included in the model. G_{12} is assumed to be equal to $E_2/(1+v_{23})$ [18]. C is called the compliance matrix.

2.4 Transformation Relation

The stress-strain relationship in Equation (3) is defined with respect to local material coordinates. In the finite element method, we need to represent the stress-strain relation of all elements in the same global coordinates. Since the fiber orientation varies in different regions of the left ventricle [2], we need to transform the local stress-strain relation into a global stress-strain relation. For the coordinates shown in Fig. 3, the stress in local fiber coordinates *(1,2,3)* can be transformed into global element coordinates *(x,y,z)* by [19]:

$$\underline{\sigma_{123}} = \begin{bmatrix} \sigma_1 \\ \sigma_2 \\ \sigma_3 \\ \tau_{23} \\ \tau_{31} \\ \tau_{12} \end{bmatrix} = \begin{bmatrix} \cos^2\theta & \sin^2\theta & 0 & 0 & 0 & 2\sin\theta\cos\theta \\ \sin^2\theta & \cos^2\theta & 0 & 0 & 0 & -2\sin\theta\cos\theta \\ 0 & 0 & 1 & 0 & 0 & 0 \\ 0 & 0 & 0 & \cos\theta & -\sin\theta & 0 \\ 0 & 0 & 0 & \sin\theta & \cos\theta & 0 \\ -\sin\theta\cos\theta & \sin\theta\cos\theta & 0 & 0 & 0 & \cos^2\theta-\sin^2\theta \end{bmatrix} \begin{bmatrix} \sigma_x \\ \sigma_y \\ \sigma_z \\ \tau_{yx} \\ \tau_{zx} \\ \tau_{xy} \end{bmatrix} = T\underline{\sigma_{xyz}} \quad . \quad (4)$$

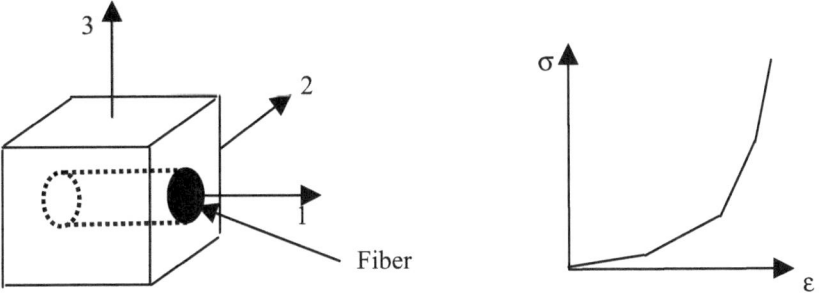

Fig. 1. Elastic moduli evaluation of composite   **Fig. 2.** Piece-wise linearity of Young's modulus

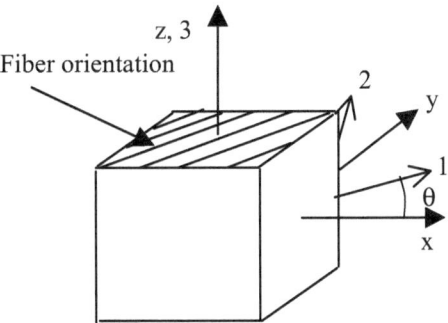

Fig. 3. Coordinate transformation

The strain transformation can be expressed in tensor form as:

$$\underline{\varepsilon_{123}} = T\,\underline{\varepsilon_{xyz}} \quad . \tag{5}$$

By inverting equations (4) and (5) and combining with equation (3), we have the stress-strain relation represented in global coordinate x-y-z as:

$$\underline{\varepsilon_{xyz}} = T^{-1}CT\,\underline{\sigma_{xyz}} = C_1\underline{\sigma_{xyz}} \quad . \tag{6}$$

3   Finite Element Formulation

3.1 Model Dynamics

The finite element equation is derived by using energy minimization and the variational formulation [16]:

$$\dot{q} + K(q - q_0) = P \quad . \tag{7}$$

where q represents the displacement, q_0 represents the displacement generated by residual strain, K is the stiffness matrix and

$$K = \sum_e (\int_{V^e} B^T D B dV), \; P = P_p + P_a = \sum_e (\int_{S^e} N^T f_p dS + \int_{V^e} N^T f_a dV),$$

where f_p is the boundary force mainly generated by the pressure of the blood in the cavity, and f_a is the active force generated by the fiber. D is the elasticity matrix; it is the inverse of the compliance matrix given in equation (3). B is the strain-displacement matrix that relates nodal strain, ε, to nodal displacements, q, as $\varepsilon = Bq$. N is the shape function matrix.

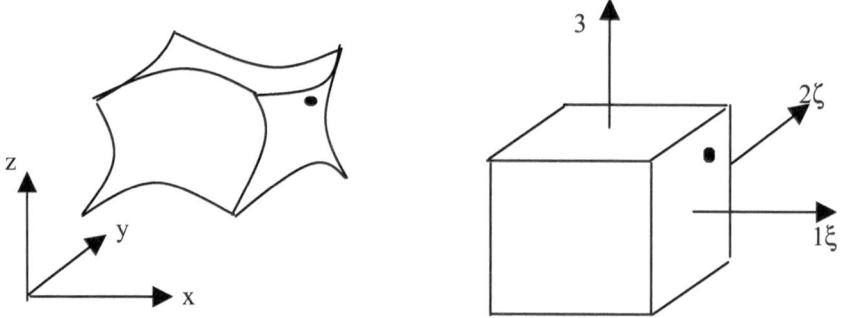

Fig. 4. Isoparametric Finite Element

3.2 Isoparametric Finite Elements

The finite elements reconstructed from MRI images have irregular geometry. We used isoparametric finite elements to map each element to a regular geometry element (Fig. 4).

The transformation is expressed in terms of shape functions:

$$x = \sum_{i=1}^{n} N_i(\xi,\zeta,\eta)x_i, \quad y = \sum_{i=1}^{n} N_i(\xi,\zeta,\eta)y_i \quad z = \sum_{i=1}^{n} N_i(\xi,\zeta,\eta)z_i \quad . \tag{8}$$

where (x_i, y_i, z_i) is the position of i th node in the element numbering system. The shape functions, N_i, depend on the node's local coordinates. The Jacobian of the transformation will determine whether the material is incompressible.

3.3 Residual Strain and Stress

To measure the residual strain, researchers have taken cross-sectional equatorial slices from potassium-arrested rat left ventricles [27]. They were then cut radially and immediately became curved arcs, with open angles which quantified the residual strain. In our model, both the circumferential residual strain and the radial residual strain are assumed to vary linearly from the epicardium to the endocardium. We assume the circumenferential residual strain is 0.05 at the epicardium and –0.05 at the endocardium, while the radial residual strain is -0.05 at the epicardium and 0.05 at the endocardium [28].

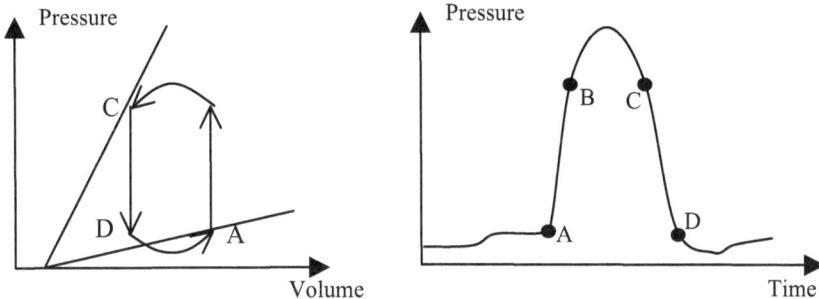

Fig. 5. Schematic ventricle pressure-volume loop   **Fig. 6.** Ventricle pressure changes over time

3.4 Boundary Conditions

In our finite element model, the principal boundary condition is the blood pressure in the left ventricle, which changes over time. The left ventricle receives blood from the left atrium and pumps the blood through the aorta to the systemic circulation. As shown schematically in Fig. 5 and Fig. 6 [1,21], the mitral valve closes at A and the left ventricle undergoes isovolumic contraction with rapidly rising pressure until B, when the left ventricular pressure exceeds the aortic pressure and the aortic valve opens, blood is ejected and the left ventricle's volume begins to decrease. The aortic valve closes at end systole of C, due to the decreasing pressure which falls below the aortic pressure. The left ventricle then undergoes isovolumic relaxation from C to D. The mitral valve reopens at D when the pressure of left ventricle is lower than that of left atrium. In our implementation, we updated the blood pressure at each time step from the end of diastole to the end of systole.

3.5 Stiffness and Active Force Estimation

In Equation 7, the elasticity matrix D and the active force f_a are unknown. The in-vitro elasticity parameters of canine heart have been measured [22]; we used this data as initial values because of the general similarity between human and canine hearts. We then used the EM algorithm [23] to estimate the stiffness and active force at each time step. EM algorithm is typically used to compute maximum likelihood estimates given incomplete samples. Define $J(\theta|\theta_0)$ as:

$$J(\theta \mid \theta_0) \equiv E_{\theta_0}\left(\log \left.\frac{p(X,\theta)}{p(X,\theta_0)}\right| S(X) = s\right) . \tag{9}$$

where X is the random variable, θ is the parameter to be estimated, S(X) is the sufficient statistics on X, $p(X, \theta)$ is the probability density function. The EM algorithm works as following:

1. Initialize $\theta_{old} = \theta_0$
2. Compute $J(\theta|\theta_{old})$ for many values of θ
3. Maximize $J(\theta|\theta_{old})$ as a function of θ
4. Set $\theta_{new} = \arg \max J(\theta|\theta_{old})$, if $\theta_{old} \neq \theta_{new}$, set $\theta_{old} = \theta_{new}$ and go to step 2, otherwise, return $\theta = \theta_{new}$.

where Step 2 is often referred to as the expectation step and Step 3 is called the maximization step.

In our experiment, the parameters are $\theta = (E_1, E_2, f_a)$. The displacement error is defined as:

$$d(x,\hat{\theta}) = \frac{1}{n}\sum_{i=1}^{n}\left[(x_i - x_{it})^2 + (y_i - y_{it})^2 + (z_i - z_{it})^2\right] . \tag{10}$$

where (x_p, y_p, z_p) is the computed global coordinate based on estimation $\hat{\theta}$, and (x_{it}, y_{it}, z_{it}) is the reconstructed global coordinate from MRI tagging. Since the smaller the displacement error, the better the estimation is, we put more weight on estimations with less displacement error. Our normalized density function is then defined as:

$$p(x,\hat{\theta}) = \frac{1/d(x,\hat{\theta})}{\sum_{\theta \in \Theta}(1/d(x,\theta))} . \tag{11}$$

As shown in Fig. 2, since E_1 and E_2 are approximated as piece-wise linear, we need to estimate θ in each time interval. We assume E_1 and E_2 are linearly related. The implementation algorithm is:

1. Initialize t = 1
2. In the t th time interval, calculate $d(\theta)$ and $p(x, \theta)$ for all $\theta \in \Theta$
3. Initialize $(E_{1, old}, E_{2, old}) = (E_{1, 0}, E_{2, 0})$, where $E_{1, 0}$ and $E_{2, 0}$ are calculated from the experimental data given in [22]
4. Fix $(E_{1, old}, E_{2, old})$, using the EM algorithm to get optimal estimation $f_{a, new}$
5. Fix $f_{a, new}$, using the EM algorithm to get optimal estimation $(E_{1, new}, E_{2, new})$
6. If $(E_{1, old}, E_{2, old}) \neq (E_{1, new}, E_{2, new})$, set $(E_{1, old}, E_{2, old}) = (E_{1, new}, E_{2, new})$, go to step 4, otherwise, return $(E_1, E_2) = (E_{1, new}, E_{2, new})$, $f_a = f_{a, new}$
7. t = t+1, if t < nt, go to step 2, otherwise, stop.

where nt is the number of time steps.

4 Results

4.1 Finite Element Model

The left ventricle reconstructed from MRI tagging has 8 levels of parallelepiped elements and 1 level of wedge-shaped elements. The wedge-shaped element has better geometric approximation to the object and is used at the left ventricle apex. Each parallelepiped level has 2 layers radially and 12 elements for each layer. Each wedge-shaped level has 72 elements. Totally we have 264 elements and 327 nodes. To model the fiber orientation change from epicardium to endocardium and improve the precision of computation, we interpolated each parallelepiped elements into 27 subelements. Then we have 5256 elements and 6303 nodes. The cycle from the end of diastole to the end of systole is divided into 4 time intervals. Time 1 corresponds to the end of diastole and time 5 corresponds to the end of systole.

4.2 Fiber Orientation

In Figure 7, we have shown the fibers in epicardium and endocardium at time 1. In (a), the epicardium elements were shown in wireframe with green color, the tiny bar in each element shows the orientation of fiber within that element. In (b), we have shown the fiber orientations of 6 elements from epicardium to endocardium. In (c), the corresponding fiber orientation of endocardium was shown similarly. The fiber orientation of each element changes over time, we can compute current fiber orientation at each time step by calculating the current coordinates of representative points on the fiber. The fiber orientation data was then plugged into Equation 6 to compute each element's current compliance matrix.

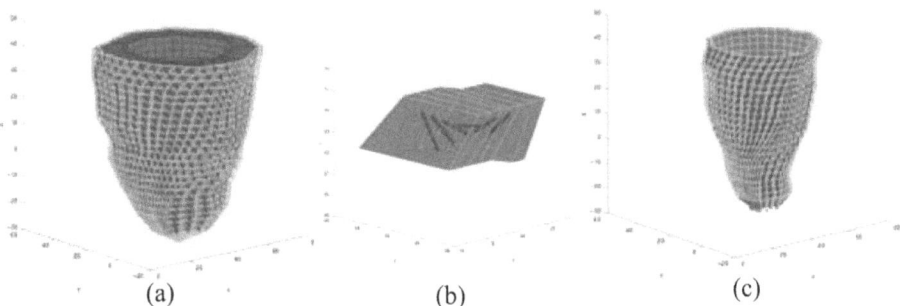

(a)   (b)   (c)

Fig. 7. Fibers: (a) in epicardium (b) from epicardium to endocardium (c) in endocardium

4.3 Blood Pressure

The blood pressure in the left ventricle cavity changes over time and it has small variance between base and apex because of hydrostatic pressure [1]. The blood pressure values over time are given in Table 1.

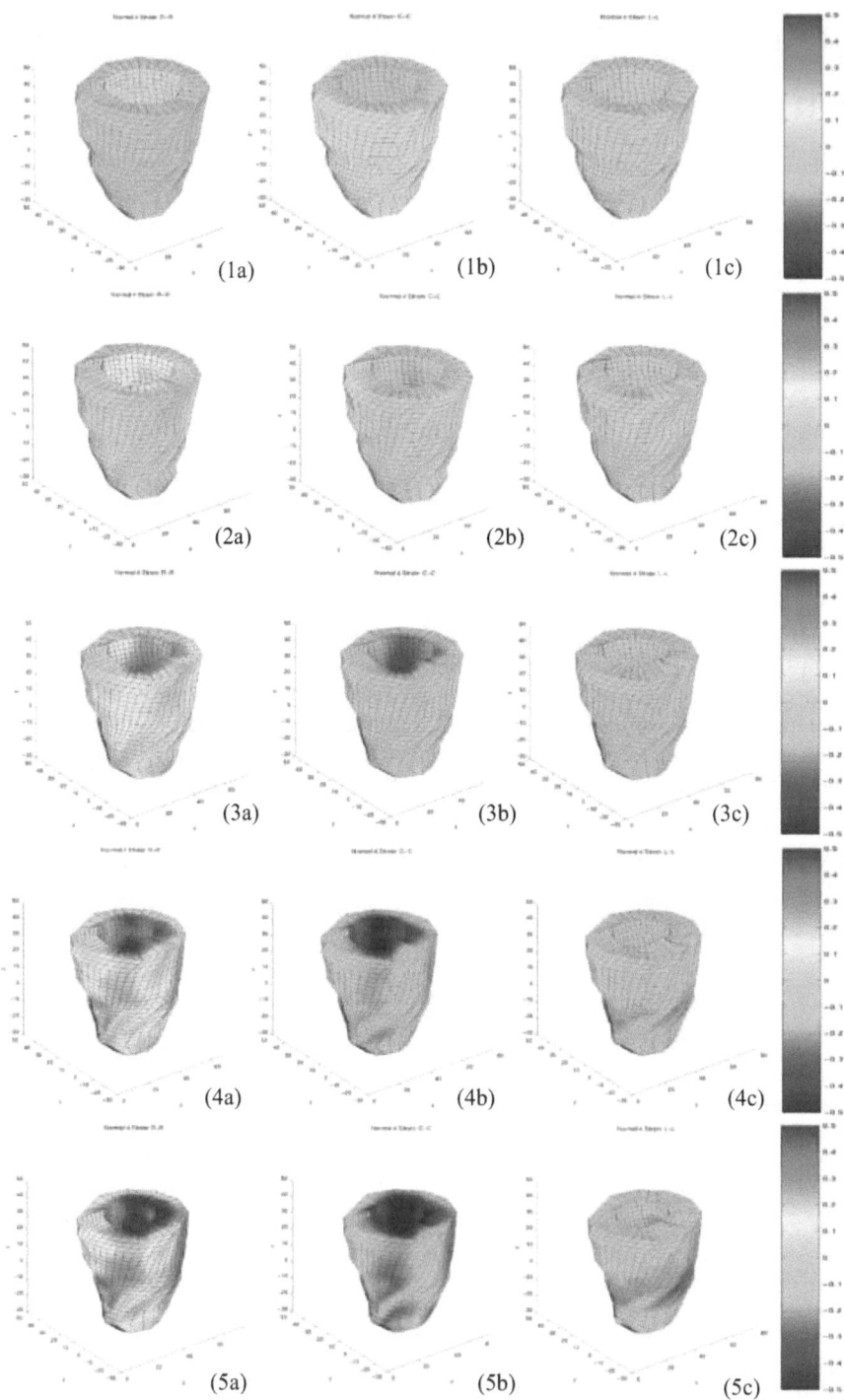

Fig. 8 (a) Radial (b) Circumferential (c) Longitudinal components of strain in 5 time steps (1-5)

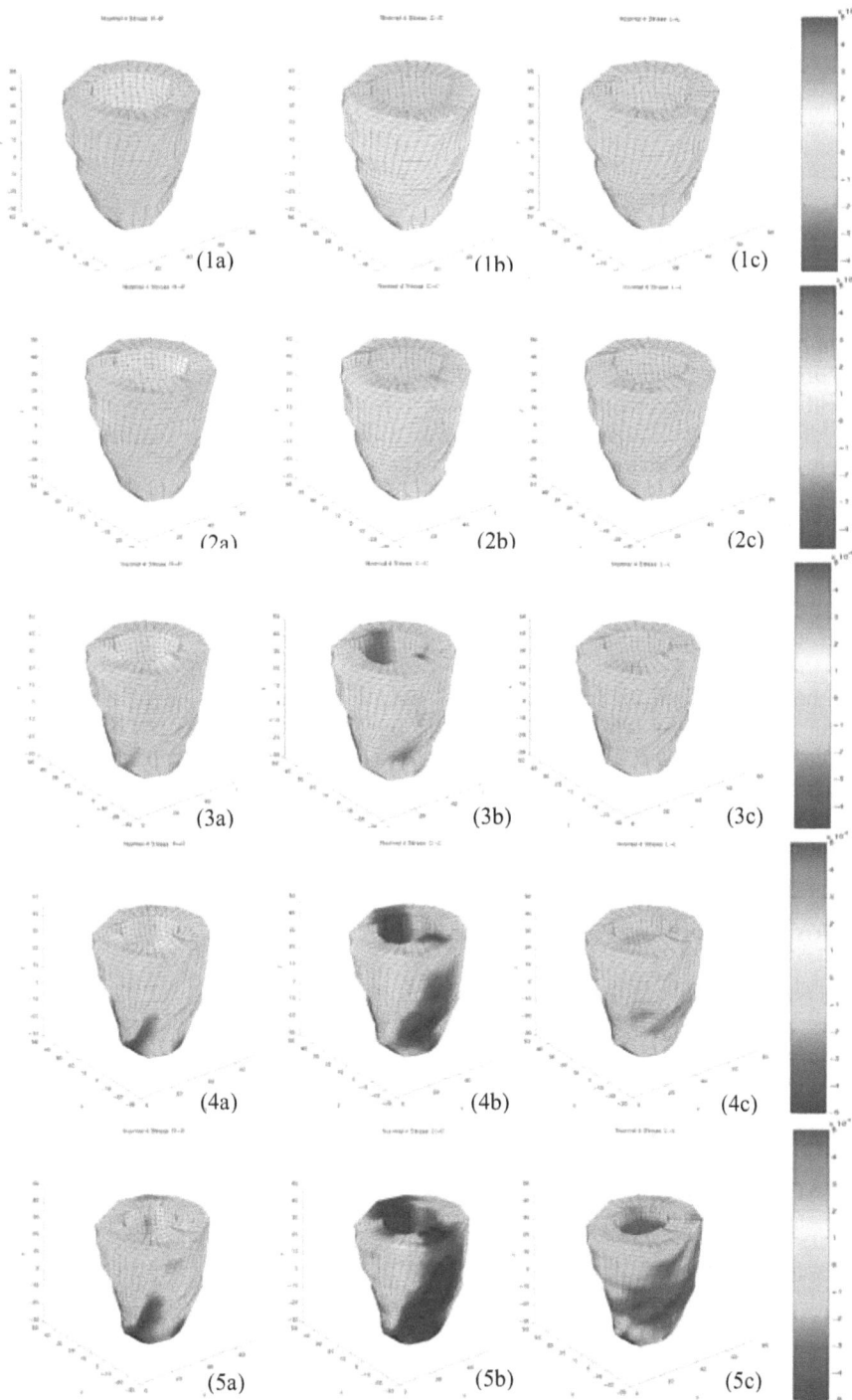

Fig. 9 (a) Radial (b) Circumferential (c) Longitudinal components of stress in 5 time steps (1-5)

Table 1. Left ventricle blood pressure in each time sitep

Time	1	2	3	4	5
Blood Pressure (KPa)	1.1 ~ 1.6	6.1 ~ 6.6	11.1 ~ 11.6	15.1 ~ 15.6	13.1~13.6

4.4 Strain and Stress Distribution

The strain and the stress distributions in one contraction cycle (5 time steps) are shown in Fig. 8 and Fig. 9, respectively. In general, the left ventricle becomes thicker radially, shorter circumferentially and longitudinally when it contracts. As we can see from the images, most radial stresses are positive while most circumferential and longitudinal stresses are negative. However some elements don't follow this rule because of either boundary condition constraints or adjacent elements' influence.

The numerical computation code was written in C and the output display was implemented in Matlab. The program was run on a DELL Precision 330 with 4 CPUs (1.5GHz each) and 1,048MB RAM. For each time step, it took around 15 minutes to estimate the strain and stress. The analysis for one cycle from the end of diastole to the end of systole requires around 60 minutes.

5 Conclusions

We have developed an improved novel composite material model to estimate the strain and stress of left cardiac ventricle by using accurate displacements reconstructed from MRI-SPAMM tagging and a deformable model. Compared to the traditional strain energy function method, our model gives more intuitive and understandable parameters. The results are consistent with earlier studies. They may be clinically useful in the future.

Acknowledgements

This research has been funded by grants from the NIH.

References

1. L. Glass, P. Hunter, A. McCulloch. Theory of Heart: Biomechanics, Biophysics, and Nonlinear Dynamics of Cardiac Function. Springer-Verlag, 1991.
2. J.B. Caulfield, T.K. Borg. The collagen networks of the heart. Lab. Invest., 40:364-371,1979.
3. D. D. Streeter Jr., W. T. Hanna. Engineering mechanics for successive states in canine left ventricular myocardium: I. Cavity and wall geometry. Circulation Research, 33:639-655, 1973.
4. I.J. LeGrice, B.H. Smaill, L.Z. Chai, S.G. Edgar, J.B. Gavin, P.J. Hunter. Laminar structure of the heart: ventricular myocyte arrangement and connective tissue architecture in the dog. Am. J. Physiol. 269(2 Pt 2):H571-82, 1995.
5. I.J. LeGrice, P.J. Hunter, B.H. Smaill. Laminar structure of the heart: a mathematical model. Am. J. Physiol. Heart Circ. Physiol. 272(5 Pt 2):H2466-76, 1997.

6. J. G. Pinto, Y. C. Fung. Mechanical properties of the heart muscle in the passive state. Journal of Biomechanics, 6:597-616,1973.
7. Y. C. Pao, G. K. Nagendra, R. Padiyar, E. L. Ritman. Derivation of myocardial fiber stiffness equation on theory of laminated composite. Journal of Biomechanical Engineering, 102:252-257, 1980.
8. L. L. Demer, F.C.P. Yin. Passive biaxial properties of isolated canine myocardium. Journal of Physiology, 339:615-630, 1983.
9. F.C.P. Yin, R. K. Strumpf, P.H. Chew, S.L. Zeger. Quantification of the mechanical properties of non-contracting myocardium. Journal of Biomechanics, 20:577-589, 1987.
10. J.D. Humphrey, F.C.P. Yin. Biomechanical experiments on excised myocardium: Theoretical considerations. Journal of Biomechanics, 22:377-383, 1989.
11. J.D. Humphery, F.C.P. Yin. On constitutive relations and finite deformations of passive cardiac tissue: I. A pseudostrain-energy function. Journal of Biomechanical Engineering, 109:298-304, 1987.
12. L. Axel, L. Dougherty. Heart wall motion: Improved method of spatial modulation of magnetization for MR imaging. Radiology, 272:349-50, 1989.
13. D. N. Metaxas. Physics-based deformable models: applications to computer vision, graphics, and medical imaging. Kluwer Academic Publishers, Cambridge, 1996.
14. I. Haber, D. N. Metaxas, L. Axel. Three-dimensional motion reconstruction and analysis of the right ventricle using tagged MRI. Medical Image Analysis, 4, 2000.
15. I. Haber. Three dimensional motion reconstruction and analysis of the right ventricle from planar tagged MRI. Ph.D. Dissertation, University of Pennsylvania, Philadelphia, PA, 2000.
16. K. Bathe. Finite element procedures in engineering analysis. Prentice Hall, 1982.
17. F.J. Vetter, A.D. McCulloch. Three-dimensional analysis of regional cardiac function: a model of rabbit ventricular anatomy. Progress in Biophysics & Molecular Biology, 69:157-183,1998.
18. A.K. Kaw. Mechanics of Composite Materials. CRC press, 1997.
19. M. W. Hyer. Stress Analysis of Fiber-Reinforced Composite Materials. McGraw-Hill, 1998.
20. A.A. Amini, Y. Chen, R. W. Curwen, V. Manu, J. Sun. Coupled B-Snake grides and constrained thin-plate splines for analysis of 2D tissue deformations from tagged MRI. IEEE Transaction on Medical Imaging 17(3),344-356, 1998.
21. F.C. Hoppensteadt, C.S. Peskin. Modeling and Simulation in Medicine and the Life Sciences. Springer, 2002.
22. T. P. Usyk, R. Mazhari, A. D. McCulloch. Effect of laminar orthotropic myofiber architecture on regional stress and strain in the canine left ventricle. Journal of Elasticity, 61, 2000.
23. P. J. Bickel, K. A. Doksum. Mathematical statistics: basic ideas and selected topics, Vol. I. Prentice Hall, 2001.
24. Z. Hu, D.N. Metaxas, L. Axel. In-vivo strain and stress estimation of the left ventricle from MRI images. Medical Image Computing and Computer-Assisted Intervention (MICCAI'02), 2002.
25. Y.C. Fung. Biodynamics: Circulation. Springer-Verlag, New York, 1984.
26. J.H. Omens. Left Ventricular Strain in the No-load State due to the Existence of Residual Stress. PhD thesis, University of California, La Jolla, CA, 1988.
27. J.H. Omens, Y.C. Fung. Residual strain in rat left ventricle. Circ. Res., 66:37-45,1990.
28. K. Costa, K. May-Newman, D. Farr, W.G. O'dell, A.D. McCulloch, J.H. Omens. Three-dimensional residual strain in midanterior canine left ventricle. Am. J. Physiol. 273:H1968-76,1997.
29. X. Papademetris, E.T. Onat, A.J. Sinusas, D.P. Dione, R.T. Constable, J.S. Duncan. The Active Elastic Model. Information Processing in Medical Imaging, 2001.

Preliminary Validation Using *in vivo* Measures of a Macroscopic Electrical Model of the Heart

Maxime Sermesant[1,*], Owen Faris[2], Franck Evans[2], Elliot McVeigh[2],
Yves Coudière[3], Hervé Delingette[1], and Nicholas Ayache[1]

[.] Epidaure Research Project, INRIA Sophia Antipolis
2004 route des Lucioles, 06902 Sophia Antipolis, France
[.] National Institutes of Health, National Heart Lung and Blood Institute,
Laboratory of Cardiac Energetics, Bethesda, Maryland
[.] Mathematics Laboratory, CNRS (UMR 6629), Nantes University

Abstract. This article describes an experimental protocol to obtain *in vivo* macroscopic measures of the cardiac electrical activity in a canine heart coupled with simulations done using macroscopic models of the canine myocardium. Electrical propagation simulations are conducted along with preliminary qualitative comparisons. Two different models are compared, one built from dissection and highly smoothed and one measured from Diffusion Tensor Imaging (DTI). Validating a macroscopic model with *in vivo* measurements of the electrical activity should allow a future use of the model in a predictive way, for instance in radiofrequency ablation planning.

1 Introduction

Cardiac electrical activity disorders are involved in many pathologies, often visible through the mechanical deficiency of the heart. Simulating the heart electromechanical activity could help to better understand these pathologies, guide diagnosis or plan interventions, like radio-frequency ablation.

Understanding and modeling cardiac electro-physiology, studying the inverse problem from body surface potentials and direct measurement of heart potentials are active research areas [4,8,11].

To completely validate an electromechanical model, direct *in vivo* measures are necessary to improve knowledge of the phenomena . Macroscopic measures of the electrical and mechanical activities on the same heart become now possible [12]. But these measures are difficult to obtain and invasive. Once validated, a possible outcome is to use the model in a predictive way and to interpolate data where needed.

This article presents a protocol to measure electrical potentials on the epicardium compatible with mechanical measurements and the construction of a macroscopic model of the myocardium from different sources along with simulated results of electrical propagation qualitatively compared to isolated and *in vivo* measures.

---

* Corresponding Author: `Maxime.Sermesant@inria.fr`

N. Ayache and H. Delingette (Eds.): IS4TM 2003, LNCS 2673, pp. 230–243, 2003.

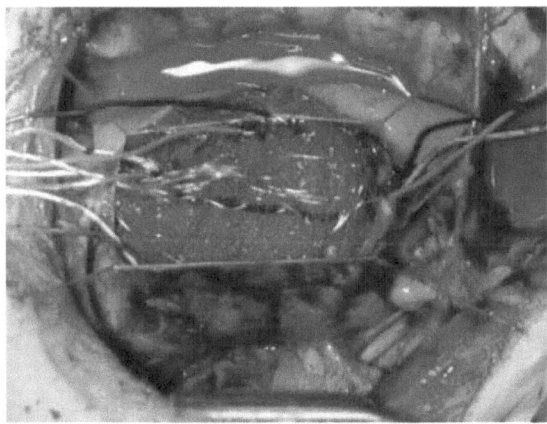

Fig. 1. Multi-electrode epicardial sock on the canine heart studied, *in vivo*.

The chosen model is simple enough so that the parameters can be adjusted from macroscopic measures, at first qualitatively. As the number of unknowns is more important than the number of measures, a more complex model would not be tractable. Moreover, the goal is more studying the differences between the pathological and normal functions than precisely studying quantitatively a given function.

The final goal is to use the complete electromechanical model in a predictive way and to simulate pathologies, therefore, it must be efficient enough to be used interactively. In this paper, a qualitative adjustment of the electrical model to the data is first performed. More quantitative adjustment is a work in progress.

2   Materials and Methods

2.1   Surgery and Experimental Layout

An adult male mongrel dog was used in this measure study. Anesthesia was induced with an initial intravenous injection of thiopental (25 mg/ml at 0.5ml/kg) and maintained after endotracheal intubation with isoflurane (0.8-2%, Siemens ventilator, 900D). A median sternotomy was performed, and a pericardial cradle was fashioned.

A multi-electrode epicardial sock consisting of a nylon mesh and 128 copper electrodes attached in an ordered fashion was then placed over the ventricular epicardium. The sock was placed in a consistent and pre-determined orientation and secured with several sutures (see fig. 1). Bipolar epicardial twisted-pair pacing electrodes were sewn onto the right atrium (RA). Similar electrodes were sewn onto the RV free-wall. A ground reference electrode was sewn onto the fat pad at the root of the aorta.

All sock and pacing wires were run directly out of the chest and connected, via a customized connection box, to two 64-channel analog to digital converter

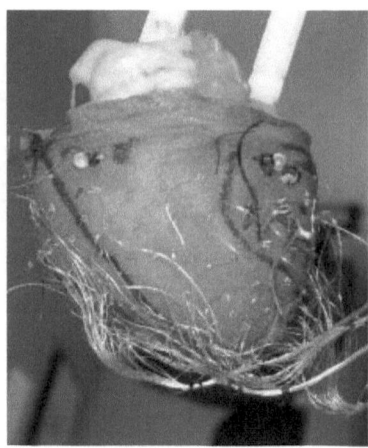

Fig. 2. Sock on the canine heart studied, once excised and polysiloxane-filled.

(A/D) boards (Hewlett-Packard, now Agilent, E1413C). All A/D boards were connected via FireWire (IEEE 1394) to a computer (Windows NT, 4.0) running data acquisition software (Hewlett-Packard, VEE 5.0). This procedure is MR compatible and fully detailed in [6].

2.2 Data Acquisition

RV pacing capture was established at a pacing rate (110-125 bpm) approximately 10-20% above intrinsic rate. Pacing current was set to approximately 20% above that needed for capture. Intrinsic electrical activation was suppressed by simultaneously pacing the RA and unipolar epicardial electrical recordings were obtained.

Electrical recordings were obtained at an acquisition rate of 1000-1450 Hz for a duration of approximately 10 seconds immediately prior to and following MR scans. Unipolar signals were electrically referenced to the aortic ground electrode.

After all *in vivo* image data and electrical recordings were obtained, the animal was heparinized and then euthanized with a bolus of potassium chloride while still under general anesthesia. The heart was excised with the sock still in place. The coronary arteries were then perfused from the aorta with isotonic saline at 50-60 mmHg to induce tissue turgor, and the heart was submerged in an isotonic saline bath to reduce body force deformation.

With the excised heart therefore in an approximate end-diastolic configuration, the LV and RV were then filled with vinyl polysiloxane by injection through the corresponding atria and atrioventricular valves to fix the shape. After approximately ten minutes, the vinyl polysiloxane solidified (see fig. 2). Using a 3D digitizer, the sock electrodes and localization markers were localized (digitizer coordinates). Additionally, locations of anatomical landmarks such as inter-ventricular sulcus, apex and aortic root were recorded.

2.3   Data Analysis

For analysis of electrical activation, epicardial readings from each electrode were averaged over approximately 20 heartbeats. The five-point finite difference estimate of the derivative of the recorded voltage, v, as a function of time, t, was used:

$$\frac{dv(t)}{dt} = \frac{1}{12\Delta t}[-v(t+2\Delta t) + 8v(t+\Delta t) - 8v(t-\Delta t) + v(t-2\Delta t)]$$

Electrical activation times, referenced to the pacing stimulus, were chosen as the point of the most negative derivative, indicating the time of local depolarization. Due to pacing artifact, the first ten milliseconds after pacing were not used for the detection of activation times.

3   Heart Myocardium Modeling

The whole process used to build a model of the myocardium from different imaging modalities is more precisely detailed in [14]. As the anisotropy created by the muscle fibers intervenes in the electrical wave propagation (and in the mechanical contraction), these directions are included in the mesh.

For the presented simulations, two different models were built. One from a canine heart dissection done in the Bioengineering Institute, Auckland University, New Zealand[1], where position and fiber directions were measured. These fiber directions from Auckland were interpolated and smoothed by the Cardiac Mechanics Research Group[2], UCSD, United States, and the latest version, which is available as a 3D vectorial image of the fiber directions, is used here.

The other model is built from a canine heart Diffusion Tensor Imaging done in the Duke University Medical Center[3], North Carolina, United States, which directly gives an estimate of the fiber directions for each voxel of the image [10].

3.1   Geometry of the Myocardium

For both models, the geometry is in a 3D image format. To build the mesh from a 3D image, the triangulated surface from the thresholded image is extracted, after using preprocessing like anisotropic diffusion and/or mathematical morphology to obtain a smooth binary mask.

As the quality of the tetrahedra depends on the quality of the surface (the surface stay unchanged when the inner volume is meshed), we improve the triangulation quality, using the INRIA YAMS[4] software, by optimizing the triangles positions depending on the curvature of the mesh.

---

[1] http://www.bioeng.auckland.ac.nz/home/home.php
[2] http://cmrg.ucsd.edu/
[3] http://wwwcivm.mc.duke.edu/civmPeople/HsuEW/EWHsu.html
[4] http://www-rocq.inria.fr/gamma/yams/

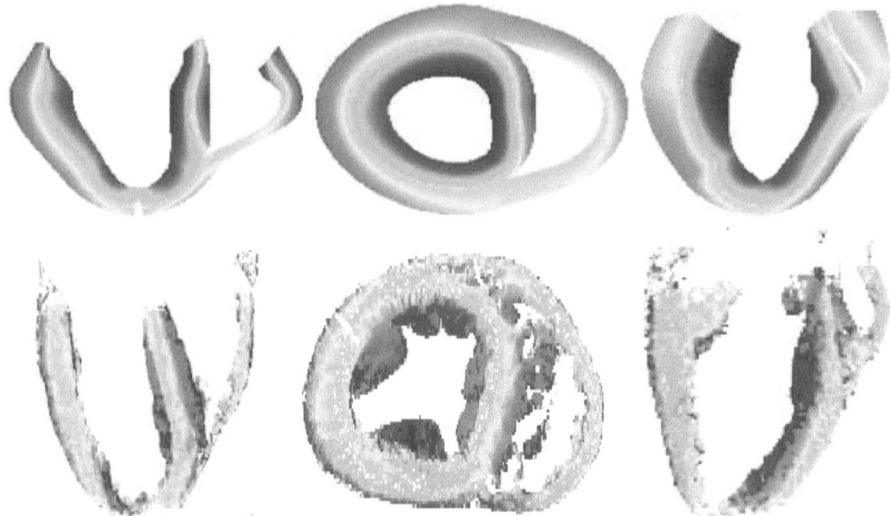

Fig. 3. Elevation angle of the fiber directions. (Top) 3 slices from a 3D image of the interpolated values representing the cardiac muscle fiber directions (data courtesy of A. McCulloch *et al.*). (Bottom) 3 slices from a 3D diffusion tensor MRI representing the cardiac muscle fiber directions (DTI data courtesy of Dr. Hsu *et al.*).

Finally, the volume is meshed with regularly sized tetrahedra to prevent smaller tetrahedra from reducing the stability of the time integration, using the INRIA GHS3D[5] software. The presented models have around 8000 vertices and 40 000 tetrahedra.

To adapt the shape of the model to the geometry of the electrodes basket, an affine registration is computed between the surface nodes of the model and electrode points positions obtained from 3D digitizer, as described in section 4.3.

3.2   Fiber Directions

The data from reduced-encoding MR Diffusion Tensor Imaging (DTI) of a canine heart is quite noisy compared to the data from UCSD, see fig. 3. Such studies can help compare the results obtained with this modality with the results obtained with a smoothed model. Fiber directions are assigned to the tetrahedral mesh using the rasterization procedure detailed in [14] (see fig. 4).

4   Electrical Model

4.1   Cell Level

At the cell level, the main idea is to study the relationship between the transmembrane ionic currents and the ionic potentials inside and outside the cell. The models concerning this relation improve while the number of phenomena observed at the cell level increases [3].

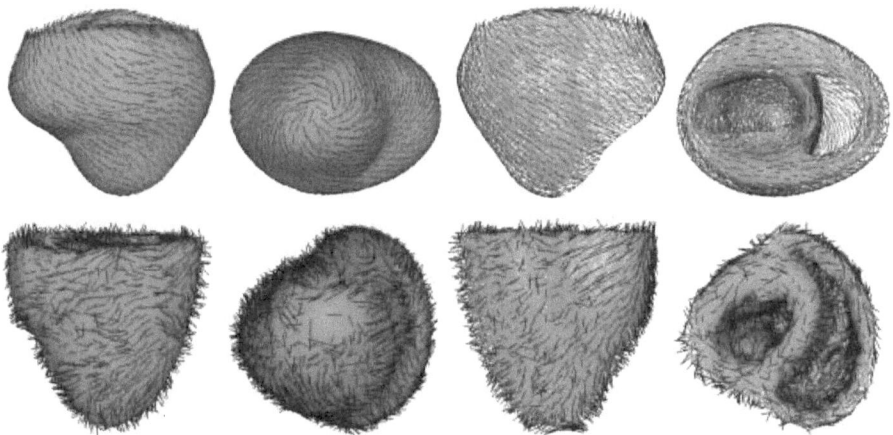

Fig. 4. Fiber directions assigned to the myocardium mesh from the UCSD interpolated data (top) and DTI data (bottom), anterior and posterior views.

At the beginning, the model is only expected to account for the most important biological phenomena:

- a cell is activated only for a stimulus larger than a certain threshold;
- the shape of the action potential does not depend on the stimulus (it is only model-dependent);
- there is a refractory period during which the cell cannot be excited;
- a cell can act as a pacemaker.

Different models are available for such a simulation. Luo-Rudy type models include many ionic currents to model precisely the evolution of the potentials. To be able to adjust the model from macroscopic measures, a more global model, FitzHugh-Nagumo like, where the variables are directly macroscopic potentials is chosen.

In so-called bidomain models, extra-cellular *and* intra-cellular potentials are included. As an objective of computing this electrical potential is also to control an electro-mechanical coupling, only a mono-domain model is used. Indeed, the contraction is controlled by the action potential, which is the difference between the two previously cited potentials.

A FitzHugh like model [7] seems to correctly capture these behaviors, and yields fast 3D computations. Aliev and Panfilov developed a modified version of the FitzHugh-Nagumo equations adapted to the dynamic of the cardiac electrical potential [1]. The model is simplified, as the complete ε term is mainly useful to model the influence of changes in pacing frequency and this behavior is not needed for the moment. Here is the set of differential equation studied:

$$\partial_t u = ku(1-u)(u-a) - uz$$
$$\partial_t z = -\varepsilon(ku(u-a-1) + z))$$

(1)

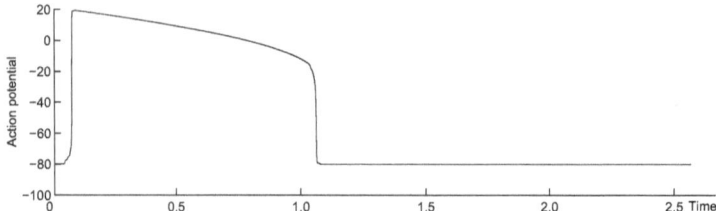

Fig. 5. 1D measure of action potential simulation with simplified Aliev and Panfilov model.

u is a normalized potential and z is a dynamic variable modeling the repolarization. k and ε control the repolarization, and a the reaction phenomenon. Parameters values are taken from [1].

With an excitation above the initialization threshold, the simulated action potential with this system is rather similar to measurements of cardiac action potentials (fig. 5).

4.2   Whole Ventricle Level – Anisotropy

At the macroscopic scale, the ventricles are considered as a conducting continuum, where the local potentials are undergoing at the same time the diffusion and the reaction phenomena described by the models above. Hence, (1) becomes:

$$\partial_t u = \text{div}\,(D\nabla(u)) + ku(1 - u)(u - a) - uz$$
$$\partial_t z = -\varepsilon(ku(u - a - 1) + z)) \tag{2}$$

On a physiological point of view, these equations are understood either as a mathematical approximation of the dynamical system introduced by Hodgkin and Huxley [9], as in [7], or as the result of some equilibrium equations that govern the conducting continuum, like in the bidomain model [15].

The anisotropy of the ventricles is taken into account through the diffusion tensor D: $D = d_0.diag(1, \rho, \rho)$, in a local orthonormal basis $(\mathbf{i}, \mathbf{j}, \mathbf{k})$ where $\mathbf{i}$ is parallel to the fiber. d_0 is a scalar conductivity and ρ the anisotropy ratio between the transverse and the axial conductivities. It is typically said that the electrical propagation goes two times faster in the fiber direction, so the used value is $\rho = 0.5$ (and $d_0 = 1.0$, as the system is adimensioned).

Once adimensioned with x and t between 0 and 1, the system is dimensioned spatially to the max dimension of the mesh and temporally so that the action potential duration is around 0.3 s ($\tau = 0.26t$, with τ the dimensioned time and t the computational time). The other parameters are: $\varepsilon = 0.01$, $k = 8.0$, $a = 0.15$.

To initiate the action potential propagation, the extremities of the Purkinje network, which are the specialized system that conducts the depolarization from the atrio-ventricular node to the myocardium, have to be located.

The Purkinje network is hardly visible by dissection and by imaging. We used the measures from Durrer *et al.* [5] to locate the Purkinje network extremities on the endocardia of both the left and right ventricles (the version of [5] presented here (fig. 6) was found on the web[6]).

A first validation of the 3D computation was comparing the resulting action potential isochrones with the measures from Durrer *et al.* (see fig. 6). The temporal integration is done with explicit Euler scheme and the spatial integration is done with linear tetrahedral elements. The computation time step is 10^{-5} and a 3D simulation takes around half an hour on a standard PC with 40 000 elements.

From the qualitative comparison between the resulting simulations, we can observe that both models seem to reproduce the activation patterns visible on the measures, with maybe a closer result for the smoother model. But as the activation initialization is not exactly at the same place, due to anatomical differences, a more detailed comparison is difficult. This is another reason why measures with precisely located pacing electrodes are very valuable.

To try to evaluate the effect of anisotropy, simulations were conducted using an isotropic diffusion tensor (see fig. 7). As the fibers have quite a small radial component, the diffusion in the wall is much slower in the anisotropic case. The anisotropy effect seems more important with the model built from DTI. As the fiber directions are noisier, there are more discontinuities, so the propagation is more affected by the anisotropy.

4.3 Simulation of the *in vivo* Measurements

Model Geometry. The 3D position of the electrodes plus different points on the left ventricle endocardium are known. Surfacic meshes are fitted to these point clouds to interpolate the positions and have more 3D data to register. These surfaces were transformed into binary 3D images. Then an affine transformation is computed between the models built and the myocardium image created from the measures, using the hierarchical method described in [14].

Ultimately, the model geometry will be fitted to the data geometry directly in a 3D image of the measured heart, using the method described in [14] to obtain local deformations.

Simulated Pacing. In order to complete our anatomical model, data about electrical onset areas of the ventricular depolarization is needed. In the presented measures, the location of the pacing electrodes is well known, which is very interesting for simulating propagation.

Pacing electrodes are simulated by imposing an initial action potential in the epicardium location corresponding to the localization of the pacing electrodes in the measure protocol, see fig. 9.

---

[6] http://butler.cc.tut.fi/~malmivuo/bem/bembook/

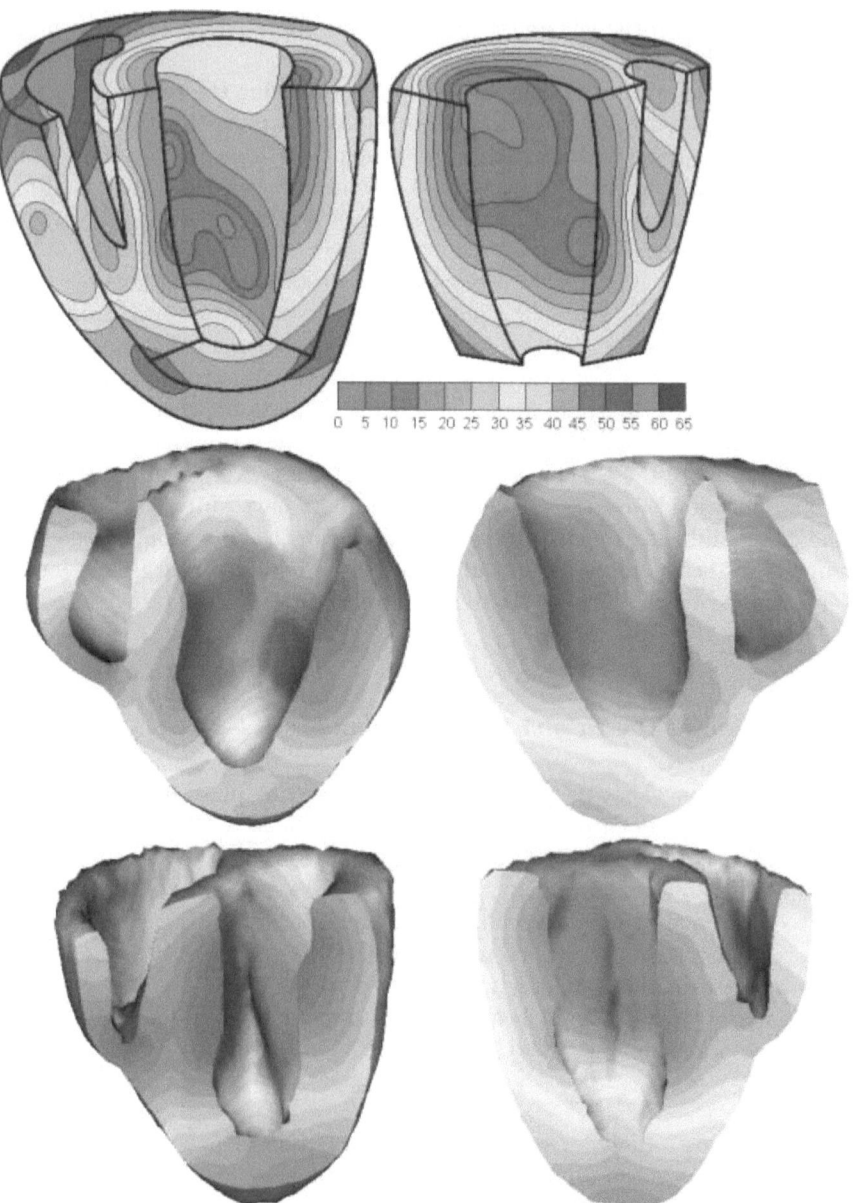

Fig. 6. Action potential isochrones measured by Durrer *et al.* (top row) compared with the simulated ones using the model built from UCSD data (middle row) and from DTI (bottom row).

4.4   Results of the Wave Propagation

The propagation of the action potential is simulated in the 3D meshes of the myocardium constructed, initiating the wave at the location defined in previous section. Fig. 10 presents resulting simulated isochrones of activation.

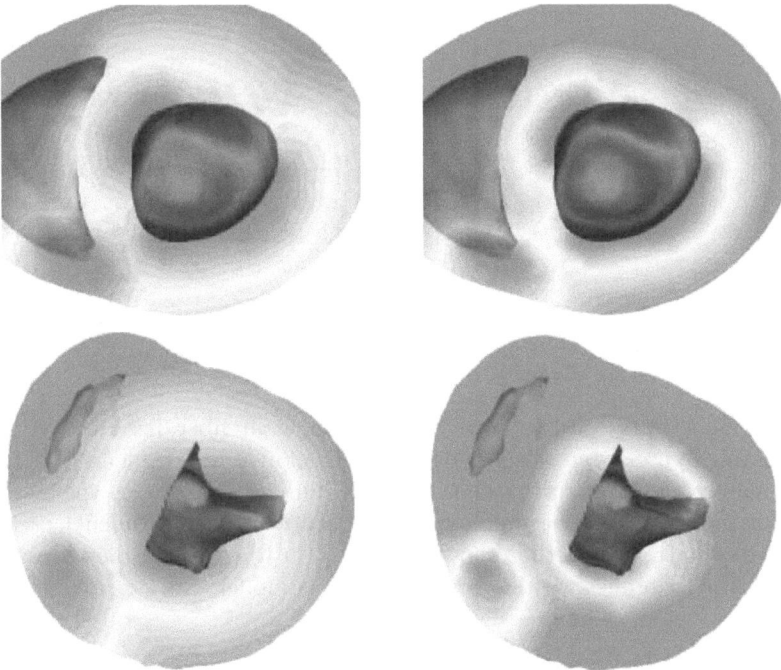

Fig. 7. Comparison between isotropic (left) and anisotropic (right) propagation for the model built from UCSD data (top row) and from DTI (bottom row). The fibers are mainly circumferential, so the radial diffusion is much slower with anisotropy.

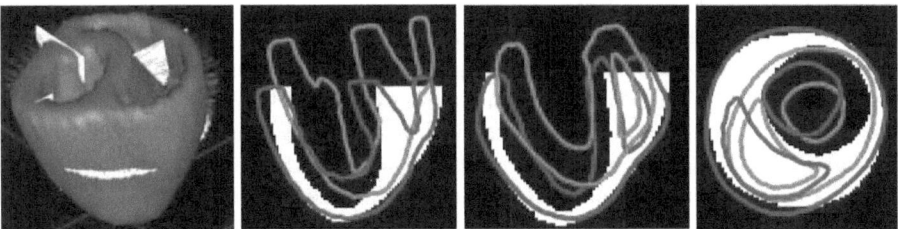

Fig. 8. Affine registration between the two models built and a 3D Image created from the position of the electrodes and points on the left ventricle endocardium. (left) matching between the model and the image voxels. (others) initial position in red and final position in blue.

As the action potential is computed and not both intra-cellular and extra-cellular potentials, the electrical measures cannot be compared directly with the simulations, only the activation times, through the isochrones, can be compared. The results look qualitatively good and using a 3D propagation model allows also to visualize the simulated potential inside the myocardium wall (see fig. 11). It appears that in the model from DTI, the discontinuities in the fiber directions in the right ventricle create a delay in the electrical propagation, and the asymmetry in the propagation does not seem to be present in the measures. It is likely that

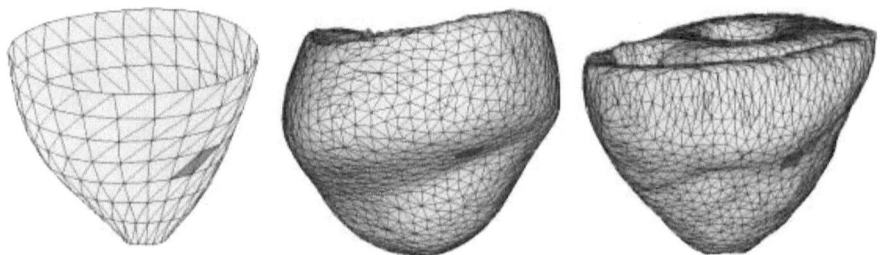

Fig. 9. Red triangles: real (left) and simulated pacing electrode location (middle: UCSD model, right: DTI model).

due to the noise in the current *in vivo* measurements in fiber directions, it would be preferable to use a smoothed model rather than measures from another heart.

Ultimately, the DTI image will be from the same canine heart as the measures (there are ongoing DTI measurements on the heart of the experiment), so it will be able to determine whether the simulations with a specific model can give better results than ones with a general smoothed model.

5   Conclusion and Perspectives

This article presents first qualitative results comparing measures and simulations of the electrical activity of a canine heart. Different data sources were used to build models of the myocardium. The simulations show that the model well captures the action potential propagation.

Quantitative adjustment of the parameters of the model are an ongoing work, using data assimilation techniques based on Kalman filtering but no existing method is really suited to adjust such a model with this kind of data. Ultimately, such a model could allow to find electrical activity patterns directly from ECG measures.

This time-dependent computed potential can also be used as an excitation entry to the system describing the mechanical behavior of the myocardium, as a model of the electro-mechanical coupling is developed [13]. Tagged cine images acquired during RV pacing in this study are very useful for this task.

Indeed, nodes displacements can be computed from the electro-mechanical coupling, and then compared with the displacements extracted from MRI. Once the whole model has been validated, it could be used in a predictive way to study pathologies and intervention planning.

Acknowledgements

The modeling and data assimilation work is a part of the multidisciplinary project ICEMA[7] (standing for Images of the Cardiac Electro-Mechanical Activ-

---

* http://www-rocq.inria.fr/who/Frederique.Clement/icema.html

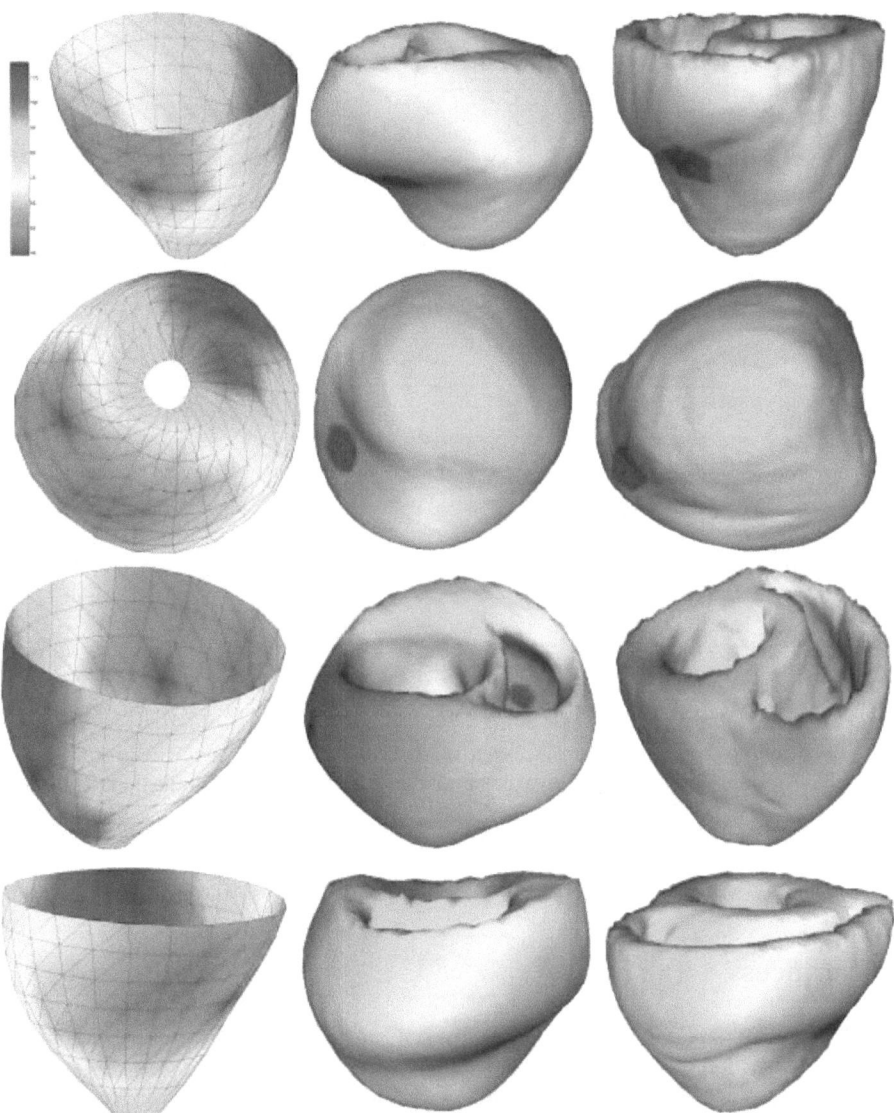

Fig. 10. Measured and Simulated Action Potential isochrones. Left: measures, middle: simulation with the UCSD model, right: simulation with the DTI model.

ity) and ICEMA-2[8], which are collaborative research actions between different INRIA projects and Philips Research France [2].

All aspects of this study were conducted in accordance with the guidelines of the Animal Care and Use Committee of the National Heart, Lung, and Blood Institute.

---

* http://www-rocq.inria.fr/sosso/icema2/icema2.html

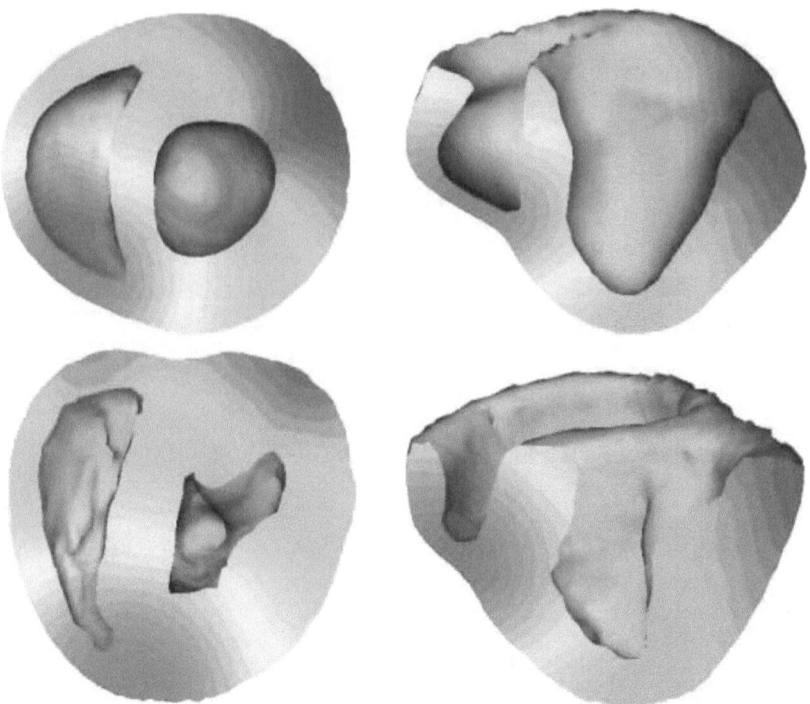

Fig. 11. Simulated Action Potential isochrones, visualized in the myocardium wall, for the UCSD model simulations (top row) and for the DTI model simulations (bottom row).

Additional color version of the images and videos can be found on the web[9].

References

1. R. Aliev and A. Panfilov. A simple two-variable model of cardiac excitation. *Chaos, Solitons & Fractals*, 7(3):293–301, 1996.
2. N. Ayache, D. Chapelle, F. Clément, Y. Coudière, H. Delingette, J.A. Désidéri, M. Sermesant, M. Sorine, and J. Urquiza. Towards model-based estimation of the cardiac electro-mechanical activity from ECG signals and ultrasound images. In *Functional Imaging and Modeling of the Heart (FIMH'01)*, number 2230 in LNCS, pages 120–127. 2001.
3. A. L. Bardou, P. M. Auger, P. J. Birkui, and J.-L. Chassé. Modeling of cardiac electrophysiological mechanisms: From action potential genesis to its propagation in myocardium. *Critical Reviews in Biomedical Engineering*, 24:141–221, 1996.
4. J. Burnes, B. Taccardi, and Y. Rudy. A noninvasive imaging modality for cardiac arrhythmias. *Circulation*, 102(17):2152–2158, 2000.
5. D. Durrer, R. van Dam, G. Freud, M. Janse, F. Meijler, and R. Arzbaecher. Total excitation of the isolated human heart. *Circulation*, 41(6):899–912, 1970.

---

[9] http://www-sop.inria.fr/epidaure/personnel/Maxime.Sermesant/gallery.php

6. O. Faris, F. Evans, D. Ennis, P. Helm, J. Taylor, A. Chesnick, M. A. Guttman, C. Ozturk, and E. McVeigh. A novel technique for cardiac electromechanical mapping with MRI tagging and an epicardial electrode sock. *Annals of Biomedical Engineering*, 31, 2003.
7. R.A. FitzHugh. Impulses and physiological states in theoretical models of nerve membrane. *Biophysical Journal*, 1:445–466, 1961.
8. P. Colli Franzone, L. Guerri, M. Pennacchio, and B. Taccardi. Myocardial anisotropy and multiphasic electrograms - simulation study in a monoventricular model. In *Workshop CardioModel. Models of the Heart: Theory and Clinical Application*, 2000.
9. A. Hodgkin and A. Huxley. A quantitative description of membrane current and its application to conduction and excitation in nerve. *Journal of Physiology*, 177:500–544, 1952.
10. E. Hsu and C. Henriquez. Myocardial fiber orientation mapping using reduced encoding diffusion tensor imaging. *Journal of Cardiovascular Magnetic Resonance*, 3:325–333, 2001.
11. R. MacLeod, B. Yilmaz, B. Taccardi, B. Punske, Y. Serinagaolu, and D. Brooks. Direct and inverse methods for cardiac mapping using multielectrode catheter measurements. *Journal of Biomedizinische Technik*, 46:207–209, 2001.
12. E. McVeigh, O. Faris, D. Ennis, P. Helm, and F. Evans. Measurement of ventricular wall motion, epicardial electrical mapping, and myocardial fiber angles in the same heart. Same as [2], pages 76–82.
13. M. Sermesant, Y. Coudière, H. Delingette, N. Ayache, J. Sainte-Marie, D. Chapelle, F. Clément, and M. Sorine. Progress towards model-based estimation of the cardiac electromechanical activity from ECG signals and 4D images. In *Modelling & Simulation for Computer-aided Medicine and Surgery (MS4CMS'02)*, 2002.
14. M. Sermesant, C. Forest, X. Pennec, H. Delingette, and N. Ayache. Biomechanical model construction from different modalities: Application to cardiac images. In *Medical Image Computing and Computer-Assisted Intervention (MICCAI'02)*, volume 2208 of LNCS, pages 714–721. 2002.
15. K. Simelius, J. Nenonen, and B.M. Horácek. Simulation of anisotropic propagation in the myocardium with a hybrid bidomain model. Same as [2], pages 140–147.

An Augmented Reality Approach Using Pre-operative Patient Specific Images to Guide Thermo-Ablation Procedures

Stijn De Buck[1], Frederik Maes[1], Wim Anné[3], Jan Bogaert[2], Steven Dymarkowski[2], Hein Heidbuchel[3], and Paul Suetens[1]

[1] Faculties of Medicine and Engineering, Medical Image Computing
(ESAT and Radiology)
University Hospital Gasthuisberg
Herestraat 49, B-3000 Leuven
stijn.debuck@uz.kuleuven.ac.be
[2] Department of Radiology, University Hospital Gasthuisberg
Herestraat 49, B-3000 Leuven
[3] Department of Cardiology, University Hospital Gasthuisberg
Herestraat 49, B-3000 Leuven

Abstract. We present a system to assist in the treatment of tachycardia patients by catheter ablation.

In an augmented reality framework we combine a patient specific pre-operative MR model, constructed from a set of transverse, coronal and sagittal images, with intra-cardial voltage potential measurements and fluoroscopic imaging to guide the electrophysiologist. The registration of the model and the fluoroscopic images, which is done by a visual matching technique, enables an easy transfer of the measurement to the pre-operative model. By visualizing annotations of different tissue types and of the measurements, extra insight is gathered about the problem, resulting in improved patient care.

Because of its low cost and similar advantages we believe our approach can compete with existing commercial solutions, which rely on dedicated hardware and costly catheters. First clinical evaluation on 31 patients indicate a considerable advantage in the diagnosis and treatment.

Our future work will consist of improving 2D-3D registration and further automating the measurement procedure.

1 Introduction

The purpose of augmented reality is the integration of real and virtual entities to attain a certain goal. In medical applications this goal is increasing procedure quality and possibly reduce its time thereby improving patient care.

Patients which suffer from tachycardia –too fast and irregular beating of the heart– can currently be treated by cardiac ablation. The origin of this tachycardia is the presence of abnormal conduction paths within the heart of these patients which causes an irregular activation and thus beating. Treatment by cardiac ablation consists of the following steps:

N. Ayache and H. Delingette (Eds.): IS4TM 2003, LNCS 2673, pp. 244–252, 2003.
© Springer-Verlag Berlin Heidelberg 2003

- First of all, the activation times are recorded by means of electrodes which are attached to catheters inserted in the heart and which measure the voltage potential as a function of time. The procedure is guided by bi-plane fluoroscopic images such that the electrophysiologist gets feedback about the location where he is measuring.
- Secondly the electrophysiologist mentally constructs a 3D representation of the patients' heart based upon the bi-plane fluoroscopic images and his anatomical knowledge. Subsequently he maps the measured activation times on this mental model and deducts from these the faulty conduction paths.
- The third phase and actual treatment consists of burning these abnormal conduction paths. Here the electrophysicist has to back-project his mental model in the real scene to know where he has to burn.

One of the major challenges for the electrophysicist is the combining of his anatomical knowledge with information from both the fluoroscopic images and the catheter measurements. In this problem augmented reality is well suited to provide a solution since it can create a framework in which several sources of information are integrated. We present here such a system which not only integrates the fluoroscopic images and the activation times but also includes a patient specific model of the heart, offering pre-operatively very detailed anatomical information.

Other approaches, of which some are commercially available ([1],[2]), solve the problem by creating a virtual environment only. Technical descriptions can be found in [3], [4], [5] and [6]. They make use of dedicated electro-magnetic tracking systems and specialized catheters. The position of these is known with respect to a fixed coordinate system and is recorded repeatedly when touching the cardiac wall. This way a point cloud of cardiac wall points is gathered and a surface model is generated. At the same time voltage potentials are stored and activation times are marked. The advantages of such systems with respect to a conventional approach are the use of a patient specific model and the display of electro-anatomical data onto it. Disadvantages include that the model has a spatial accuracy limited by the number of places visited by the catheter and that there is no 3D knowledge about the heart prior to surgery. Also the high installation cost and the costly catheters, which are to be used only once, have to be taken into account.

The approach we present here resolves these disadvantages while retaining most of the advantages.

2   Material & Methods

Our approach can be subdivided in the following parts:

1. Acquisition of pre-operative MR-datasets of the patients' heart.
2. Delineation of the MR-images and construction of a surface model.
3. Registration of surface model with the fluoroscopic images.
4. Annotation of the patient model with the activation timing measurements.
5. Visualization of the activation times on the surface model.

Part 1 and 2 are completed pre-operatively while parts 3 until 5 have to be used during the intervention and thus imply a time constraint on the processing time. These parts are therefore constructed such that processing time and acquisition time of the fluoroscopic images is kept very low and almost real-time.

2.1   Acquisition & Surface Model Generation

Cardiac MR images (256 x 256 matrix, 20 slices, 1.25 x 1.25 mm pixel size, 6 mm slice spacing) are acquired in a single breathhold at end- systole of the heart cycle (i.e. atrial diastole) using a balanced fast- field-echo sequence (Philips Intera, TR: 2.62 ms, TE: 1.31 ms, Flip Angle: 55, slice thickness: 6 mm). Three subsequent orthogonal scans are made in the patient's transverse, coronal and sagittal orientation. The images are digitally transferred in DICOM format over the PACS network from the scanner to a workstation and subsequently converted to 3-D stacks. 3-D transverse, coronal and sagittal images are first co-registered by an affine coordinate transformation in order to correct for global patient motion and differences in breathhold between different acquisitions. The registration is initialized from the information about the position and orientation of each image within the space of the scanner as provided by the DICOM header of the images and further refined using an intensity-based automated procedure based on maximization of mutual information [7]. The right atrium is delineated by manual contouring in each slice of the three orthogonal scans. The 2-D contours delineated in one image are transformed into the other images using the previously derived registration transformations to allow verification and adjustment of the position of all contour points in 3-D. All 3 orthogonal sets of contours are then combined and a single surface is constructed through all contour points using a scattered data interpolation algorithm based on fitting radial basis functions. The resulting surface consists of a dense grid of points.

2.2   Registration

The fluoroscopes of which the orientation parameters are input to a software camera, are modeled as orthographic cameras. This assumption can be made because the size of the heart (± 10 cm diameter) is small with respect to the focal distance of a fluoroscope (± 100 cm).

Registration is currently implemented by a modified version of the visual matching technique studied in [8]. Orientation of the model with respect to the fluoroscopes is determined by means of information from both the fluoroscopes and the MR-equipment. The remaining degrees of freedom, which are the position in the viewing plane and the zoom factor are left to the user to determine by visual matching.

For this purpose, contrast enhanced fluoroscopic images are made (cfr figure 1) and imported by a framegrabber in the system. The surface model is rendered transparently in overlay while the electrophysiologist uses the mouse to align the model with the images. Once good alignment is reached that state is fixed (cfr figure 1).

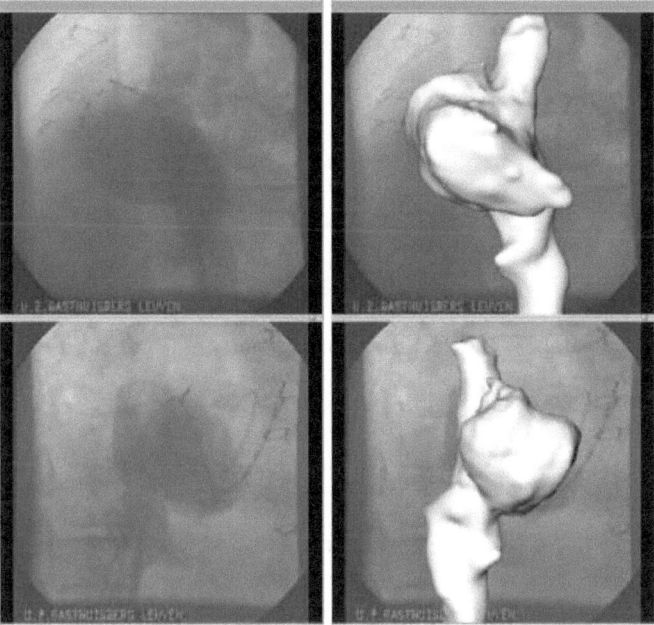

Fig. 1. On the left the bi-plane contrast enhanced fluoroscopic images and on the right the same image with the model in overlay after visual matching.

The use of a visual matching technique can be justified by a number of arguments. First of all the contrast enhanced images might differ from the projections of the model in a number of ways due to for instance errors in the manual segmentation, approximations in the fusion of contour points (cfr section 2.1) or due to the movement of the atrium in one heartbeat. By using visual matching the electrophysiologist can apply his knowledge of anatomy and fluoroscopic imaging to identify these inconsistencies and compensate for their presence. Secondly time of registration will vary around 2-3 min [8], which is a negligible percentage of procedure time. Thirdly we can make an extrapolation of the accuracy results of visual matching obtained in [8]. Recomputing the x and y movement (these are the movements parallel to the viewing plane), we obtain a result of $\pm\frac{1}{60}cm$ which is negligibly small. The influence of the zoom estimation is computed as $\pm 0.5 cm$ which is larger but still acceptable for our goal, certainly if we compare this error with the one due to the beating of the heart. These considerations indicate strongly the suitability of the visual matching approach for this application because of a sufficient registration accuracy and flexibility to incorporate user knowledge.

The registration procedure makes an easy transfer possible from the intra-operative scene to the pre-operative model as described in the following section. On the other hand, it also facilitates the transfer in the other direction: once the locations to burn are identified on the model, they can easily be seen in the augmented scene and thus on the registered fluoroscopic images.

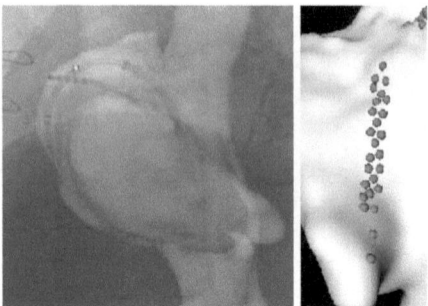

Fig. 2. The patient model on which one annotation point is attached by visual transfer from the fluoroscopic images (left) and an example of the spraying capability used to mark regions of injured tissue or valves (right).

2.3 Annotation

Annotation is the process of marking the surface with characteristics. In this case they are derived from the catheter measurements within a conventional electrophysiological setup. Currently, we implemented three types of annotation:

- The marking of activation times
- The marking of injured tissue: because of their non-conducting property, these tissue regions are often the centers of circular currents which cause the tachycardia. They originate for instance of previous open heart surgery or a specific pathology which causes microcracks in the cardiac wall. Therefore it is essential to be able to manually mark these regions based on previous surgery reports or derived from the catheter measurements
- The marking of holes and valves on the model. These natural non-conducting spaces were not modeled in the segmentation step (cfr 2.1), but can be entered by annotation. They can also cause circular currents in combination with the previous type.

Individual measurement points can be marked by manually clicking the surface model and entering delay values or tissue types. Using the two fluoroscopic views the transfer of measurement locations onto the model is quite intuitive: by making the model transparent the electrophysiologist just indicates the electrode visible through the surface and a new point is added to the model. Problems like on which side he should add a point, can be resolved by the biplane fluoroscopic images, which are mostly set at about 90°. Also the ability to mirror both the model and the fluoroscopic images, is an important feature in this matter.

Injured tissue and holes can be entered manually one by one as in the previous case or they can be sprayed onto the surface. By spraying, points are added to the model in a given, modifiable radius and at a certain resolution (cfr figure 2).

Tools were added to change the properties of previous measurements or to relocate them on the surface. The latter feature is mainly used when no fluoroscopic images are present or to compensate possible registration inaccuracies.

Since patients with complex pathologies often have several different activation patterns sequentially, the possibility is provided to add points to a certain type of arithmia while the injured and hole surface types are kept invariant over all arithmias. Even when previous measurements or tissue types are altered, this invariance is maintained.

2.4   Visualization

To offer a proper insight into the patients anatomy, three viewing windows are constructed: two "augmented" views, which show the fluoroscopic images aligned with the model and a third one allowing free manipulation of the pre-operative model. This enables the electrophysiologist to get a good 3D perception of the atrium shape before and during the intervention.

Once the model is annotated, we can visualize this information by color coding the surface. Since the outcome of our model generation step is constructed upon a dense grid of points, we can colorize the surface by assigning a color to each vertex. The method to accomplish this proceeds as follows:

1 First of all, the annotation values are assigned to vertices in a predetermined radius r. This radius is chosen as a validity radius of the annotation measurement.
2 Together with the assignment of annotation values a weight factor is assigned to each vertex which is given by:

$$w_i = e^{-\frac{a d_i}{r}^2} \tag{1}$$

with d_i equal to the distance between the annotation point i and the vertex and a a fixed factor.
3 For each vertex the resulting value is computed as

$$\sum_{i=1}^{k} w_i M_i \tag{2}$$

for k equal to the number of annotation points that influence this vertex. The resulting value is used to modulate the hue component of the vertex color. The M_i values represent activation times of the heart contraction but can without modification be used for voltage potentials or other local characteristics.

For the special types of tissue, we assign special weights such that the normal measurements are suppressed.

By following this procedure we make sure that annotation measurements are extrapolated across the surface as far as they are valid. On the other hand, they are interpolated in overlapping regions. This results in a simulation model as depicted in figure 3. In the case of activation time measurements, the electrophysiologist can clearly observe regions of early and late activation from the local color component and doing so, he can identify the activation flow path.

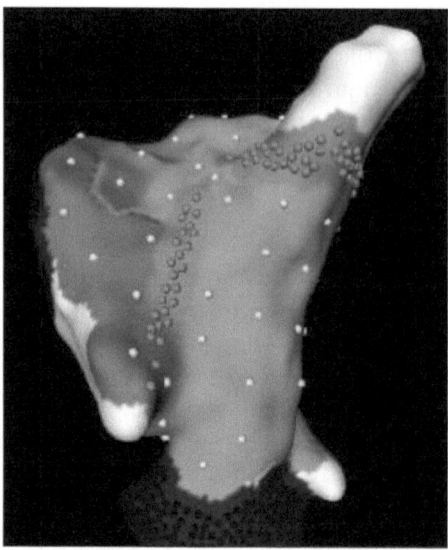

Fig. 3. A patient atrium model, after annotation with activation timings, injured tissue and a valve which was followed by color coding the surface. The color codes represent the activation times of the potential.

2.5   Hardware Requirements

To use our software a common personal computer (Pentium III) with OpenGL graphics card is sufficient. Currently the fluoroscopic images are digitized by means of a DataTranslation framegrabber. Therefore hardware costs are reduced to the minimum and we can acquire fluoroscopic images at video rate. We also constructed a dual screen setup such that the physiologist can immediately view the model while manipulating the catheters. In that case the program is controlled by his assistent which can manipulate and annotate the model.

3   Qualitative Intra-operative Evaluation

The system has already been used in 31 cases and found to be very useful and offering a substantial amount of added value with respect to the conventional approach: the practicing electro-physiologist stated that the fact only of having pre-operatively a patient specific model at hand provides extra insight and sometimes leads to the choice of a different approach for the intervention than conventionally would be assumed. The color coded surface also provides an intuitive visualization and simplifies the mental reconstruction problem of the electrophysiologist where he has to fuse electrical, anatomical and imaging information to come up with a treatment plan.

One drawback in the current implementation is the fact that all annotation measurements have to be entered manually, which is rather time consuming. This will be dealt with in our future work.

4   Discussion

In this paper we presented an augmented reality tool to assist an electrophysiologist in the treatment of tachycardia. Three sources of information are combined, which offer a substantial amount of added value: a patient specific model of the heart incorporating detailed anatomy, intra-operative fluoroscopic, which give immediate feedback and activation measurements from the cardiac wall.

Since the system can be used as an extension to a conventional electrophysiological setup, hardware costs were reduced to the minimum and its advantages are comparable to other commercial approaches, we believe our approach to be competitive to these. Nevertheless, a number of drawbacks remain unsolved: although the registration accuracy between pre- and intra-operative images seems fairly good from the analysis provided in section 2.2, we believe it might be further improved by an automated method. Furthermore we would like to automate recording of cardiac potentials and activation timing to reduce procedure time.

The first problem we intend to solve by an intensity based registration. In that case the procedure will consist of a manual alignment as implemented in the current approach followed by the intensity based approach. Several criteria can be used like correlation, mutual information, etc. [9].

In order to automate the recording, we will design virtual catheters, which can be aligned with the real ones thereby enabling fast localization of the electrodes on multi-polar catheters. The latter can contain 20 or more electrodes. This way the activation of a complete chamber can be measured by placing the catheters at 3 to 5 locations.

Acknowledgements

This work is supported by the GOA/99/05-project, entitled "Variabiliteit in menselijke vorm en spraak".

References

1. http://www.biosensewebster.com/US/products_cartonav.htm.
2. http://www.endocardial.com.
3. Dipen C. Shah, Pierre Jais, Michel Haissaguerre, Salah Chouairi, Atsusi Takahashi, Meleze Hocini, Stephane Garrigue, and Jacques Clementy. Three-dimensional Mapping of the Common Atrial Flutter Circuit in the Right Atrium. *Circulation*, 96(11):3904–3912, 1997.
4. Lior Gepstein, Gal Hayam, and Shlomo A. Ben-Haim. A Novel Method for Nonfluoroscopic Catheter-Based Electroanatomical Mapping of the Heart : In Vitro and In Vivo Accuracy Results. *Circulation*, 95(6):1611–1622, 1997.
5. Charles C. Gornick, Stuart W. Adler, Brian Pederson, John Hauck, Jeffrey Budd, and Jeff Schweitzer. Validation of a New Noncontact Catheter System for Electroanatomic Mapping of Left Ventricular Endocardium. *Circulation*, 99(6):829–835, 1999.

6. Richard J. Schilling, Nicholas S. Peters, and D. Wyn Davies. Simultaneous Endocardial Mapping in the Human Left Ventricle Using a Noncontact Catheter Comparison of Contact and Reconstructed Electrograms During Sinus Rhythm. *Circulation*, 98(9):887–898, 1998.
7. Frederik Maes, Andre Collignon, Dirk Vandermeulen, Guy Marchal, and Paul Suetens. Multimodality image registration by maximization of mutual information. *IEEE transactions on Medical Imaging*, 16(2):187–198, 1997.
8. S. De Buck, J. Van Cleynenbreugel, G. Marchal, and P. Suetens. Evaluation of a visual matching registration technique for intraoperative surgery support. In Seong K. Mun, editor, *Medical Imaging, Visualization, Image-guided Procedures, and Display*, pages 492–499, Februari 2002.
9. Graeme P. Penney, Jürgen Weese, John A. Little, Paul Desmedt, Derek L. G. Hill, and David J. Hawkes. A comparison of similarity measures for use in 2D–3D medical image registration. *IEEE Transactions on Medical Imaging*, 17(4):586–595, August 1998.

Modeling of Cardiac Electro-Mechanics in a Truncated Ellipsoid Model of Left Ventricle

Frank B. Sachse[1] and Gunnar Seemann[2]

[1] Nora Eccles Harison Cardiovascular Research and Training Institute, University of Utah, Salt Lake City, USA
[2] Universität Karlsruhe (TH), Institut für Biomedizinische Technik, 76128 Karlsruhe, Germany

Abstract. Modeling of cardiac electro-mechanics enables and simplifies understanding of physiology and pathophysiology of the heart. In this work a model is presented, which allows the reconstruction of macroscopic electro-mechanical processes in the left ventricle of small mammals. The model combines a three-dimensional model of left ventricular anatomy represented as truncated ellipsoid with an integrated electro-mechanical model. The integrated model includes electrophysiological, force development and elastomechanical models of myocardium. The model is illustrated by simulations, which reflect the behavior of an extracorporated heart. These simulations yield temporal distributions of electrophysiological parameters as well as descriptions of electrical propagation and mechanical deformation. The simulations show the connection between cellular electrophysiology, electrical excitation propagation, force development, and mechanical deformation.

1 Introduction

Basic mechanisms and complex phenomena of the heart can be described and reconstructed by mathematical modeling and computer-aided simulation [1]. These simulations deliver physical and physiological quantities, which are partly not available by measurements due to technical limitations and ethical objection. Thereby achieved insights can be applied to improve clinical diagnosis and therapy as well as education in cardiology and heart surgery. A description of cardiac electro-mechanics necessitates a combination of different models, e.g. anatomical, electrophysiological, and mechanical models. An anatomical model defines the tissue distribution. An electrophysiological model of single myocytes reconstructs e.g. transmembrane voltage, intra- and extracellular ion concentrations, and ion flows through membranes. The combination with a model of electrical current flow allows the reconstruction of electrical excitation propagation. A tension development model quantifies the tension developed by contractile units outgoing from cellular electrophysiologic and mechanic states. Another mechanical model specifies the stress-strain relationship of myocardium.

In this work a three-dimensional model of a left ventricle was developed, which allows the coupled simulation of electrical excitation propagation, force

N. Ayache and H. Delingette (Eds.): IS4TM 2003, LNCS 2673, pp. 253–260, 2003.
© Springer-Verlag Berlin Heidelberg 2003

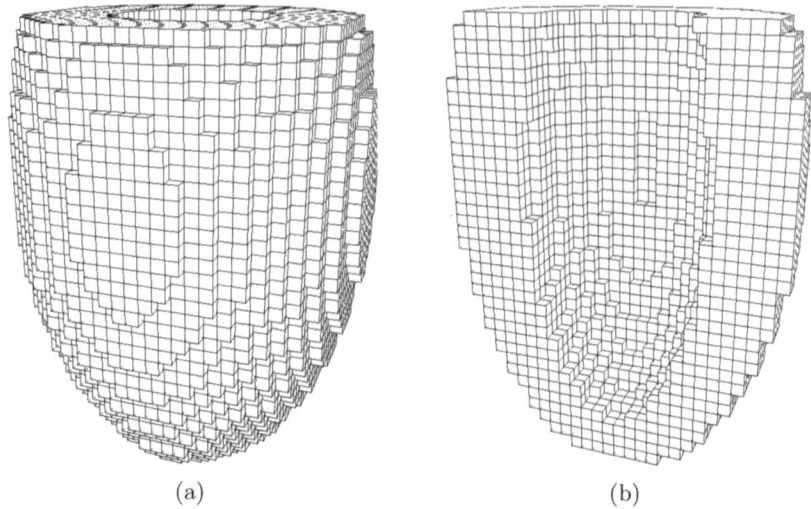

Fig. 1. Model of ventricular anatomy. The (a) full and (b) sectioned model is shown in wire frame representation. The model consists of 30 x 30 x 38 cubic elements.

development and deformation. The anatomy was represented by a truncated ellipsoid model with varying fiber orientation from endo- to epicardial myocardium. The model was derived from the Noble-Varghese-Kohl-Noble model of the electrophysiology of a ventricular myocyte [2], a monodomain model of intercellular current flow [1], a hybrid model of force development [3,4,5], and a deformation model [6], which bases on the strain energy density function of Guccione-McCulloch-Waldman for passive myocardium [7]. The force development is controlled by the concentration of intracellular calcium. The calculation of the deformation was performed in an incremental Lagrangian formulation with displacement-based isoparametric finite elements [8]. The integrated model was used to investigate the distribution of physical and physiological quantities in the left ventricle. The results of simulations, e.g. patterns of transmembrane voltage, intracellular calcium concentration, electrical propagation and mechanical displacement, were analyzed quantitatively.

2   Modeling of Ventricular Anatomy

The geometry of left ventricle can be approximated by crop of two confocal truncated ellipsoids [9]. The ellipsoid's focus length d is defined as $d = \sqrt{a^2 - b^2}$ with the ellipsoid's minor radius b and major radius a. The truncation of an ellipsoid is quantified by a truncation factor f_b specified by $f_b = l_{ab}/l_{ea}$ with the length from apex to base l_{ab} and the length from equator to apex l_{ea}. Commonly, a truncation factor f_b of 0.5 is chosen. The orientation of myocytes was included by interpolation starting from boundary conditions in three depths of the myocardium. The orientation was set subepicardial to $-70°$, midwall $0°$, and subendocardial 70° reflecting knowledge from anatomical studies [10]. The size of the ellipsoids was chosen in such a manner, that they approximate the size of small animal's left ventricle. The ventricle's geometry and fiber orientation was rendered in lattices of 30 x 30 x 38 and 16 x 16 x 20 cubic elements with a length of 200 and 400 μm, respectively (figure 1).

3   Modeling of Electrophysiology

Cellular electrophysiology. Variant models of myocyte's electrophysiology were published in the last 40 years. In this work the Noble-Varghese-Kohl-Noble model was applied, which describes the electrophysiology of ventricular myocytes of guinea-pig. The model includes effects on ionic channels by the concentration of adenosine triphosphate (ATP) and acetylcholine (ACh) as well as by stretching. Furthermore, a tension generation model and a description of the diadic space was incorporated. Different variants and configurations of the model exist. The variant applied in this work is based on [2,11,12] neglecting ATP and ACh activated ionic channels as well as using only the electrophysiological part of the model. The model includes dependencies of electrophysiological parameters on the length or tension of the sarcomere. The mechano-electric feedback is realized by introducing stretch-activated ion conductances, a modulation of calcium binding to troponin C, and a modulation of sarcoplasmic leak current. The usage of the mechanisms in the integrated model presented in this work is restricted to the incorporation of length dependencies of the electrophysiological parameters. Stretch activated ion channels were deactivated.

Excitation propagation. Different modeling approaches of the electrical excitation propagation in the myocardium can be distinguished depending on the representation of the microscopic and macroscopic anatomy as well as depending on the approximation of the cellular electrophysiology [1]. In this work a monodomain model was used [13], which describes the electrical current flow in the intracellular domain by the generalized Poisson's equation:

$$\nabla \cdot (\sigma_i \nabla \phi_i) = \beta I_m - I_{si}$$

with the intracellular potential ϕ_i, the intracellular conductivity tensor σ_i, the intracellular current source density I_{si}, the transmembrane current density I_m, and the surface to volume ratio β of myocytes. The intracellular conductivity σ_i consists of conductivities for the intracellular components and for the gap junctions. The hereby arising 2nd order differential equations were solved on deformable grids with the finite element method. A deformation dependent transformation of the intracellular conductivity tensor was performed [14].

4   Modeling of Tension Development

Overview. Of special interest for biophysically motivated modeling are descriptions of cellular tension development, which base on electrophysiological quan-

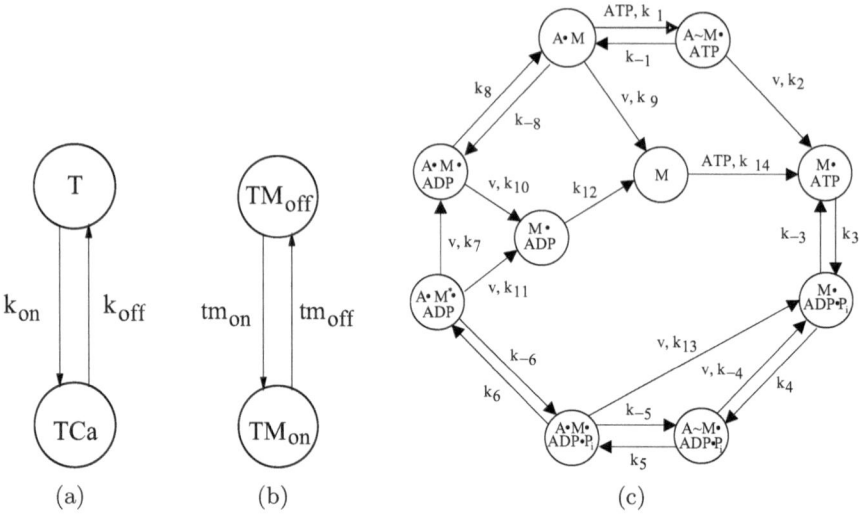

Fig. 2. State diagram of hybrid model. (a) Two state variables quantify the calcium binding to troponin C. (b) Two further state variables describe the configuration of tropomyosin. (c) Ten state variables detail the interaction of actin and myosin as well as the hydrolysis of adenosine triphosphate. M and A symbolize myosin and actin, respectively. ATP, ADP and P_i represent adenosine triphosphate, adenosine diphosphate and phosphate, respectively. The transition between states is depicted by an arrow, strong binding by a closed circle, and weak binding by a tilde. The arrows are labeled with constants k_x, ATP and stretch velocity v indicating that these are influencing rate coefficients functions.

tities delivered e.g. by electrophysiological cell models [15]. Commonly, the concentration of intracellular calcium $[Ca^{2+}]_i$ is used to modulate rate coefficients, which depict the interaction between states of actin and myosin. The state variables describe e.g. the binding of intracellular calcium Ca^{2+} to the troponin complex and the cross bridge cycling. Further parameters influencing rate coefficients are the sarcomere length and the state variables.

Hybrid model. The hybrid model combines a description of the binding of intracellular calcium $[Ca^{2+}]_i$ to troponin C, the configuration change of tropomyosin, and the interaction of actin and myosin. The calcium binding to troponin C is similarly described as in the 3rd model of Rice et al. [15]. The interaction of actin and myosin is adopted from Gordon et al. [16], Bers et al. [17], and Spudich [18]. The model uses 14 state variables, which are coupled by rate coefficients. Two state variables, T and TCa, detail the binding of intracellular calcium Ca^{2+} to troponin C (figure 2a). The state variable T describes the normalized concentration of troponin C with no bound calcium, TCa the normalized concentration of troponin C with bound calcium. Two further state variables, TM_{on} and TM_{off}, quantify the configuration of tropomyosin (figure 2b). The state variable TM_{on} describes the normalized concentration of tropomyosin in permissive

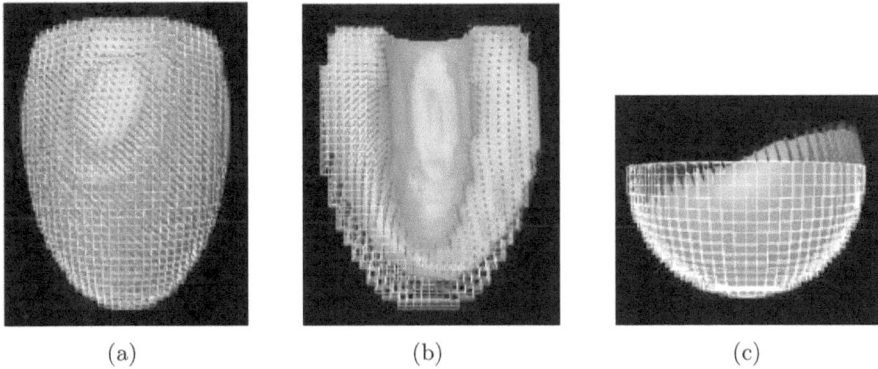

(a)   (b)   (c)

Fig. 3. Deformation at time $t = 290$ ms in electro-mechanical model of left ventricle. (a) All and (b,c) half of the ventricle's model is shown with a bright wire-frame as reference configuration and arrows indicating displacements.

configuration, TM_{off} in non permissive. Ten state variables are used to quantify the interaction between actin and myosin, particularly the cross-bridge cycling (figure 2c), by describing normalized concentrations of myosin. The interaction between the states of the model is described by a system of 1st order differential equations. The differential equations describe the change of states controlled by rate coefficients. Partly, the rate coefficients are dependent on the sarcomere stretch velocity v, the sarcomere stretch λ and the concentration of intracellular calcium $[Ca^{2+}]_i$. The sum of tension generating states T_{AM} is given by:

$$T_{AM} = A \bullet M + A \bullet M \bullet ADP + A \bullet M^* \bullet ADP$$

The normalized tension T is determined by

$$T = \alpha \frac{T_{AM}}{T_{max}}$$

with the sarcomere overlap function $\alpha = \alpha(\lambda)$, which is tissue and species specific [17,15], and maximal tension T_{max}, which is dependent on the rate coefficients. The normalized tension T can be multiplied by a tissue and species specific factor f_{T0} to quantify tension development of myocardium.

5   Results

The integrated ventricular model was investigated by exemplary simulations, whereby the tension factor f_{T0} was varied. The tension factor f_{T0} was set to 5, 10, and 20 kPa. Two different resolutions were chosen for the electrophysiological and mechanical model, i.e. 200 and 400 μm per voxel length. The basal positions were fixed, i.e. the displacements were set to zero. An exemplary simulation started by applying an electrical stimulus at the apex at time $t = 0$ ms.

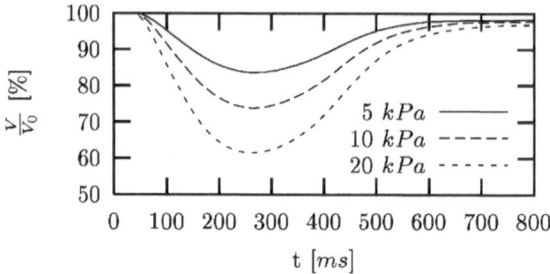

Fig. 4. Ratio of endocardial volume. The ratio between the volume of left ventricular cavity in deformed and undeformed configuration is dependent on the tension produced by contractile units. The course of the volume ratio is shown calculated by simulations with a tension factor f_{T0} of 5, 10 and 20 kPa.

The simulation had a duration of 800 ms. The displacements were determined every 5 ms. Every 20 μs a calculation of each voxel was performed in the electro-physiological, excitation propagation, and force development model.

The simulations with the integrated model showed a rapid spread of electrical excitation and delayed force development (figure 5). The simulations lead to an inhomogeneous force development and deformation, which is characterized by the fiber orientation (figure 3). Significant torsions were appearing in regions near to the apex (figure 3 c). The transmembrane voltages and intracellular calcium concentrations differ only slightly for different tension factors and measurement positions (figure 5 a,b). Significant differences are found for tension development (figure 5 c). Normalized developed tensions T were larger for small tension factors. The differences can be attributed to the decrease in stretch by increase of tension factors. This decrease leads to reduced calcium-troponin C binding and small values in the sarcomere overlap function. The endocardial volume was decreased during the contraction to maximal 84, 74, and 61 % of its reference volume for a tension factor of 5, 10, and 20 kPa, respectively. The course of volume decrease was similar for the different tension factors (figure4).

6   Discussion

A integrated model of cardiac electro-mechanics and its application in a left ventricular model was presented. Simulations in an environment similar to the classical Langendorff experiments [19] were performed to illustrate the properties of the model. Different limitations of the simulations can be assigned, e.g. neglect of transmural inhomogeneities of myocardium [20] as well as simplified mechanical and electrical boundary conditions. Nevertheless, the methods presented illustrate a strategy, which can be adapted e.g. for clinical cardiologic diagnosis and therapy planning on patient specific data sets.

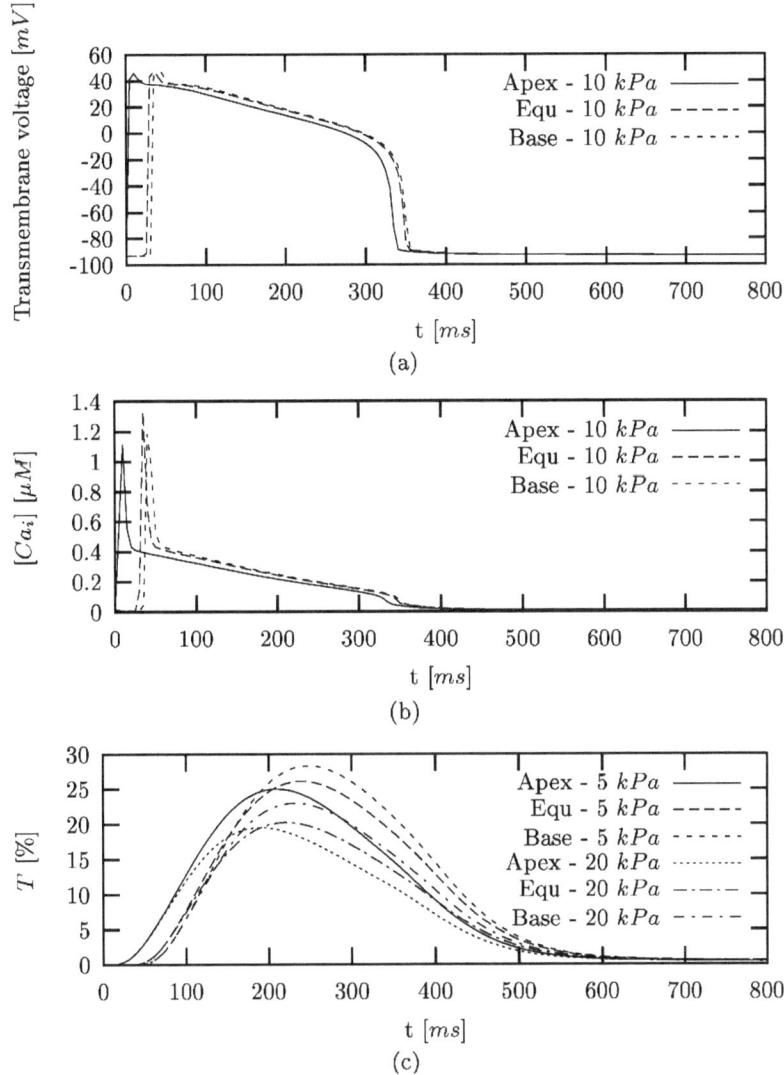

Fig. 5. (a) Transmembrane voltage, (b) concentration of intracellular calcium, and (c) normalized tension at different positions and for different tension factors f_{T0}. Tension factors of 5, 10 and 20 kPa were assigned to the contractile units. Positions at the epicardial apex, equatorial midwall, and basal midwall were selected.

References

1. Sachse, F.B.: Modeling of the mammalian heart, Universität Karlsruhe (TH), Institut für Biomedizinische Technik (2002) Habilitationsschrift.
2. Noble, D., Varghese, A., Kohl, P., Noble, P.: Improved guinea-pig ventricular cell model incorporating a diadic space, I_{Kr} and I_{Ks}, and length- and tension-dependend processes. Can. J. Cardiol. **14** (1998) 123–134

3. Glänzel, K.: Kraftentwicklung im Sarkomer unter Berücksichtigung elektromechanischer Kopplung. Diploma Thesis, Institut für Biomedizinische Technik, Universität Karlsruhe (TH) (2002)
4. Sachse, F.B., Glänzel, K., Seemann, G.: Modeling of protein interactions involved in cardiac tension development. IJBC (2003) submitted.
5. Sachse, F.B., Glänzel, K., Seemann, G.: Modeling of electro-mechanical coupling in cardiac myocytes: Feedback mechanisms and cooperativity. LNCS (2003) submitted.
6. Sachse, F.B., Seemann, G., Werner, C.D.: Modeling of electro-mechanics in left ventricle. In: Proc. Computers in Cardiology. (2002) 705–708
7. Guccione, J.M., McCulloch, A.D., Waldman, L.K.: Passive material properties of intact ventricular myocardium determined from a cylindrical model. J. Biomechanical Engineering **113** (1991) 42–55
8. Bathe, K.J.: Finite-Elemente-Methoden. Springer, Berlin, Heidelberg, New York (1990)
9. McCulloch, A.D.: Cardiac biomechanics. In Bronzino, J.D., ed.: The Biomedical Engineering Handbook. 2 edn. CRC Press (2000) 28–1–28–26
10. Streeter, jr., D.D., Bassett, D.L.: An engineering analysis of myocardial fiber orientation in pig's left ventricle in systole. Anatomical Record **155** (1966) 503–512
11. Kohl, P., Day, K., Noble, D.: Cellular mechanisms of cardiac mechano-electric feedback in a mathematical model. Can. J. Cardiol. **14** (1998) 111–119
12. Noble, P.: -. personal communication (2000)
13. Sachse, F.B., Seemann, G., Riedel, C., Werner, C.D., Dössel, O.: Modeling of the cardiac mechano-electrical feedback. Int. J. Bioelectromagnetism **2** (2000)
14. Sachse, F.B., Seemann, G., Riedel, C.: Modeling of cardiac excitation propagation taking deformation into account. In: Proc. BIOMAG 2002. (2002) 839–841
15. Rice, J.J., Winslow, R.L., Hunter, W.C.: Comparison of putative cooperative mechanisms in cardiac muscle: length dependence and dynamic responses. Am. J. Physiol. Circ. Heart. **276** (1999) H1734–H1754
16. Gordon, A., Regnier, M., Homsher, E.: Skeletal and cardiac muscle contractile activation: Tropomyosin "rocks and rolls". News Physiol. Sci. **16** (2001) 49–55
17. Bers, D.M.: Excitation-Contraction Coupling and Cardiac Contractile Force. Kluwer Academic Publishers, Dordrecht, Netherlands (1991)
18. Spudich, J.A.: TIMELINE: The myosin swinging cross-bridge model. Nature Reviews Molecular Cell Biology **2** (2001) 387–392
19. Langendorff, O.: Untersuchungen am überlebenden Säugetierherzen. Pflügers Arch. ges. Physiologie **61** (1895) 291–332
20. Liu, D.W., Gintant, G.A., Antzelevitch, C.: Ionic bases for electrophysiological distinctions among epicardial, midmyocardial, and endocardial myocytes from the free wall of the canine left ventricle. Circ Res. **72** (1993) 671–687

Physical Modeling of Airflow-Walls Interactions to Understand the Sleep Apnea Syndrome

Yohan Payan[1], Xavier Pelorson[2], and Pascal Perrier[2]

[1] TIMC Laboratory, UMR CNRS 5525 and Universite Joseph Fourier, Institut
Albert Bonniot 38706 La Tronche, France
Yohan.Payan@imag.fr
http://www-timc.imag.fr/Yohan.Payan/
[2] Institut de la Communication Parlee, INPG and UMR CNRS Q5009, 46 Ave. Felix
Viallet, F-38031 Grenoble Cedex, France
{pelorson,perrier}@icp.inpg.fr

Abstract. Sleep Apnea Syndrome (SAS) is defined as a partial or total
closure of the patient upper airways during sleep. The term "collapsus"
(or collapse) is used to describe this closure. From a fluid mechanical
point of view, this collapse can be understood as a spectacular example of fluid-walls interaction. Indeed, the upper airways are delimited in
their largest part by soft tissues having different geometrical and mechanical properties: velum, tongue and pharyngeal walls. Airway closure
during SAS comes from the interaction between these soft tissues and
the inspiratory flow. The aim of this work is to understand the physical
phenomena at the origin of the collapsus and the metamorphosis in inspiratory flow pattern that has been reported during SAS. Indeed, a full
comprehension of the physical conditions allowing this phenomenon is a
prerequisite to be able to help in the planning of the surgical gesture that
can be prescribed for the patients. The work presented here focuses on
a simple model of fluid-walls interactions. The equations governing the
airflow inside a constriction are coupled with a Finite Element biomechanical model of the velum. The geometries of this model is extracted
from a single midsagittal radiography of a patient. The velar deformations induced by airflow interactions are computed, presented, discussed
and compared to measurements collected onto an experimental setup.

1 Introduction

Sleep apnea is a disorder in which a person stops breathing during the night,
usually for periods of 10 seconds or longer. In most cases the person is unaware of
it. This disorder results from collapse and obstruction of the throat pharyngeal
airway (figure 1).

It is accepted that this occurs due to both a structurally small upper airway
and a loss of muscle tone, as there is loss of the wakefulness stimulus to upper
airway muscles with sleep onset. This results in airway collapse, increased resistance to airflow, decreased breathing, and increased breathing effort ([8]). In
most subjects, the narrowest airway cross-section area occurs behind the palate

N. Ayache and H. Delingette (Eds.): IS4TM 2003, LNCS 2673, pp. 261–269, 2003.

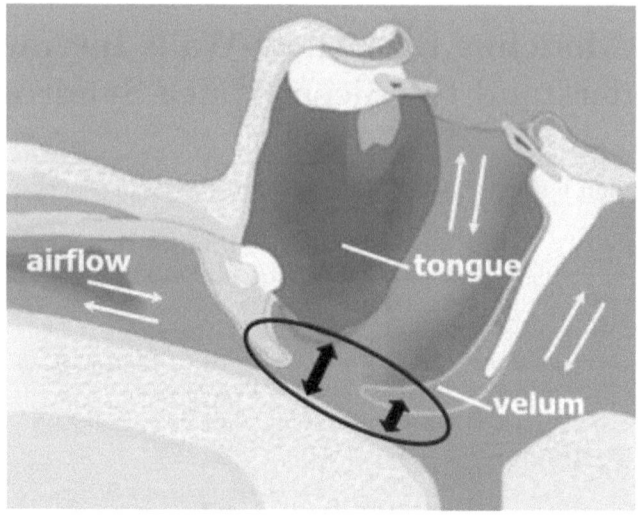

Fig. 1. Sleep Apnea Syndrome: fluid (airflow) / wall interactions.

and uvula. This area is the most vulnerable to obstruction from loss of muscle tone during sleep. On physical exam, a long and wide velum, large tonsils, and redundancy of pharyngeal walls may be found. Lower throat findings may include a large tongue and lingual tonsils. An estimated 5 in 100 people, typically overweight middle-aged men, suffer from sleep apnea. In addition to a strong daily fatigue, several chronic cardiovascular complications have been found to be related to sleep apnea syndrome (SAS), such as systemic arterial and pulmonary hypertension, heart failure or arrhythmias. The Apnea Index (AI) measures the number of apneas per hour. Hypopnea is defined as a decrease in airflow of 50% or more (without complete cessation) accompanied by a drop in oxygen saturation. The Hypopnea Index (HI) is the number of hypopneas per hour. The Apnea/Hypopnea Index (AHI) is the sum of AI and HI. The definition of sleep apnea in terms of the AHI is assumed to be the following: normal is AHI 0-5, mild 5-15, moderate 15-30, and severe 30+. Surgical techniques used for the treatment of the SAS can either reduce the volume of the tongue, stiffen the velum, or try to have a more global and progressive action on the entire upper airways. In order to fully understand the metamorphosis occurring in the inspiratory flow pattern that has been reported during SAS, some mechanical models of the upper airway have been developed, assuming that upper airways can be represented by a single compliant segment ([1], [10]), or by series of individual segments representing singularities ([2]). In this aim, a complete biomechanical model of the upper airways appears thus to be interesting, to describe and explain - at the physical point of view - the upper airway obstruction. The work presented here focuses on a simple model of fluid-walls interactions. A 2D Finite Element (FE) biomechanical model of the velum is introduced, and coupled with an analytical approximation of the equations that govern the airflow inside a constriction. The simulations of airflow / walls interactions are then compared with in vitro measurements collected onto an experimental setup.

2   Finite Element Model of the Velum

Lingual, velar and pharyngeal soft tissues are partly responsible for the SAS as their deformations can even lead to a total closure of the upper airways. In a first step, we only focus on the velo-pharyngeal region, at the intersection between the nasal and the oral cavity. In this perspective, a 2D sagittal continuous model of the velum was elaborated and discretized through the Finite Element Method. The codes developed for this model assume no displacement in the transverse direction (*plane strain hypothesis*) as well as a small deformation hypothesis. Velar tissues being assumed as quasi-incompressible (because mainly composed of water), a value close to 0.5 was chosen for the Poisson ratio. A 10 kPa value was taken for the Young modulus, which seems coherent with values reported for tongue ([5]) and vocal folds ([4]). The geometry of the model was extracted from a single midsagittal radiography of a patient (figure 2).

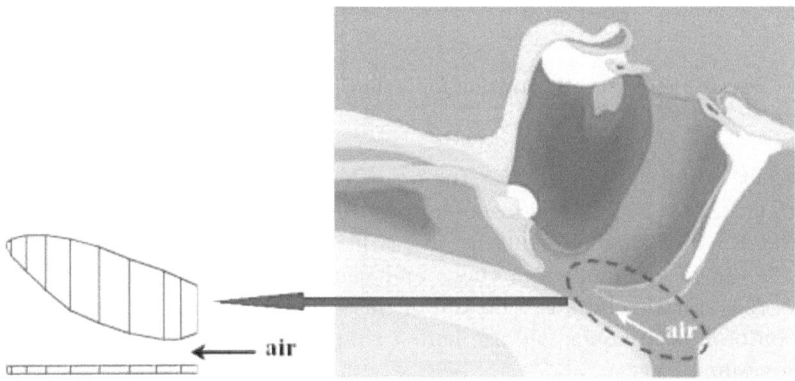

Fig. 2. Midsagittal view of the upper airways (right) and FE model of the velum (left).

The upper part of the model represents the velar tissues. Geometrical dimensions were close to values reported in the literature ([7]) ([11]): the total length is of order of 30 mm while thickness varies from one extremity to the other (with a mean of 5-mm). Only the velum deformations were taken into account for the current simulations. The pharyngeal walls (the lower part of the model in figure 2) were assumed to be rigid. The two points located onto the right part of the velum were also considered as fixed in order to model the velar attachment to the hard palate. Finally, simulations were limited to the 2D midsagittal plane, but for the computation of pressure forces (integrated along the 3D geometry), a 30-mm value was taken for the velar thickness in the frontal plane.

3   Physical Modeling of the Airflow

From a fluid mechanical point of view, the partial or the total collapse of the
upper airway, as observed during sleep hypopnea or apnea, can be understood
as a spectacular example of fluid-walls interaction. While the most important
parameters influencing this effect *in vivo* are well known, this phenomenon is
still difficult to model and thus to predict. Figure 3 presents in a simple way a
constriction inside the upper airways.

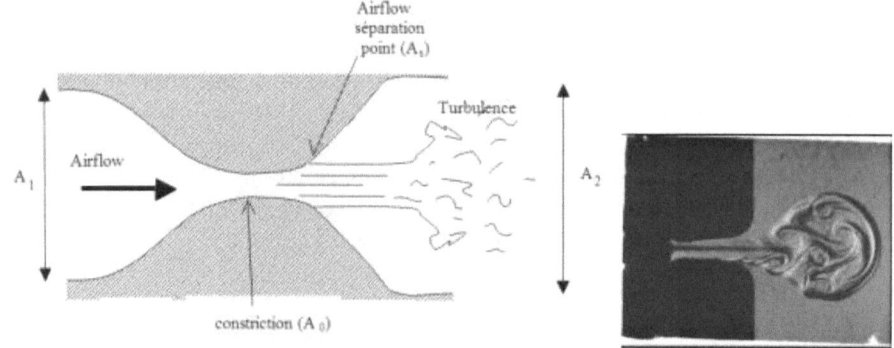

Fig. 3. Schematic Illustration for airflow inside the constriction.

An exact analytical solution for the flow through such a constriction is not
available. Further, full numerical simulations of the unsteady three-dimensional
flow through a deformable structure are still, at present time, very complex
even using the recent numerical codes and using powerful computers [3]. For
these reasons, and also because the aim of this paper is to provide a qualitative
description of a sleep apnea, we use here a simplified flow theory based on the
following assumptions:

 – As the airflow velocity in the upper airways is, in general, much smaller than
   the speed of sound, it can be assumed that the flow is locally incompressible.
 – It can be reasonably assumed that the time needed for the constriction to
   collapse (of order of a second) is large compared with typical flow convection
   times (the time needed for the flow to pass the constriction is of order of a
   few milliseconds). Therefore, it will be assumed that the flow is quasi-steady.

The principle of mass-conservation thus yields the following relationship:

$$\Phi = constant \tag{1}$$

where $\Phi = \nu \cdot A$ is the volume flow velocity, ν and A are respectively the (local)
flow velocity and upper-airways area. As a third assumption, it is considered
that all viscous effects can be neglected. This assumption can be rationalized

partially by considering that typical Reynolds numbers involved are of order of 1000 and therefore that viscous forces are negligible compared with convective ones. This leads to the Bernoulli law:

$$p + \frac{1}{2}\rho\nu^2 = constant \tag{2}$$

where p is the local pressure and ρ the (constant) air density. Equations (1) and (2) must be corrected in order to take into account a spectacular viscous effect: flow separation. Indeed, it is expected that the strongest pressure losses are due to the phenomenon of flow separation at the outlet of the constriction. This phenomenon is due to the presence of a strong adverse pressure gradient that causes the flow to decelerate so rapidly that it separates from the walls to form a free jet (see right part of figure 3). Very strong pressure losses, due to the appearance of turbulence downstream of the constriction, are associated with flow separation. As a matter of fact, the pressure recovery past the flow separation point is so small that it can in general be neglected. In the following, it is assumed that the flow separates from the walls of the constriction at the point where the area reaches 1.2 times the minimum area A_0 (see figure 3). This approximated value was empirically proposed and constitutes an acceptable approximation of the phenomena [6]. To summarize, for a given pressure drop $(p_1 - p_2)$, and for a given geometry of the constriction, the volume flow velocity Φ is:

$$\Phi = A_S\sqrt{\frac{2(p_1 - p_2)}{\rho}} = 1.2A_0\sqrt{\frac{2(p_1 - p_2)}{\rho}} \tag{3}$$

and the pressure distribution $p(x)$ within the constriction is predicted by:

$$p(x) = p_1 + \frac{1}{2}\rho\Phi^2\left(\frac{1}{A_1^2} - \frac{1}{A(x)^2}\right) \tag{4}$$

where $A(x)$ is the transversal area at the x abscissa (figure 3). Therefore, the force exerted by the airflow onto the walls of the constriction can be computed by integrating the pressure along the x axis up to the flow separation point. This force induces a deformation of the upper airways soft tissues, thus modifying the airways geometry, and therefore changing the pressure distribution along the airways.

4   Coupling the Airflow with the Model of the Velum

The coupling between the airflow pressure forces computation and the deformations of the FE model of the velum was iteratively processed. An adaptive Runge Kutta algorithm was used to solve the dynamical equations that govern the deformations of the velum. At each integration time step of the algorithm, the new deformed geometry of the velum is used to calculate the pressure forces distribution along the constriction. This new pressure force distribution is then defined as new boundary conditions for the Finite Element computation of the velum

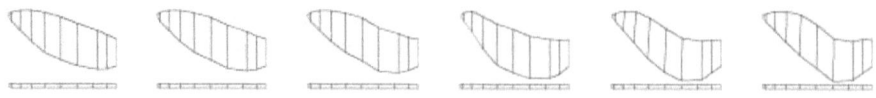

Fig. 4. FE model of the velum coupled with the airflow: from initial (left) to final (right) positions. One frame each 0.025 second.

deformations. As a first very qualitative approximation of the respiratory cycle, the pressure drop $(p_1 - p_2)$ was taken as a sinusoidal: $p_1 - p_2 = p_{max} \cdot sin(4\pi \cdot t)$. Figure 4 shows simulations of the airflow / velum interactions for a half-period pressure drop command with an 800 Pa maximal value. A clear reduction of the constriction can be observed, thus simulating a hypopnea [9].

This decrease in the size of the constriction is associated with a limitation of the volume flow velocity. Left part of figure 5 plots this phenomenon, and shows also that this limitation can be avoided if the stiffness of the velum (the Young modulus value in our FE model) is increased. This point is qualitatively coherent with the clinical practice of uvuloplasty that consist in burning the velum in order to stiff the velar tissues. Similarly, the right part of figure 5 shows how a decrease in the size of the constriction can increase the volume flow velocity limitation.

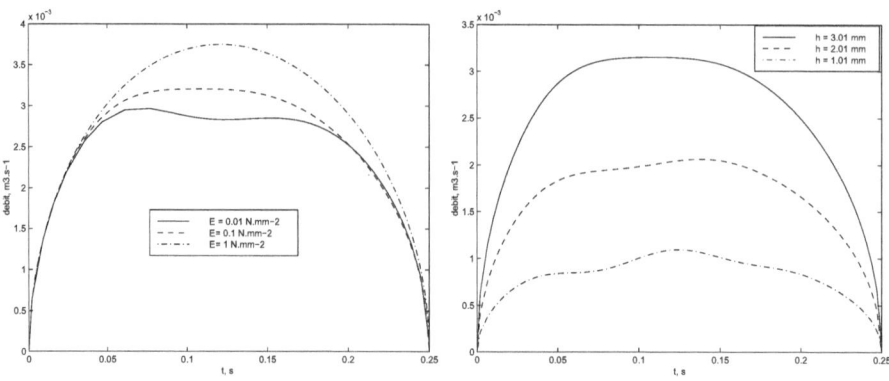

Fig. 5. Simulated limitation of the volume flow velocity: influence of the velum stiffness (left) and the constriction size (right).

5   Experimental Setup

Our investigations are based on an experimental exploitation of a setup specially designed using classical dimensions obtained through in-vivo data acquisition. Although in vivo data serve as a reference for model development, they are highly non reproducible and don't allow quantitative validation of theory, as many parameters are not controlled, or even not reachable. This underscores the interest of using an experimental setup where most of the parameters are under control. In order to investigate precisely the interaction phenomenon, the elastic characteristics of the upper airway soft tissues must be reproduced with sufficient realism. To fulfill this requirement, the soft tissues were modeled in a first approximation by a latex cylinder filled with water under pressure. This latex is assumed to represent to some extent the rounded backward part of the tongue, also responsible for a collapse during SAS. The latex is placed inside a squared rigid pipe (figure 6) that represents the larynx. It is assumed that larynx walls are rigid compared to the tongue elasticity. Extremities of the latex are fixed and glued to the pipe with silicone. Pressure measurements can be performed at different positions : before (P_{sub}), after and at constriction (P_{gu}, P_{gd}). Complementary velocity measurements can also be made using a hot film (TSI, model 1210). Various parameters can also be controlled: air flow conditions in order to simulate different aerodynamic conditions, constriction height hc.

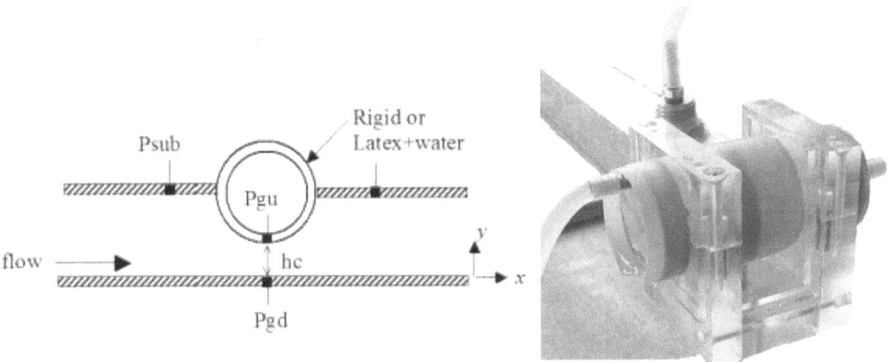

Fig. 6. Schematic view and photography of the experimental setup used to study airflow/tongue interaction.

Figure 7 plots the pressure drop and volume flow velocity values measured at the constriction. When the pressure drop command has a low value (low respiratory effort), the volume flow velocity profile is similar to the pressure drop temporal evolution (figure 7, left). On the contrary, when the pressure drop command increases (close to 1000 Pa), the latex tends to collapse and the area of the constriction decreases. A limitation of the volume flow velocity is then observed (right lower part of figure 7).

6   Discussion

Despite the limitations of our modeling hypotheses, preliminary interesting simulations were carried out. Indeed, the airflow model coupled with a 2D Finite Element model of the velum provides a decrease of the size in the constriction

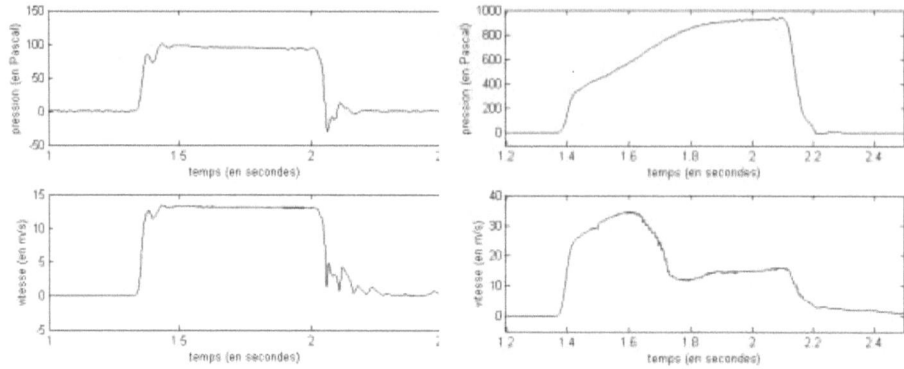

Fig. 7. Pressure drop (top) and volume flow velocity (down) measured for two conditions. Left: low pressure drop; normal conditions. Right: high pressure drop; pathological conditions with sleep hypopnea.

area with the increase of the pressure drop. This result is well known by the clinicians and is described as the airflow limitation phenomenon (hypopnea). Moreover, this phenomenon is observed in the *in vitro* experimental set up that we have developed. If they differ from a quantitative point of view, figure 5 and figure 7 (right lower part) both illustrate the same phenomenon: the volume flow velocity limitation. Finally, it is also interesting to note that an increase of the velum stiffness (modeled with an increase of the Young modulus value) or an increase in the size of the velo-pharyngeal constriction both tend to limit the hypo-apnea syndrome. Those results are consistent with some surgery techniques that try to modify mechanical properties of the velum (by burning tissues, thus increasing their stiffness) or try to have a more global and progressive action on the entire upper airways (in order to increase the size of the upper-airway constriction).

References

1. Y. Auregan and N. Meslier. Modélisation des apnées obstructives du sommeil. *C.R. Acad. Sci.*, 316:1529–1534, 1993.
2. R. Fodil, C. Ribreau, B. Louis, F. Lofaso, and D. Isabey. Interaction between steady flow and individualised compliant segments: application to upper airways. *Medical & Biological Engineering & Computing*, November 1997.
3. C. Hirsch. *Numerical computation of internal and external flows*. John Wiley & sons, 1994.
4. Y. Min, I. Titze, and F. Alipour. Stress-strain response of the human vocal ligament. *NCVS Status and Progress*, 7:131–137, 1994.
5. Y. Payan and P. Perrier. Synthesis of v-v sequences with a 2d biomechanical tongue model controlled by the equilibrium point hypothesis. *Speech Communication*, 22(2–3):185–205, 1997.
6. X. Pelorson, A. Hirschberg, A.P.J. Wijnands, and H. Bailliet. Description of the flow through the vocal cords during phonation. *Acta Acustica*, 3:191–202, 1995.

7. J.L Pépin, G. Ferreti, D. Veale, P. Romand, M. Coulomb, C. Brambilla, and P. Lévy. Somnofluoroscopy, computerised tomography and cephalometry in the assesment of the airway in obstructive sleep apnoea. *Thorax*, 47:150–156, 1992.
8. J.L Pépin, P. Lévy, D. Veale, and G. Ferreti. Evaluation of the upper airway in sleep apnea syndrome. *Sleep*, 15:s50–s55, 1992.
9. D.M. Rapoport. Methods to stabilize the upper airway using positive pressure. *Sleep*, 19:123–130, 1996.
10. P.L. Smith, R.A. Wise, A.R. Gold, A.R. Schwarts, and S. Permutt. Upper airway pressure-flow relationship in obstructive slepp apnea. *J. Appl. physiol*, 64:789–795, 1998.
11. F.J. Trudo, W.B., Gefter, K.C. Welch, K.B. Gupta, G. Maislin, and R.J. Schwab. State-related changes in upper airway caliber and surrounding soft-tissue structures in normal subjects. *Am. J. Respir. Care Med.*, 158:1259–1270, 1998.

Evaluation of a New 3D/2D Registration Criterion for Liver Radio-Frequencies Guided by Augmented Reality

Stéphane Nicolau[1,2], Xavier Pennec[1], Luc Soler[2], and Nicholas Ayache[1]

[1] INRIA Sophia, Epidaure, 2004 Rte des Lucioles, F-06902 Sophia-Antipolis Cedex
{Stephane.Nicolau,Xavier.Pennec,Nicholas.Ayache}@sophia.inria.fr
http://www-sop.inria.fr/epidaure/Epidaure-eng.html
[2] IRCAD-Hopital Civil, Virtual-surg, 1 Place de l'Hopital, 67091 Strasbourg Cedex
{stephane.nicolau,luc.soler}@ircad.u-strasbg.fr

Abstract. Our purpose in this article is to superimpose a 3D model of the liver, its vessels and tumors (reconstructed from CT images) on external video images of the patient for hepatic surgery guidance. The main constraints are the robustness, the accuracy and the computation time. Because of the absence of visible anatomical landmarks and of the "cylindrical" shape of the upper abdomen, we used some radio-opaque fiducials. The classical least-squares method assuming that there is no noise on the 3D point positions, we designed a new Maximum Likelihood approach to account for this existing noise and we show that it generalizes the classical approaches. Experiments on synthetic data provide evidences that our new criterion is up to 20% more accurate and much more robust, while keeping a computation time compatible with real-time at 20 to 40 Hz. Eventually, careful validation experiments on real data show that an accuracy of 2 mm can be achieved within the liver.

1 Introduction

The treatment of liver tumors by radio-frequencies is a new technique which begins to be widely used by surgeons. However, the guidance procedure to reach the tumors with the electrode is still made visually using per-operative 2D cross-sections of the patient using either Ultra-Sound (US) or Computed Tomography (CT). Because of the difficulty to locate in 3D the tumor's center, surgeons estimate that the tumor size has to exceed 2 cm to perform a reliable intervention. Our purpose is to build an augmented reality system that could superimpose reconstructions of the 3D liver and tumor onto a video image in order to improve the surgeon's accuracy during the guidance step. In such a system, the overall accuracy has to be less than 5 mm to provide a significant help to the surgeon.

Just before the intervention, a CT-scan of the patient is acquired and an automatic 3D-reconstructions of his skin, his liver and the tumors is performed [17]. Two cameras (jointly calibrated) are viewing the patient's skin from two different points of view. The patient is intubated during the intervention, so the

N. Ayache and H. Delingette (Eds.): IS4TM 2003, LNCS 2673, pp. 270–283, 2003.

volume of gas in his lungs can be controlled and monitored. Then, it is possible to fix the volume at the same value during a few seconds repetitively and to perform the CT and the electrode's manipulation almost in the same volume's condition. Thus, we assume that a rigid registration is sufficient to register accurately the 3D-model extracted from the CT with the 2D video images. Consequently we are confronted to the classical rigid problem of *3D/3D and 3D/2D Object Registration.*

Surface and iconic registration using mutual information have been used to register the 3D surface of the face to either video images [18] or another 3D surface acquired with a laser range scanner [5]. In both cases, thanks to several highly curved "edges" on the model (nose, ears, eyes), the reported accuracy was under 5 mm. We believe that in our case, the "cylindrical" shape of the human abdomen is likely to lead to much larger uncertainties along the cranio-caudal axis.

Landmarks 3D/3D or 2D/3D registration can be performed when several precisely located points are visible both in the 3D-model and in the video images. Since the landmarks are really homologous, the "cylindrical" geometry of the underlying abdomen surface is not any more a problem. As there are no visible anatomical landmarks in our case, we chose to stick to the patient skin some radio-opaque markers that are currently localized and matched interactively.

This problem was largely considered in a wide variety of cases. Closed-form solution for few points [2,1] and linear resolution [3] were proposed in the last decades to find the registration as quickly as possible to fulfill real-time constraints. Others [9,15] determine the object pose by minimizing the classical projective least-squares (LSQ) error function. In these cases, the methods differ by the minimization procedure and by the rotation parameterization. Haralick [6] and Or [10] turn the 2D/3D problem into a 3D/3D points registration problem in which they estimate the depth of the points seen in the image by minimizing a 3D LSQ Euclidean distance criterion. Finally, Yuan and Liu [19,8] solve the problem by separating the rotational components from the translational one.

Linear and closed-form solution provide a direct resolution, but they are very sensitive to noise because they assume that data points are exact. As the accuracy is crucial in our application, we cannot afford using them. An alternative approach that separates the rotation from the translation was examined by Kumar [7], and he shows that it leads to worse parameters estimation in the presence of noise than the classical LSQ estimation.

Therefore, we think that a LSQ criterion has a definite advantage among the other methods because it can take into account the whole information provided by the data. However, all of the existing methods implicitly consider that 2D points are noisy, but that 3D points are exact. In our case, this assumption is definitely questionable, which lead to the development of a new maximum likelihood (ML) criterion generalizing the standard 3D/2D LSQ criterion. Last but not least, LSQ criterion enables the prediction of the noise influence on the final registration. This would be very useful as it would allow us to detect an inaccurate registration. However, this aspect will be considered in a future work.

In the sequel, we first recall that the classical 2D/3D registration criterion for points correspondences can be considered as a *Maximum Likelihood* estimation if the 3D points are exact. Then, modifying the statistical assumptions to account for an existing noise on the 3D point measurements, we derive a new and original criterion that generalizes the classical ML one. Section 3 is devoted to the comparative performance evaluation of both criteria with synthetic data while Section 4 focuses on a careful validation with real data.

2   Maximum Likelihood 2D/3D Registration

Let $M_i = [x_i, y_i, z_i]^\top$ be the 3D points that represent the exact localization of the radio-opaque fiducials in the CT-scan reference frame and $m_i^{(l)} = [u_i^{(l)}, v_i^{(l)}]^\top$ be the 2D points that represent its exact position in the images of camera (l). In this article, we assume that correspondences are known. To account for occlusion, we use a binary variable ξ_i^l equal to 1 if M_i is observed in camera (l) and 0 otherwise. We denote by $T \star M$ the action of the rigid transformation T on the 3D point M and by P_l ($1 \leq l \leq M$) the camera's projective functions from 3D to 2D such that $m_i^{(l)} = P^{(l)}(T \star M_i)$ (we used in our implementation the calibration algorithm of [20]). In the following sections, $\hat{A}$ will represent an *estimation* of a perfect data A, and $\tilde{A}$ an *observed measure*. Thus, the 3D points measured by the user will be written $\tilde{M}_i$, and the measured video 2D points $\tilde{m}_i$.

2.1   Standard Projective Points Correspondences (SPPC) Criterion

Assuming that the 3D points are exact ($\tilde{M}_i = M_i$) and that the 2D points only are corrupted by an isotropic Gaussian noise η_i of variance σ_{2D}^2, we have:

$$\tilde{m}_i^{(l)} = m_i^{(l)} + \eta_i = P^{(l)}(T \star M_i) + \eta_i \quad \text{with} \quad \eta_i \sim N(0, \sigma_{2D})$$

The probability of measuring the projection of the 3D point M_i at the location $\tilde{m}_i^{(l)}$ in image (l), knowing the transformation parameters $\theta = \{T\}$ is given by:

$$p(\tilde{m}_i^{(l)} \mid \theta) = \frac{1}{2\pi\sigma_{2D}^2} \cdot \exp\left(-\frac{\| P^{(l)}(T \star M_i) - \tilde{m}_i^{(l)} \|^2}{2 \cdot \sigma_{2D}^2} \right)$$

Let χ be the data vector regrouping all the measurements, in this case the 2D points $\tilde{m}_i^{(l)}$ only. Since the detection of each point is performed independently, the probability of the observed data is $p(\chi \mid \theta) = \prod_{l=1}^{M} \prod_{i=1}^{N} p(\tilde{m}_i^{(l)} \mid \theta)^{\xi_i^l}$. In this formula, unobserved 2D points (for which $\xi_i^l = 0$) are implicitly taken out of the probability. Now, the *Maximum likelihood* transformation $\hat{\theta}$ maximizes the probability of the observed data, or equivalently, minimizes its negative log:

$$C_{2D}(T) = \sum_{l=1}^{M} \sum_{i=1}^{N} \xi_i^l \cdot \frac{\left\| P^{(l)}(T \star M_i) - \tilde{m}_i^{(l)} \right\|^2}{2 \cdot \sigma_{2D}^2} + \left(\sum_{l=1}^{M} \sum_{i=1}^{N} \xi_i^l \right) \cdot \log[2\pi\sigma_{2D}^2] \quad (1)$$

Thus, up to a constant factor, this ML estimation boils down to the classical LSQ criterion in the 2D coordinates. Of course, this criterion assumes that there is no noise on the 3D points, or that this noise could be distributed over the 2D measurements. This simple hypothesis makes the criterion very easy to optimize because there are only 6 parameters to estimate and lead to a very low calculations cost. However, from a statistical point of view, distributing the 3D error on 2D measurements leads to correlated noises, which does not agree with the independence assumption used to derive the ML estimation.

2.2 Extended Projective Points Correspondences (EPPC) Criterion

To introduce a more realistic statistical hypothesis on the 3D data, it is thus safer to consider that we are measuring a noisy version of the exact points:

$$\tilde{M}_i = M_i + \varepsilon_i \quad \text{with} \quad \varepsilon_i \sim N(0, \sigma_{3D}).$$

In this case, the exact location M_i of the 3D points is considered as a parameter, just as the transformation T. In statistics, this is called a *latent or hidden variable*, while it is better known as an *auxiliary variable* in computer vision. Thus, knowing the parameters $\theta = \{T, M_1, \ldots M_N\}$, the probability of measuring respectively a 2D and a 3D point is:

$$p(\tilde{m}_i^{(l)} \mid \theta) = G_{\sigma_{2D}} \left(P^{(l)}(T \star M_i) - \tilde{m}_i^{(l)} \right) \quad \text{and} \quad p(\tilde{M}_i \mid \theta) = G_{\sigma_{3D}} \left(M_i - \tilde{M}_i \right).$$

One important feature of this statistical modeling is that we can safely assume that all 3D and 2D measurements are independent. Thus, we can write the probability of our observation vector $\chi = (\tilde{m}_1^1, \ldots, \tilde{m}_N^1, \ldots, \tilde{m}_1^M, \ldots, \tilde{m}_N^M, \tilde{M}_1, \ldots, \tilde{M}_N)$ as the product of the above individual probabilities. The ML estimation of the parameters is still given by the minimization of $-\log(p(\chi|\theta))$:

$$C(T, M_1, \ldots M_N) = \sum_{i=1}^{N} \frac{\parallel \tilde{M}_i - M_i \parallel^2}{2 \cdot \sigma_{3D}^2} + \sum_{l=1}^{M} \sum_{i=1}^{N} \xi_i^l \cdot \frac{\parallel \tilde{m}_i^{(l)} - m_i^{(l)} \parallel^2}{2 \cdot \sigma_{2D}^2} + K \quad (2)$$

where K is a normalization constant depending on σ_{2D} and σ_{3D}.

The obvious difference between this criterion and the simple 2D ML is that we now have to solve for the hidden variables (the exact locations M_i) in addition to the previous rigid transformation parameters. An obvious choice to modify the optimization algorithm is to perform an alternated minimization w.r.t. the two groups of variables. Starting from a transformation initialization T_0, we initialize the M_i with the $\tilde{M}_i$ and perform a first minimization on T (this corresponds to optimizing the simple 2D ML criterion). In a second step, we keep the transformation T fixed and we optimize for the M_i. We then continue to alternatively update the transformation $\hat{T}$ (given the lastly estimated values $\hat{M}_i$ of the exact M_i) and the exact positions $\hat{M}_i$ (given the lastly estimated transformation $\hat{T}$). The algorithm is stopped when the distance between the two last estimation of the parameters become negligible. The convergence is insured since we minimize the same positive criterion at each step.

2.3   Dealing with 3D and 2D Anisotropic Noise

Up to now, we considered isotropic 2D and 3D noises. However, most of the
CT-scan acquisition are not isotropic (the slice thickness is often larger than the
pixel size within a slice). In that case, the markers 3D localization error will
most probably be anisotropic:

$$\varepsilon_i \sim N(0, \Sigma_{3D}) \qquad \text{with} \qquad \Sigma_{3D} = \begin{pmatrix} \sigma_{3D_x}^2 & 0 & 0 \\ 0 & \sigma_{3D_y}^2 & 0 \\ 0 & 0 & \sigma_{3D_z}^2 \end{pmatrix}$$

This induces very small changes in our ML formulation: we just have to replace
in our criterion (Eq. 2) the first term $\frac{\|\tilde{M}_i - M_i\|^2}{2 \cdot \sigma_{3D}^2}$ by half of the Mahalanobis
distance $(\tilde{M}_i - M_i)^\top \cdot \Sigma_{3D}^{-1} \cdot (\tilde{M}_i - M_i)$. The same kind of modifications obviously
holds for the second term of the equation in the case of a 2D anisotropic noise.

2.4   Link with Reconstruction and 3D/3D Registration

Let us consider that the exact 3D points are measured in the reference frame of
the cameras (instead of the CT frame as previously) and that we are looking for
the transformation from the camera world to the CT (instead of the reverse as
previously). The 3D/2D ML criterion is then rewritten:

$$C(T, M_1, \ldots M_N) = \sum_{i=1}^{N} \frac{\| \tilde{M}_i - T * M_i \|^2}{2 \cdot \sigma_{3D}^2} + \sum_{l=1}^{M} \sum_{i=1}^{N} \xi_i^l \cdot \frac{\| \tilde{m}_i^{(l)} - P^{(l)}(M_i) \|^2}{2 \cdot \sigma_{2D}^2} + K.$$

As explained in the section 2.2, this criterion can be optimized iteratively by
successively estimating the 3D coordinates and the transformation T. Moreover,
it has to be noticed that the change of variable does not affect the transformation
to be found.

Now, assuming that we are performing an estimation with a largely overestimated σ_{3D} and a correct estimation of σ_{2D}. Around the optimal transformation
$\hat{T}$, we will have

$$\frac{1}{N} \sum_{i=1}^{N} \| \tilde{M}_i - T * M_i \|^2 \simeq \frac{1}{N} \sum_{i=1}^{N} \| \tilde{M}_i - \hat{T} * M_i \|^2 = \hat{\sigma}_{3D} \ll \sigma_{3D}$$

Thus, the first term of the criterion is negligible with respect to the second term
(since σ_{2D} is assumed to be correctly estimated): optimizing for the exact 3D
point positions boils down to the minimization of

$$C_{Rec}(M_1, \ldots M_N) = \sum_{l=1}^{M} \sum_{i=1}^{N} \xi_i^l \cdot \| \tilde{m}_i^{(l)} - P^{(l)}(M_i) \|^2$$

This criterion is in fact one of the more widely used reconstruction criterion.
Then, after the determination of the 3D coordinates of the points in the cameras

frame, the next step consists in minimizing the criterion with respect to T. As the second term does not depends on T, this corresponds to the minimization of

$$C_{3Dreg}(T) = \sum_{i=1}^{N} \| \tilde{M}_i - T * M_i \|^2$$

which is no more than the standard 3D LSQ criterion.

As a conclusion, the method consisting in reconstructing the position of 3D points from the cameras and then registering in 3D can be viewed as a limit case of our 3D/2D ML criterion where the noise on 3D points is largely *overestimated* (with respect to the noise on 2D points). On the other hand, we have already seen that the standard 2D ML criterion for 3D/2D registration is a limit case of our 3D/2D ML criterion where the noise on 3D points is largely *underestimated* or really very small (still with respect to the noise on 2D points).

Thus, we may expect our criterion to perform better than these methods when there is effectively some noise on the 3D points and if we have a good estimation of relative 3D and 2D variances.

3   Performances Evaluation and Criteria Comparison

The goal of this section is to assess on synthetic data the comparative effectiveness of the SPPC and EPPC criteria in terms of computing cost, accuracy and robustness. Experiments are realized with two synthetic cameras with a default angle of 45 degrees and jointly calibrated in the same reference frame. The two cameras are focusing on the same 15 points M_i representing the markers localizations, distributed in a volume of about $10 \times 10 \times 10$ cm^3. The ratio of the distance cameras/points with the cameras focal length is 25. We modeled the fiducial localization error by a Gaussian noise with different standard deviations on both the 3D and 2D data (default values are $\sigma_{3D} = \sigma_{2D} = 2$ which corresponds to a SNR of 75 dB[1]. The optimization procedure used is the Powell algorithm. The registration error is evaluated using 9 control points C_i different from the M_i to assess a *Target Registration Error* (TRE) instead of a *Fiducial Localization Error* (FLE) that does not reflect correctly the real accuracy. For each experiment, we give the mean computation time, the mean RMS TRE and the mean relative error (TRE_{SPPC}/TRE_{EPPC}) over 10000 registrations (a value above one means that EPPC is more accurate than SPPC).

3.1   Accuracy Performances

Focusing on the accuracy and not on the robustness, we kept the initial and sought transformation fixed and close enough so that both algorithms do converge correctly. The two following tables present the performances w.r.t. a varying 3D/2D noise ratio (mean TRE values are meaningless and then not reported

---

[1] $SNR_{dB} = 10 \log_{10}(\frac{\sigma_s}{\sigma_n})$ where σ_s (resp. σ_n) is the variance of the signal (resp. noise).

since the noises do vary) and a varying angle between the cameras. In these tables, we should only consider the relative values of the computation times as the absolute value depends on the initialization.

Noise ratio σ_{3D}/σ_{2D}		4	2	1	0.5	0.25
Computation	SPPC	0.0024	0.0023	0.0024	0.0023	0.0024
time (sec.)	EPPC	0.119	0.057	0.025	0.024	0.021
Mean relative error		1.17	1.15	1.09	1.03	1.006

Cameras angle		90	60	45	30	10
Computation	SPPC	0.0025	0.0034	0.0026	0.0025	0.0031
time (sec.)	EPPC	0.015	0.021	0.025	0.028	0.022
Mean TRE	SPPC	1.76	1.94	2.14	2.50	4.12
(mm)	EPPC	1.75	1.83	1.94	2.22	3.40
Mean relative error		1.008	1.051	1.089	1.114	1.188

One can see that EPPC always provides a better TRE (up to 20%) than SPPC, at the cost of a 10 to 20 times larger computational time. The gain in accuracy is all the more sensitive that the angle between the cameras is small and the 3D noise is important w.r.t. the 2D noise. As the amount of 3D information depends on these two parameters, this results was expected since EPPC better captures than the SPPC the noisy nature of information on 3D data.

Eventually, the following table presents the influence of the number of points on the computation times and accuracy performances.

Number of points		30	15	8	4
Computation	SPPC	0.0042s	0.0026s	0.0016s	0.0011s
time (sec.)	EPPC	0.067s	0.025s	0.011s	0.006s
Mean TRE	SPPC	1.55	2.16	2.73	4.45
(mm)	EPPC	1.42	1.97	2.51	3.94
Mean relative error		1.080	1.090	1.079	1.117

The first observation is that the computation times are roughly proportional to the number of points. This was foreseeable since the computational complexity is linear in the number of data points. The second observation is that the measured TRE seems inversely proportional to the square root of the number of points (multiplying 2 times the number of points decrease the RMS error by a factor $\sqrt{2}$). This is also in accordance with the standard accuracy improvements in statistics. One interesting consequence is that we need about 15% less points with EPPC than with SPPC to reach the same accuracy on this example.

3.2 Robustness Evaluation

To evaluate the robustness w.r.t the initial transformation, we select a random initial and/or sought transformation (uniform rotation and translation in a range of the order of the cameras' field of view), random 2D (resp. 3D) noises from 1 to 3 pixels (resp. mm), and a random angle between the cameras (from 10 to 90 degrees). In the following table, we display the performance results when the transformation is initialized using the identity or randomly. The first case represents the usual situation where we have no prior information, and the second one simulates a very bad initialization.

		Random T with identity initialization	Random T with random initialization
Computation time	SPPC	0.0223s	0.0250s
	EPPC	0.138s	0.342s
Ratio of wrong convergence (RWC)	SPPC	20.07%	31.55%
	EPPC	0.09%	0.67%
Mean relative error		1.107	1.139

In both cases, one can see that the SPPC converges wrongly in broadly 20% of the case, whereas the EPPC almost always converges toward the optimal transformation (RWC under 1%). The difference in performances between the two columns probably comes from the optimization algorithm (Powell) that uses a research domain centered around the identity. Another optimization scheme may lead to slightly different results.

3.3   Computation Times

EPPC is a more accurate and much more robust criterion than SPPC. This is paid by a higher computation time that remains however limited to a few tenths of seconds when the initialization is unknown, and to 0.025 to 0.05 seconds when the initialization is close to the sought transformation. Thus, in view of a real-time system, one may expect to obtain with EPPC the best performances with an initialization time of say 0.3 sec. and an update rate of 20 to 40 Hz if the motions are sufficiently slow w.r.t. the video-rate. One may further improve the tracking rate by updated the "exact 3D coordinates" only once in a while or using a Kalman filter.

4   Performances Assessment with Real Data

This section is devoted to accuracy experiments with real 3D CT-scans and 2D images of a plastic mannequin with approximately 25 fiducials sticked on its surface. Ideally, the method's accuracy should be assessed by comparing each registration result with a gold-standard that relates both the CT and the camera coordinate systems to the same physical space, using an external and highly accurate apparatus. As such a system is not available (otherwise we would not have to develop a 3D/2D registration algorithm), we adapted the registration loops protocol introduced in [13,16,14], that enables to measure the TRE error for a given set of test points.

The principle is to acquire several CT scans of the mannequin that are registered using a method described below. Then, several couples of 2D images are

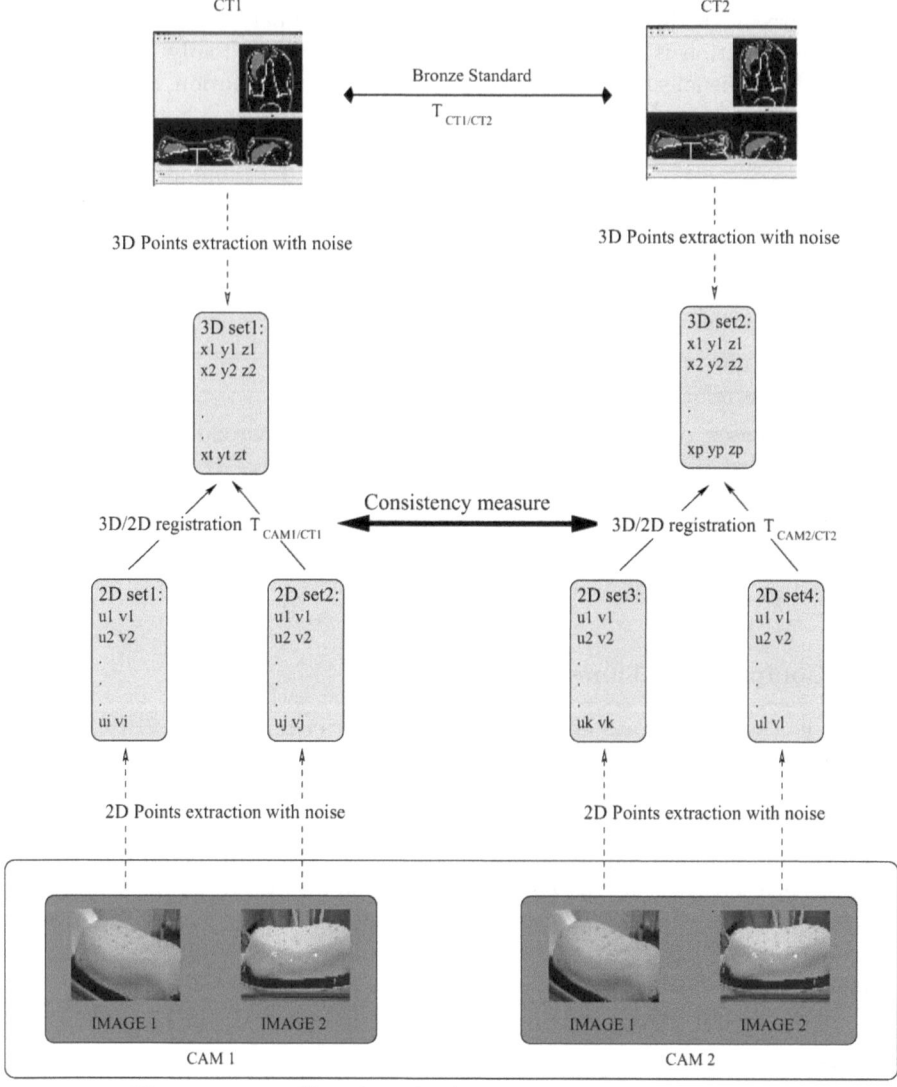

Fig. 1. Registration loops used to estimated the registration consistency.

acquired with the cameras jointly calibrated so that we can compare *independent* 3D/2D registration of the same object (different 2D and 3D images). A typical loop is sketched in Fig. 1: a test point in the faked liver of CT1 is transformed into the CAM1 coordinate system using a first 3D/2D registration, then into the coordinate system of CT2 using a second 3D/2D registration (the coordinate system of CAM1 and CAM2 are identical since cameras are jointly calibrated), and back to CT1 using the bronze standard registration. Using two different CT images allows to de-correlate the two 3D/2D transformation. Indeed, if we register the 2D points to the same set of 3D points (extracted from a single CT scan), the two transformations are similarly affected by the 3D points errors. Consequently, the variability of the 3D points extraction (and any possible bias) is hidden.

If all transformations were exact, we would obtain the same position for our test point. Of course, since the transformations are not perfect, we measure a Target Registration Error (TRE) whose variance is $\sigma^2_{loop} = 2\sigma^2_{CAM/CT} + \sigma^2_{CT/CT}$. This experiment providing only one error measurement, we still need to repeat it with different datasets to obtain statistically significant measures. In order to take into account possible calibration error and/or bias, it is necessary to repeat the experiment with different calibrations and positions of the cameras, and not only to move the object in the physical space. As a result, we are able to measure σ^2_{loop} and thus to estimate the variability due to the 3D/2D registration $\sigma^2_{CAM/CT}$, provided we know $\sigma^2_{CT/CT}$. For this purpose, we devise the following experimental procedure based on multiple CT scan registration that not only gives a very accuracy registration between CT images, but also evaluates the accuracy of this registration.

4.1   Bronze Standard Registration between the CT Images

Our goal here is to compute the $n - 1$ most reliable transformations $\bar{T}_{i,i+1}$ that relate the n (successive) CT_i images. Estimations of these transformations are readily available by computing all the possible registrations $T_{i,j}$ between the CT images using m different methods ([16]). Then, the transformations $\bar{T}_{i,i+1}$ that best explain these measurements are computed by minimizing the sum of the squared distance between the observed transformations $T_{i,j}$ and the corresponding combination of the sought transformation $\bar{T}_{i,i+1} \circ \bar{T}_{i+1,i+2} \ldots \bar{T}_{j-1,j}$. The distance between transformations is chosen as a robust variant of the left invariant distance on rigid transformation developed in [13].

The estimation $\bar{T}_{i,i+1}$ of the perfect registration $T_{i,i+1}$ is called *bronze standard* because the result converges toward $T_{i,i+1}$ as m and n become larger. Indeed, considering a given registration method, the variability due to the noise in the data decreases as the number of images n increases, and the registration computed converges toward the perfect registration up to the intrinsic bias (if there is any) introduced by the method. Now, using different registration procedures based on different methods, the intrinsic bias of each method also becomes a random variable, which is hopefully centered around zero and averaged out in the minimization procedure. The different bias of the methods are now integrated into the transformation variability. To fully reach this goal, it is important to use as many independent registration methods as possible.

In our setup, we used five CT scan of the plastic mannequin in different positions, and five different methods with different geometric features or intensity measures. Three of these methods are intensity-based: the algorithm aladin [11] has a block matching strategy where matches are determined using the coefficient of correlation, and the transformation is robustly estimated using a least-trimmed-squares; the algorithm yasmina uses the Powell algorithm to optimize the SSD or a robust variant of the correlation ratio (CR) metrics between the images [16]. For the feature-based methods, we used the `crest lines` registration described in [12], and the multi-scale `EM-ICP` algorithm of [4] on zero-crossings of the Laplacian surfaces (the images were down-sampled by a factor of 2 to limit the number of surface points to about 1.5 million...). Since none of these methods uses the 3D extracted points as registration data, we ensure the independence with respect to the marker localization noise that corrupts the 3D/2D registration.

As a side effect, we may use only four of the five methods to determine the bronze standard registration, and use that standard to determine the accuracy of the fifth method (a kind of leave-one-method-out test). This uncertainty is then propagated into the final bronze standard registration (including all methods) to estimate its accuracy. In the table below, we give the standard deviation determined this way on the rotational and translational components of each method, the uncertainty of the resulting bronze standard registration and the uncertainty of the 3D registration using standard least-squares of the fiducial markers positions (w.r.t. the bronze standard).

	σ_{rot} (deg)	σ_{trans} (mm)
Aladin	0.09	0.56
Yasmina SSD	0.02	0.41
Yasmina CR	0.06	0.41
Crest lines	0.04	0.27
EM-ICP	0.08	0.68
Bronze standard	0.01	0.07
Fiducials	0.15	0.85

One can observe that the crest lines registration is performing the best, quickly followed by the Yasmina registrations. EM-ICP is not performing very well due to the down sampling of the images. The final bronze standard registration accuracy is very good (it corresponds to 0.08 mm TRE on the test points). Finally, we point out that the lack of accuracy of the fiducials registration w.r.t. the other methods is due to the fact that the markers were stick on the "skin" of the mannequin, which is elastic and did move by about 1 to 2 mm between the acquisitions, while all other methods did focus on the rigid structure of the mannequin (this effect was checked on the images after registration).

4.2   Validation Results and Discussion

After the bronze standard registration of our 5 CT images of the mannequin, we took pictures of the mannequin in four different positions with four different cameras and three different (but joint) calibrations. As a result, we have 12 pairs of 2D images to register to 5 CT scans for each pair of cameras (the second pair of camera being used to close the registration loop on a different CT scan). To robustify the experiment, we randomized the camera used in each pair among the four available, which finally leads to 180 registration loops. Approximately

25 stick fiducials were interactively localized in each image ($\sigma_{3D} = 0.75$ mm, $\sigma_{2D} = 2.0$ pixel). This lead to the following quantitative evaluation.

Angle between the cameras		60°	40°	10°
Computations	SPPC	0.023 s	0.025 s	0.027 s
time	EPPC	0.21 s	0.21 s	0.25 s
Mean TRE RMS	SPPC	1.80	1.99	2.30
error in mm	EPPC	1.70	1.83	2.20
Relative error		1.062	1.089	1.054

One can observe that EPPC is always more accurate than SPPC but the relative error does not increase as much as denoted with synthetic data when the angle is small (10°). This may be explained by the various conditions in which the measurement were done (different local lengths, number of fiducial observed...), and the simple representation of the camera (pinhole model).

Another explanation is obviously the consistent but non-rigid motion (mentioned in the last section) of the skin markers on which the registration is done. However, these small movements realistically simulate the imperfect repositioning of the skin and the organs that is induced by our gas volume monitoring protocol. In this context, the assumption (on which are based our criteria) of independent noises on each 3D marker position may not be fulfilled. Despite these variations from the theoretical assumptions, we underline that the achieved accuracy is around 2mm, which is by far better than the 5mm needed for our medical application. One can assess the visual accuracy of our registration on one particular case in Fig. 2.

5   Conclusion

We devised in this paper a new 2D/3D Maximum Likelihood registration criterion (EPPC) based on better statistical hypotheses than the classical 3D/2D least-square registration criterion (SPPC). Experiments on synthetic and real data showed that EPPC is 5 to 20% more accurate and much more robust than SPPC, but requires higher computation times when the initialization is unknown (0.3 instead of 0.03 sec.). However, synthetic experiments provide evidences that a refreshment rate of 20 to 40 Hz is achievable in a tracking phase, where only small motions have to be detected. In the context of augmented reality for liver radio-frequencies, we showed that the proposed method provides an accuracy of about 2mm within the liver, which fits the initial goal of 5mm that was necessary to provide a significant help for radiologists and surgeons.

In the future, we will focus on the automatic detection, tracking and matching of the fiducials in the 3D and 2D images, in order to fully automate the algorithm. Incidentally, we expect to obtain a much better (probably sub-pixel) localization of the fiducials, and thus drastically improve the accuracy of the 3D/2D registration down to less than one millimeter. We will also investigate the effects of the optimization procedure and of the camera calibration algorithm. Another work in progress concerns the prediction of the uncertainty of

Fig. 2. The top left image shows the plastic mannequin with the radio-opaque fiducials. The top right image displays the augmented reality view of the surgeon, i.e. the super-imposition on the video image of the 3D reconstruction of the fiducials and interns parts (plastic frame and liver) after the registration process. To check visually the quality of the achieved registration on the liver, we put off the skin (bottom left image), and we superimposed the reconstruction of the fiducials on the liver (bottom right).

the 3D/2D registration w.r.t. the current data used. This important element of the system safety will allow the detection of bad geometric fiducials configurations (for instance not enough fiducials visible in both cameras, etc) that lead to very inaccurate registrations. Lastly, we intend to extend the current system in order to take into account the deformations due to breathing.

References

1. M. Dhome and al. Determination of the attitude of 3d objects from a single perspective view. *IEEE Trans. on PAMI*, 11(12):1265–1278, December 1989.
2. M. Fischler and R. Bolles. Random sample consensus : A paradigm for model fitting with applications to image analysis and automated cartography. *Com. of the ACM*, 24(6):381–395, June 1981.
3. S. Ganapathy. Decomposition of transformation matrices for robot vision. *Pattern Recognition Letters*, 2(6):401–412, December 1984.

4. Sébastien Granger and Xavier Pennec. Multi-scale EM-ICP: A fast and robust approach for surface registration. In A. Heyden, G. Sparr, M. Nielsen, and P. Johansen, editors, *European Conference on Computer Vision (ECCV 2002)*, volume 2353 of *LNCS*, pages 418–432, Copenhagen, Denmark, 2002. Springer.
5. W. Grimson and al. An automatic registration method for frameless stereotaxy, image-guided surgery and enhanced reality visualization. *IEEE TMI*, 15(2):129–140, April 1996.
6. R. Haralick and al. Pose estimation from corresponding point data. *IEEE Trans. on Systems, Man. and Cybernetics*, 19(06):1426–1446, December 1989.
7. R. Kumar. Determination of camera location and orientation. In *DARPA Image Understanding Workshop*, pages 870–881, Palo Alto, Calif., 1989.
8. Y. Liu and al. Determination of camera location from 2d to 3d line and point correspondences. *IEEE Trans. on PAMI*, 12(01):28–37, January 1990.
9. D. Lowe. Fitting parameterized three-dimensional models to images. *IEEE Trans. on PAMI*, 13(5):441–450, May 1991.
10. Siu-Hang Or and al. An efficient iterative pose estimation algorithm. In *Proc. of ACCV'98*, pages 559–566, 1998.
11. S. Ourselin, A. Roche, S. Prima, and N. Ayache. Block Matching: A General Framework to Improve Robustness of Rigid Registration of Medical Images. In *Proc. of MICCAI'00*, pages 557–566, Pittsburgh, Penn. USA, October 11-14 2000.
12. X. Pennec, N. Ayache, and J.-P. Thirion. Landmark-based registration using features identified through differential geometry. In I. Bankman, editor, *Handbook of Medical Imaging*, chapter 31, pages 499–513. Academic Press, September 2000.
13. X. Pennec, C.R.G. Guttmann, and J.-P. Thirion. Feature-based registration of medical images: Estimation and validation of the pose accuracy. In *MICCAI'98*, LNCS 1496, pages 1107–1114, October 1998.
14. G. Penney and al. Overview of an ultrasound to ct or mr registration system for use in thermal ablation of liver metastases. In *MIUA'01*, July 2001.
15. T. Phong and al. Object pose from 2-d to 3-d points and line correspondences. *IJCV*, 15:225–243, 1995.
16. A. Roche, X. Pennec, G. Malandain, and N. Ayache. Rigid registration of 3D ultrasound with MR images: a new approach combining intensity and gradient information. *IEEE TMI*, 20(10):1038–1049, October 2001.
17. L. Soler and al. Fully automatic anatomical, pathological, and functional segmentation from ct-scans for hepatic surgery. *Computer Aided Surgery*, 6(3), August 2001.
18. P. Viola and W.M. Wells. Alignment by maximization of mutual information. *International Journal of Computer Vision*, 24(2):137–154, 1997.
19. J. Yuan. A general photogrammetric method for determining object position and orientation. *IEEE Trans. on Robotics and Automation*, 5(2):129–142, April 1989.
20. Zhengyou Zhang. A flexible new technique for camera calibration. Technical report, Microsoft Research, December 1998.

In vitro Measurement of Mechanical Properties of Liver Tissue under Compression and Elongation Using a New Test Piece Holding Method with Surgical Glue

Ichiro Sakuma[1], Yosuke Nishimura[1], Chee Kong Chui[1], Etsuko Kobayashi[1], Hiroshi Inada[2], Xian Chen[1], and Toshiaki Hisada[1]

[1] Institute of Environmental Studies, Graduate School of Frontier Sciences,
The University of Tokyo, 7-3-1 Hongo, Bunkyo-ku, Tokyo 113-8656 Japan
isakuma@k.u-tokyo.ac.jp,
{yosuke,cheekong,etsuko}@miki.pe.u-tokyo.ac.jp
{chen,hisada}@sml.k.u-tokyo.ac.jp
[2] Department of Precision Machinery Engineering, Graduate School of Engineering,
The University of Tokyo, 7-3-1 Hongo, Bunkyo-ku, Tokyo 113-8656 Japan
inadah@miki.pe.u-tokyo.ac.jp

Abstract. There is a need to determine biomechanical properties of liver tissue to develop realistic elastic deformable liver model for computer aided surgery. In this report, we introduced a method to measure mechanical properties using surgical instant adhesive (surgical glue). The method made easier to define the mechanical boundary conditions for test pieces. It also makes it possible to conduct both compression and elongation test on the same test piece. In actual deformation of liver during surgical intervention, the tissue is subject both to compression and elongation. Identification of mechanical properties in the range where mechanical force changes from compression to elongation is important. We can identify the stress-strain relationship of liver samples in the transition range from compression to elongation. We also investigated viscoelastic properties by compressing the sample at different velocities. The obtained results can be applied to non linear FEM analysis of liver tissue.

1 Introduction

To develop realistic deformable model for liver that can be used for computer aided surgery such as surgical navigation, identification of mechanical properties of liver is important. However, no standard protocol to determine the mechanical properties is available currently. Recently deformation analyses based on biomechanics using finite element analysis (FEM) have been widely investigated to develop surgical navigation system. To identify the mechanical properties of liver, reliability of the analysis is dependent on the accuracy of mechanical properties used. Ishihara et al. recently reported results of elongation test using sheet shaped liver test pieces[1]. Yamada performed elongation tests on liver tissue[2]. Ottenmayer and Salisbury measured normal indentation fore-displacement over a frequency range from DC to

N. Ayache and H. Delingette (Eds.): IS4TM 2003, LNCS 2673, pp. 284–292, 2003.
© Springer-Verlag Berlin Heidelberg 2003

approximately 100Hz to identify solid organ mechanical properties in vivo.[4] Kauer et al [4] reported inverse finite element characterization of soft tissue based on aspiration method of soft tissue.

In actual deformation, liver tissue is subjected to both elongation and compression at different positions in the liver tissue. Thus it is important to identify mechanical properties under both of elongation and compression. In this paper, we propose a new mechanical test method applicable both for elongation and compression [5]. The proposed method can simplify the mechanical boundary condition near the interface between liver test piece and holding chuck of a test instrument. The preliminary results on stress-strain relationship, heterogeneity of mechanical properties of liver, and vico-elastic properties are demonstrated.

2 Material and Methods

2.1 Liver Test Piece Preparation and Fixation of a Test Piece

Fresh liver was collected at local slaughterhouse and was refrigerated and set at room temperature before experiments. A cylindrical test piece was obtained using a circular surgical knife (8 mm in diameter, 5-10 mm in length). In this procedure, cylindrical samples with diameter of 7 mm were obtained. The longitudinal length of the test piece free from external force was measured and used as natural length of the test piece. Both ends of the cylindrical test piece were attached to the two rubber plates using instant surgical glue as shown in Figure 1. A holding part was attached on the plate as interface between the test piece and chuck of a loading test machine. Although the bonding force of the glue is not large enough to hold the liver tissue at its breakpoint, strain as large as ±40% could be realized.

2.2 Measurement of Stress-Strain Relationship

A compact testing instrument (Eztest, Shimadzu Co.Kyoto,Japan) was used. This instrument can conduct elongation, compression, indentation, and stress relaxation test. The elongation and compression speed can be changed from 1 to 1000mm/min. Resolution of force measurement was 0.001N. A photograph of a test piece placed on the instrument is shown in Figure 2. The change in the diameter of the test piece was monitored using a video camera.

Elongation and compression forces as large as ±0.1 N were applied three times at rate of 30 mm/min as preconditioning. Throughout the tests, the temperature was kept at 22±1. Humidity was kept from 60 to 70% to prevent drying of the test pieces.

A cylindrical test piece was obtained using a circular surgical knife (8 mm in diameter, 5-10 mm in length). Both ends of the cylindrical test piece were attached to the two rubber plates using instant surgical glue.

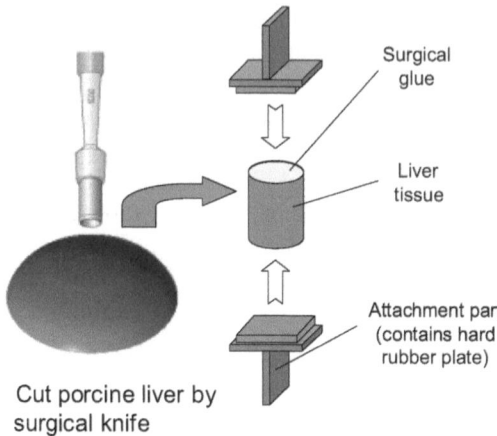

Fig.1 Preparation of a liver test piece

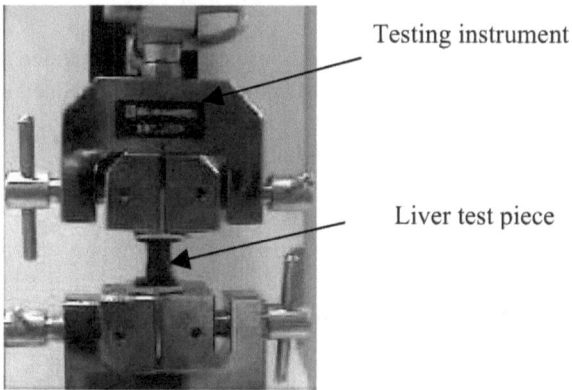

Fig.2 A liver test piece set on the testing instrument

3   Results

3.1   Effect of Compression on Mechanical Strength for Elongation

To evaluate influence of history of compression on the mechanical properties during
elongation test, we applied fixed compression force prior to the elongation test. Afterwards we measured force-displacement relationship during elongating the test
piece. The compression force was set at 1, 2, 5, 10, and 18 N respectively. The obtained relationship between applied force and displacement is shown in Figure 3. The
upper panel shows the relationship between strain and force for compression experiments. The lower panel shows those for elongation tests. When the applied compression forces were larger than 5 N, change in the relationship was observed. There was
leakage of interstitial fluid under these conditions. When the applied compression
fore was 15N, irreversible destruction of liver tissue was observed. When the applied

compression force was less than 1 N, there were no significant differences between the obtained force-displacement curves under elongation.

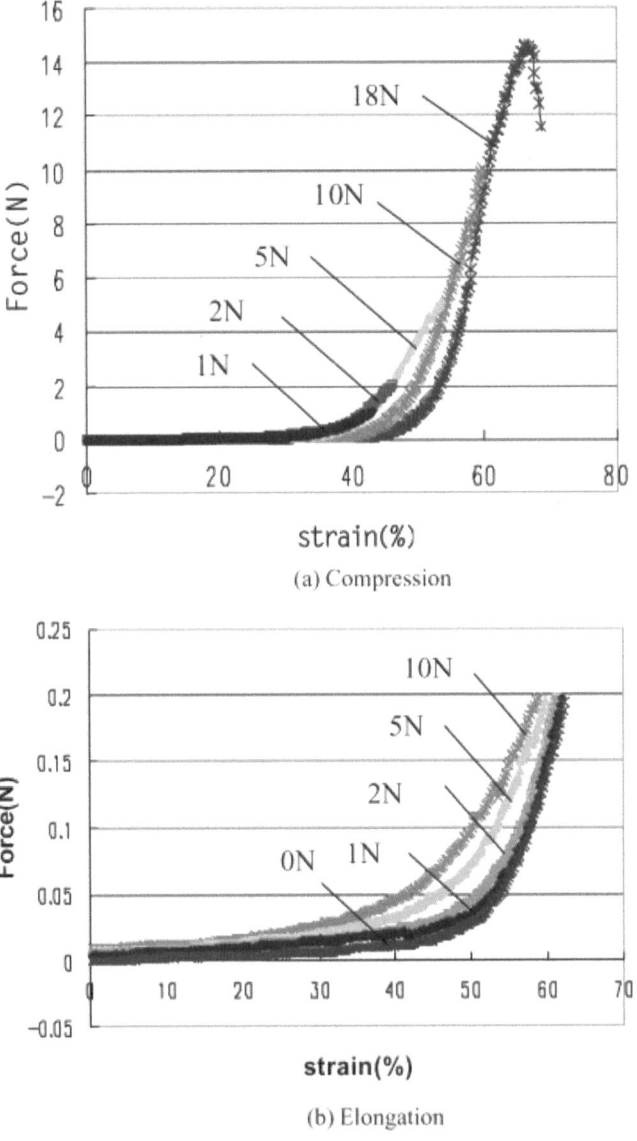

(a) Compression

(b) Elongation

Fig. 3. Influence of compression force on mechanical properties during elongation. Compression force ranging from 1 to 18 N was applied to the test pieces. Relationship between applied force and displacement was evaluated under elongation on the same test piece after initial compression with forces ranging from 1-18N. In case of application of 18N, the test piece was destroyed at approximately 15 N before reaching the maximum of the programmed compression force.

3.2 Stress and Strain Relationship Measurement

Figure 4 summarizes results of elongation and compression tests for 15 test pieces.
The average relationship between strain and stress is illustrated. Elongation and compression velocity was set as 10mm/min. The length of test pieces was 5 ±1mm. The
diameter of the test piece was 7 mm. We tested 15 test pieces. Obtained average
stress and strain relationship is shown in Fig.4. Vertical axis denotes nominal stress
obtained using the initial cross sectional area of the test piece. The data showed non-
linear relationship between stress and strain in liver tissue samples. In small strain
range (less than 10 %), reproducibility of the measured stress is good while the standard deviation of the measured stress among the test pieces were large.

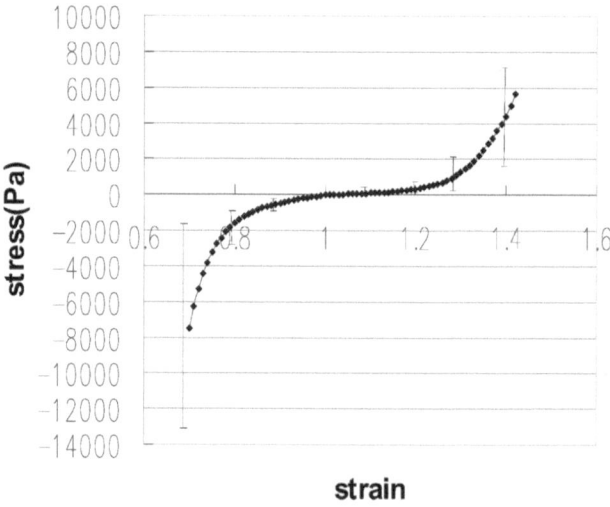

Fig. 4. Relationship between nominal stress and strain of liver test pieces

The average value of 15 measurements is shown. Error bar stands for standard deviation among 15 different test pieces.

3.3 Heterogeneity of Mechanical Properties in Liver Tissue

 Test pieces were collected from visceral side, diaphragmatic side, and edge part of
the liver. as shown in Figure 4. These portions have relatively homogeneous structures. The length of the test pieces was 10 ±1 mm. The locations of sampling points
in liver tissue and measured average specific gravity of test pieces at each point are
shown in Figure 5. Their mechanical properties were measured at leading speed of 10
mm/min. The strain-stress results are shown in Figure 6, where vertical axis denotes
nominal stress obtained using the initial cross sectional area of the test piece. Liver
tissue at visceral side showed slightly softer properties under elongation compared
with the other test pieces. The size of hepatic lobule was larger than those in the other

sides. It is considered that macroscopic stress-strain relationship is affected by these differences in microscopic structure of liver tissue. These results imply that heterogeneity of mechanical properties in liver tissue should be taken into consideration for accurate simulation of organ deformation.

Specific gravity of test samples

	diaphragmatic side	$1.095(g/cm^3)$
.	viscereal side	$1.105(g/cm^3)$
.	Edge part	$1.061(g/cm^3)$

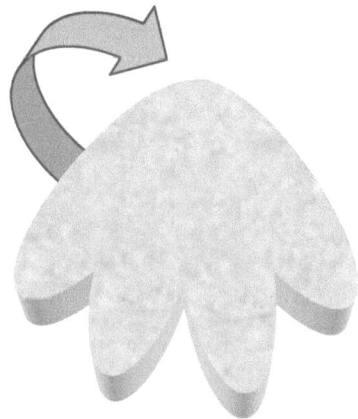

Fig. 5. Location of test piece harvest and specific gravity of the test pieces

3.4 Measurement of Visco-Elastic Properties with Stress Relaxation Tests

To identify visco-elastic properties of liver, stress relaxation tests ware conducted on the liver test pieces. The diameter of the test piece was 7 mm and its length was 5 mm.

Initially, 40 % compressive displacement was applied to the test pieces at different displacement rate (1, 2, 10, 20, and 50 mm/min) respectively. Test with displacement rate of 1 mm/min was conducted first, and test with displacement rate of 50 mm/min was conducted last. Change in applied force at fixed displacement was continuously monitored after the onset of force application.

The results are shown in Figure 7. The magnitude of observed force was highly dependent on displacement rate. With displacement rate of 50 mm/min, the maximum observed force was less than 0.35N. At 300 sec after the onset of force application, the observed forces approached to the value of 0.01-0.015 N in all cases.

Figure 8 shows the relaxation of elongation force under change during elongation of liver tissue samples. Initially, 20 % elongation applied to the test pieces at different displacement rate (1, 2, 10, 20, and 50mm/min respectively). Test with displacement rate of 1 mm/min was conducted first, and test with displacement rate of 50 mm/min was conducted last. The maximum observed force was approximately 0.03 N with displacement rate of 50 mm/min,, while that with 1mm/min was 0.025N. Several hundreds sec after the onset of force application, the observed forces approached to the value of 0.01 to 0.015 N in all cases.

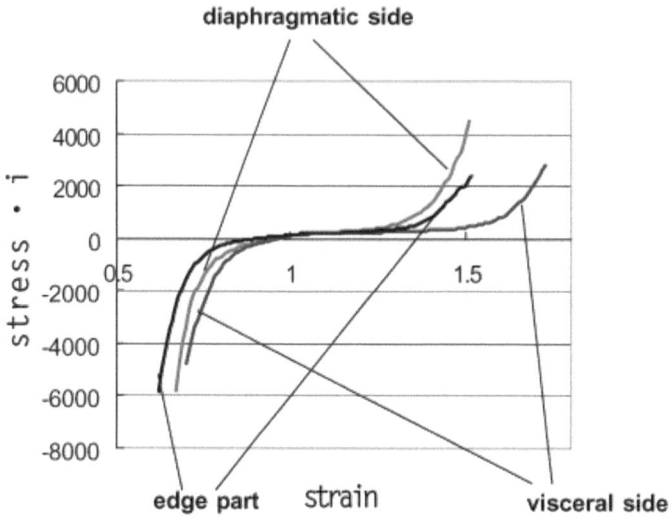

Fig. 6. Relationship between nominal stress – strain of the liver samples taken from different part of the liver

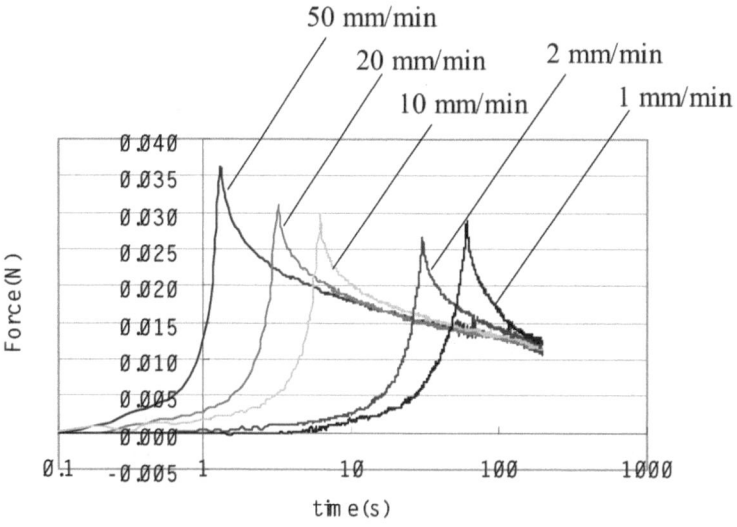

Fig. 7. Relaxation of compression force under 40 % compression at different displacement rate

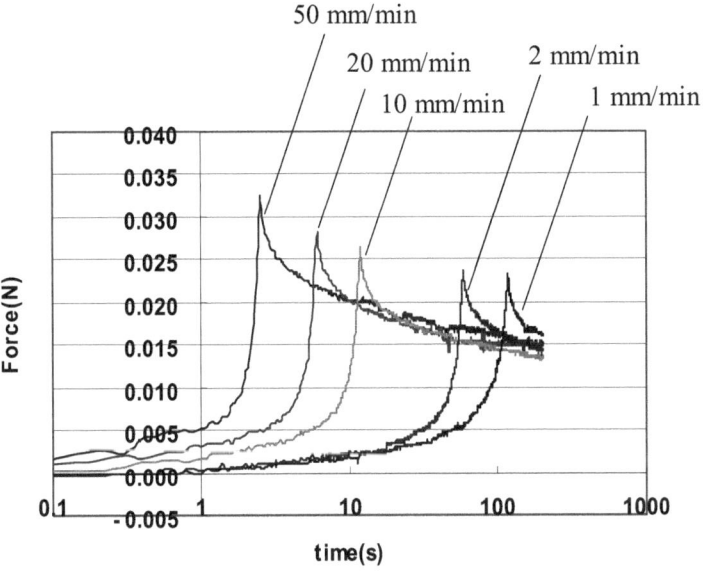

Fig. 8. Relaxation of elongation force under 20 % elongation at different displacement rate

Similar results were obtained both in compression and elongation tests. These results imply that displacement rate should be taken into consideration in developing deformable model of liver.

4 Discussion

Ishihara et al. [1]measured mechanical properties of liver using a sheet specimen under elongation. In their method, mechanical stress is applied at the interface between the sample and mechanical chuck of the testing instrument. They measured change in distance between two marker points at the central area free from the above mentioned device related mechanical stress. There is still ambiguity in determining the natural distance under no load condition.

Miller and Chinzei[6] used surgical glue to fix the brain sample on the plate and conducted stretch test of the brain tissue. The proposed method is similar to their method. One of the advantages of this holding method is simplified boundary condition at the interface between test pieces and mechanical chuck of the instrument.The advantages of the proposed testing method are as follows:

1. It can conduct both compression and elongation test continuously on the same test pieces. Thus we can estimate the origin of stress-strain curve accurately.
2. Since cylindrical test piece was attached on the rubber plate, the mechanical boundary condition is relatively simple [6].

In actual deformation of liver during surgical intervention, the tissue is subject both to compression and elongation. Identification of mechanical properties in the range where mechanical force changes from compression to elongation is important. For this purpose, the proposed method will provide valuable information. Preliminary results showed that the compression of the test pieces with less than 1 N did not cause any change in mechanical properties of the test piece under elongation. Thus the proposed method is effective in analyzing the stress-strain relationship in the transient range from compression to elongation.

The result also showed the existence of heterogeneity of mechanical properties among liver tissue even for relatively homogeneous portion in the liver tissue.

Visco-elastic property was investigated in compression test of the liver. Viscous force was three times larger than elastic force when 50 mm/min displacement rate was applied. It implies that constitutive equation of liver tissue should take the visco-elastic properties into consideration for finite element analysis of liver deformation.

Acknowledgment

This study was partly supported by research for the future program (JSPS-RFTF 99I00904) .

References

1. Toshikazu ISHIHARA, Yuko Nakahira, Kazunori Furukawa. : Measurement of Mechanical Properties of the Pig Liver and Spleen, JSME Annual Meeting (2000) 209-210
2. Hiroshi Yamada :Strength of biological materials; edited by F. Gaynor Evans.- Baltimore : Williams & Wilkins , 1970
3. Mark P. Ottensmeyer, J.Kenneth Salisbury. :In Vivo Data Acquisition Instrument for Solid Organ Mechanical Property Measurement MICCAI2001, LNCS 2208, pp.975-982,2001.
4. M. Kauer, V. Vuskovic, J. Dual, G. Szekely, M. Bajka. :Inverse Finite Element Characterization of Soft Tissues, MICCAI2001, LNCS 2208, pp.128-136,2001.
5. C Chui, Y Nishimura, E Kobayashi, H Inada and I Sakuma, "A medical simulation system with unified multilevel biomechanical model", to appear in Proc. ICBME 2002 Conference, December 2002.
6. Karol Miller, Kiyoyuki Chinzei :Mechanical properties of brain tissue in tension, :Journal of Biomechanics 35(2002)483-490

Validation of the Interval Deformation Technique for Compensating Soft Tissue Artefact in Human Motion Analysis

Rita Stagni[1,2], Silvia Fantozzi[1,2], Angelo Cappello[1], and Alberto Leardini[2]

[1] Department of Electronics, Computer Science and Systems, University of Bologna,
Viale Risorgimento 2, 40136 Bologna, Italy
acappello@deis.unibo.it
www-bio.deis.unibo.it
[2] Movement Analysis Laboratory, Istituti Ortopedici Rizzoli
Via di Barbiano 1/10, 40136 Bologna, Italy
{stagni,fantozzi,leardini}@ior.it
www.ior.it/movlab

Abstract. Soft tissue artefact is the most invalidating source of error in human motion analysis. This error is caused by the erroneous assumption that markers on the skin surface are rigidly connected to the underlying bone. Several methods have been proposed in the literature to compensate for this spurious effect by mathematical modelling of the interposed soft tissues. Validation of these methods has been performed only on simulated data or on a small sample of data acquired from subjects mounting devices which limit skin motion. In the present study, the performance of one of the most recent compensation methods was evaluated using experimental data acquired combining stereophotogrammetry and 3D video-fluoroscopy. The effectiveness of the compensation method was found strongly dependent on the modelling form assumed for the motion of the markers in the bone frame. Even when the compensation produced a significant advantage in the evaluation of bone orientation, the estimation of position was critical.

1 Introduction

The description of human joint kinematics during activities of daily living is the main aim of human motion analysis. Stereophotogrammetry allows the reconstruction of the trajectories of markers attached to the skin surface of the body segments to be analysed. These trajectories are used to calculate the pose of the underlying bony segments, with the erroneous assumption that markers and bony segments are rigidly connected. It is well known that markers on the surface of the body move significantly with respect to the underlying bones because of the interposition of soft tissues [1]. This interposition is the origin of two different sources of error: anatomical landmarks mislocation and Soft Tissue Artefact (STA). The latter has been recently recognized to be the major source of error in human motion analysis [2]. It is caused by the relative movement between the markers, necessarily attached on the skin, and the underlying bones, and it is therefore associated to the specific marker-set and experimental protocol adopted. Inertial effects, skin deformation and sliding, gravity and muscle contraction contribute to this phenomenon. This artefact has a frequency content similar to that of bone movement in space and it is therefore not possible to distinguish between this latter and STA by means of any filtering technique.

N. Ayache and H. Delingette (Eds.): IS4TM 2003, LNCS 2673, pp. 293–301, 2003.

Because of the relevance of this source of error in human motion analysis, several compensation methods have been suggested in the literature in order to model and to try to compensate STA. Cappello et al. [3] have described a method designed to reduce errors associated with non-rigid segment movement. This technique overcomes the previous constraint of rigidity between the marker cluster and the underlying bone and advises a double calibration of the anatomical landmarks, at the two extremes of the expected range of motion. Lu and O'Connor [4] expand the rigid-body model approach. Rather than seeking optimal rigid body transformation on each segment individually, they imposed model-based constraints to the joints connecting each couple of body segments. Lucchetti et al. [5] take an entirely different approach, using *ad-hoc* movements to characterize the artefact and to determine the correlation between flexion-extension angles and skin marker trajectories. Andriacchi et al. [6] described a Point Cluster Technique (PCT), which employs an overabundance of markers placed on each segment and an optimal weighting to minimize the effects of skin motion. A more recent paper by the same authors [7] illustrated a further extension of this method, the Interval Deformation Technique (IDT), including equations assumed to describe soft tissue deformation. This recent method was tested simply on simulated data, biased by the fact that generated deformation is of the same functional form as the model used for its compensation, and on one subject mounting an Ilizarov external fixator, which limits skin motion.

The purpose of the present study was to assess the performance of the proposed IDT on experimental position data obtained from the combination of stereophotogrammetry and 3D video-fluoroscopy. This test allows for the assessment of the technique in real conditions, without any restriction to skin motion, and with the knowledge of the reference motion of the underlying bone.

2   Materials and Methods

2.1   Subjects

Two female subjects, who underwent total knee replacement, were analysed. The subjects were clinically successful arthroplasty and gave informed consent to participate in this study. The age, height, weight, Body Mass Index, and follow-up of subject #1 and #2 were 67 and 64 years, 155 and 164 cm, 58 and 60 Kg, 24 and 22, and 18 and 25 months, respectively.

2.2   Set-Up and Methodology

The subjects were contemporary analysed with stereophotogrammetry and video-fluoroscopy.

In order to assess the amount of STA on the lateral aspect of the thigh, a cluster of many reflecting skin markers was stuck on this segment. The clusters were 4-5 cm spaced grids of markers with a diameter of 0.6 cm. Nineteen and twenty-five markers were uniformly attached laterally on the skin of the thigh of subject #1 and #2, respectively (Fig. 1).

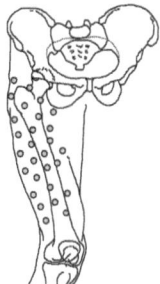

Fig. 1. Grid of markers evenly distributed on the lateral surface of the thigh.

Further five specialized reflecting and radiopaque markers, therefore visible to both systems, with a diameter of 12 mm were used. Four of these were placed on a plane parallel to the image plane of the fluoroscope for spatial registration of the two measurements. The fifth specialized reflecting and radiopaque marker was attached on the skin in correspondence of the patella for temporal synchronization.

A static up-right posture was acquired. The position of 6 anatomical landmarks (right and left anterior iliac spines, mid-point between right and left posterior iliac spines, great trochanter, lateral and medial epicondyles) was calibrated with respect to the relevant cluster of markers [8]. The two subjects were acquired during step up/down with fluoroscopy and stereophotogrammetry simultaneously. Three repetitions were collected.

The subjects performed each motor task with the knee under analysis inside the fluoroscopic 32cm field of view.

Marker trajectories were collected at 50 frames per second by means of a stereophotogrammetric system with 5 TV cameras (Smart, e-Motion, Padova, Italy). The stereophotogrammetric calibrated field of view was 1.5 x 1.5 x 1 m.

2.3 Data Analysis

The accurate 3D pose of the prosthesis components was reconstructed by means of single-plane lateral 2D fluoroscopic projections and relevant CAD models (Fig. 2).

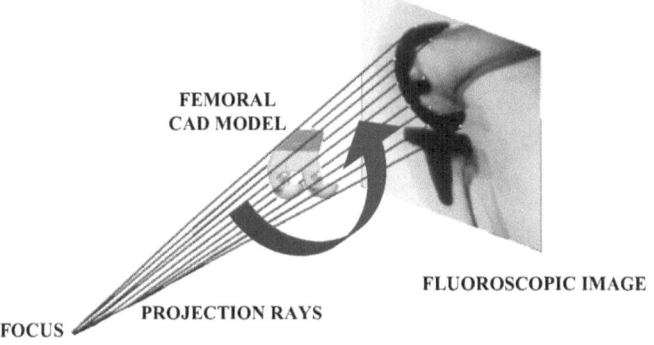

Fig. 2. Sketch of the model for fluoroscopic image generation process used to estimate component 3D poses.

Series of lateral images were acquired at the nominal frequency of 6 samples per second with a standard fluoroscope (DRS, System 1694 D, General Electric CGR, USA). The images were printed out on films and digitised by means of a scanner (Scanmaster DX, Howtek, Hudson, NH, USA). Images of a 3D cage of plexiglas with 18 tantalum balls in known positions and of a rectangular grid of lead balls 10 mm apart were collected in order to calculate respectively the position of the camera focus and the parameters necessary for image distortion correction. This was obtained using a global spatial warping technique [9]. An established technique for 3D kinematics analysis of a known object from a single view was implemented [10]. Prosthesis component poses in space, assumed to well represent position and orientation of the bones, were obtained from each fluoroscopic image by an iterative procedure using a technique based on CAD-model shape matching. Previous validation work [10] showed that orientation and position in the sagittal plane can be estimated with an accuracy better than 1 degree and 0.5 mm, respectively.

Spatial registration between the two measurement systems was obtained by defining a common absolute reference frame by means of the four radiopaque and reflecting markers. The temporal synchronization was obtained by matching the fluoroscopic with the resampled stereophotogrammetric trajectories of the fifth specialized marker. The matching was obtained by calculating the maximum cross correlation between the two trajectories considering the resampling frequency and the starting frame as the parameters to be determined. Skin marker trajectories obtained from the stereophotogrammetric system and the 3D poses of the prosthesis components obtained from 3D fluoroscopy were then reported in the same absolute reference frame.

The possible misalignment of the prosthesis components with respect to the relevant anatomical reference frame was taken into account. This misalignment was calculated in the static up-right posture, considered as reference position, and the fluoroscopy-based 3D pose of the anatomical reference frame was calculated accordingly.

According to the Calibration Anatomical System Technique (CAST) proposed by Cappozzo et al. [11], the kinematics of the thigh anatomical reference frame can be reconstructed from the trajectories of the skin markers, once the position of the anatomical landmarks relative to the corresponding technical reference frame is known from a calibration procedure (Fig. 3). The technical reference frame is defined for each segment at frame n from the position of its origin $o_T(n)$ and its rotation matrix $R_T(n)$, calculated from the trajectories of the corresponding markers in the global reference frame.

The local coordinates AL_{loc} of the anatomical landmarks in the technical reference frame are calculated from data acquired during the calibration trials. Then the global coordinates are given by the equation:

$$AL_{glo}(n) = o_T(n) + R_T(n) \cdot AL_{loc} \qquad (1)$$

Thus, the kinematics of the anatomical reference frame in terms of its origin $o_A(n)$ and rotation matrix $R_A(n)$ in the global frame, are:

$$o_A(n) = o_T(n) + R_T(n) \cdot o_A^T \qquad (2)$$

$$R_A(n) = R_A(n) \cdot R_A^T \tag{3}$$

Where o_A^T is the origin and R_A^T is the rotation matrix of the anatomical in the technical reference frame.

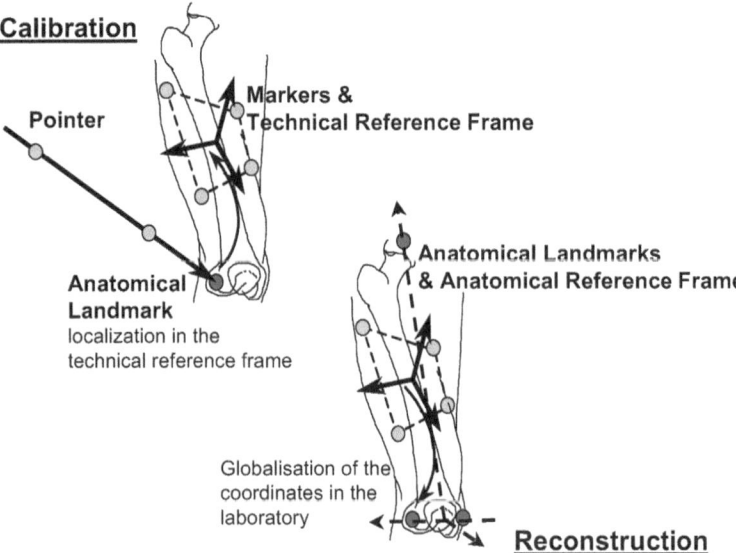

Fig. 3. Schematic representation of the calibration and reconstruction in the global reference frame of the position of the anatomical landmarks and of the relevant anatomical frame (dashed axes), through the definition of the technical reference frame (solid bold axes), according to the Calibrated Anatomical System Technique [11].

Five different methods were used for the definition of the technical reference frame and for the following reconstruction of the kinematics of the thigh anatomical frame:

PCT:
As suggested by Andriacchi et al. [6], the technical frame associated to the cluster of skin markers is defined by the eigen-vectors of the inertia tensor described by the expression:

$$I(n) = \begin{bmatrix} \sum_{i=1}^{n}[(p_{i,y})^2 + (p_{i,z})^2] \cdot m(n)_i & \sum_{i=1}^{n} p_{i,x} \cdot p_{i,y} \cdot (-m(n)_i) & \sum_{i=1}^{n} p_{i,x} \cdot p_{i,z} \cdot (-m(n)_i) \\ \sum_{i=1}^{n} p_{i,x} \cdot p_{i,y} \cdot (-m(n)_i) & \sum_{i=1}^{n}[(p_{i,x})^2 + (p_{i,z})^2] \cdot m(n)_i & \sum_{i=1}^{n} p_{i,y} \cdot p_{i,z} \cdot (-m(n)_i) \\ \sum_{i=1}^{n} p_{i,x} \cdot p_{i,z} \cdot (-m(n)_i) & \sum_{i=1}^{n} p_{i,y} \cdot p_{i,z} \cdot (-m(n)_i) & \sum_{i=1}^{n}[(p_{i,y})^2 + (p_{i,x})^2] \cdot m(n)_i \end{bmatrix} \tag{4}$$

Where p_i is the trajectory of the i marker. The mass m_i associated to each marker is estimated at each time instant t in order to isolate rigid body motion of the cluster. Being $e_1(n)$, $e_2(n)$, and $e_3(n)$ the eigen-vectors of $I(n)$:

$$R_T(n) = \begin{bmatrix} e_1(n) & e_2(n) & e_3(n) \end{bmatrix} \tag{5}$$

$o_T(n)$ is the centroid of the skin markers.

Having defined the kinematics of the technical frame according to this technique, the kinematics of the anatomical frame is reconstructed, having determined o_A^T and R_A^T in the static reference calibration condition.

Three different IDT compensations on PCT:
The change of position of the skin markers with respect of the underlying bone, caused by the skin motion artefact, is taken into account. The technical reference frame is defined according to the former PCT [6].

A compensation [7] is introduced, by minimising the cost function:

$$\chi^2 = \sum_{i=1}^{M} \sum_{j=1}^{3} \sum_{n=1}^{N} \left(\frac{(L(i,j,n) - \overline{L}(i,j,n))}{\sigma(i,j,n)} \right)^2 \tag{6}$$

where $L(i,j,n)$ is the position of marker i, at frame n, along direction j of the technical frame, and $\sigma(i,j,n)$ is the individual standard deviation of the noise. The motion of the markers in the technical frame was modelled as follows:

$$\overline{L}(i,n) = E_{cb}^{-1}(n)[\overline{R}(i,n) - C_{cb}(n)] \tag{7}$$

where $\overline{R}(i,n)$ is the position of marker i at frame n in the anatomical reference frame, $C_{cb}(n)$ and $E_{cb}(n)$ represent the position and orientation of the technical with respect to the anatomical bone frame.

This compensation was applied to calculate the kinematics of the anatomical frame by considering the (1) *real* Trajectories of the Markers in the Bone-embedded Frame (TMBF). As this knowledge is not usually available, compensation was also performed assuming a (2) *cubic* (IDT-3) and (3) *gaussian* modelling of the TMBF. The parameters of the cubic and gaussian trajectories were obtained as the minimum square approximation of TMBF, in order to evaluate the performance of the compensation method in the best possible conditions.

Singular Value Decomposition (SVD):
The Singular Value Decomposition (SVD) method [12] was also used to calculate $o_T(n)$ and $R_T(n)$. Then, o_A^T and R_A^T were obtained from the static reference calibration condition.

The position and orientation error of the bone-embedded frame estimated for each of the five different methods with respect to the fluoroscopy based gold standard was calculated. The mean and standard deviation of the error was evaluated over the six trials of the two subjects. The relative orientation was expressed in terms of Euler angles (α, β, γ), according to XYZ convention. In the bone-embedded frame, X is the antero-posterior axis, Y is the longitudinal and Z is the medio-lateral.

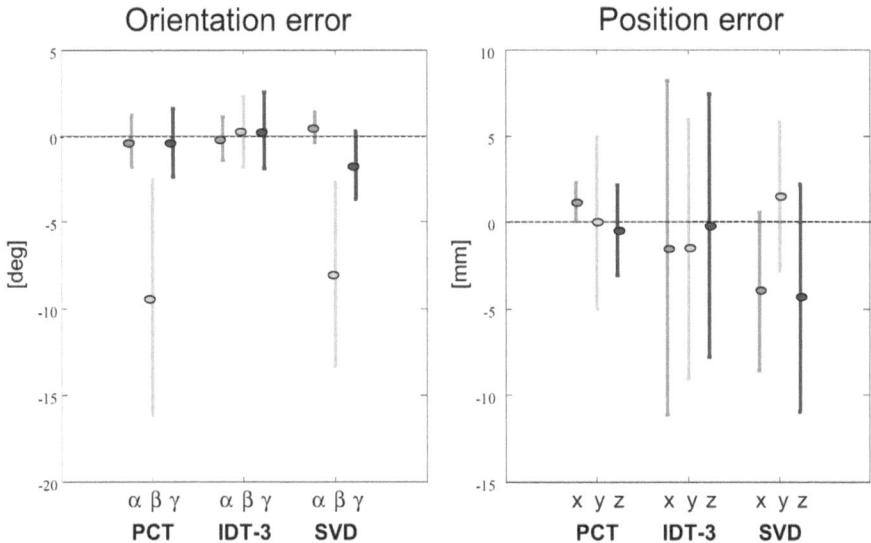

Fig. 4. Mean value ± one standard deviation bar representation of the orientation and position error of the bone frame estimated with PCT, IDT with cubic polynomial modelling of the markers trajectories (IDT-3), and CAST with SVD definition of the technical frame. Values are reported for each method for the three angles *(α, β, γ)* and directions *(x, y, z)* with respect to the bone anatomical frame of the femur.

3   Results

The root mean square error of the cubic approximation of the TMBF was 3.6 mm, whereas the gaussian was 4.5 mm.

The orientation and position error of the estimated bone frames as estimated by the PCT, the IDT with cubic polynomial approximation and the CAST based on SVD estimation of the pose of the bone anatomical frame are reported in terms of mean value ± one standard deviation of the error in Fig. 4. IDT-3 performed well and reduced significantly the orientation error with respect to PCT and SVD, with a mean value in the range of $-0.2° \div 0.3°$ and a standard deviation within $2.2°$. On the other hand, for position error, IDT-3 showed small mean position errors but a large standard deviation, up to 9.7 mm.

Results obtained from IDT with gaussian approximation and real marker trajectories were not reported in Figure 4. The gaussian approximation produced orientation and position errors much larger than any other method, with magnitudes up to several degrees and centimetres. On the other hand, the compensation based on real trajectories produced an accurate estimation of the bone frame, with negligible orientation and position error.

4 Discussion and Conclusions

IDT is the most recent method proposed in the literature to compensate for STA in human motion analysis data. It features the modelling of the effect of the soft tissues interposed between the markers and the underlying bones [7]. The compensation to the PCT in the estimation of the pose of the bone anatomical frame is excellent when the real trajectories of the markers in that frame are available, as it had already been shown [7]. In the routine experimental practice, however, the knowledge of the TMBF is not given, and the problem can be faced only by means of modelling with some degree of approximation. In this context, the IDT compensation appeared to be extremely dependant on the modelling form adopted for the TMBF approximation, as shown by the extremely different results obtained with the cubic and gaussian approximations, although the root mean square value errors of the trajectory approximation were comparable for the two methods. Moreover, the cubic approximation performed was the best possible, and did not produce a significant increase in the bone pose estimation accuracy, especially in terms of translations. Finally, it was observed [13, 14, 15] that TMBF are strongly subject and task specific, thus a generalized trajectory approximation is extremely difficult without any further subject-specific trial.

In conclusion, the performance of IDT is strongly affected by the convention adopted to represent TMBF. IDT does not introduce a significant improvement in bone pose estimation, unless TMBF are accurately known, but this is unlike in routine human motion analysis.

References

1. Cappozzo, A., Catani, F., Leardini, A., Benedetti, M.G., Croce, U.D.: Position and orientation in space of bones during movement: experimental artefacts. Clinical Biomechanics 11 (1996) 90-100.
2. Andriacchi, T.P., Alexander, E.J.: Studies of human locomotion: past, present and future. Journal of Biomechanics 33 (2000) 1217-1224.
3. Cappello, A., Cappozzo, A., La Palombara, P.F., Lucchetti, L., Leardini, A.: Multiple anatomical landmark calibration for optimal bone pose estimation. Human Movement Science 16 (1997) 259-274.
4. Lu, T.W., O'Connor, J.J.: Bone position estimation from skin markers co-ordinates using global optimisation with joint constraints. Journal of Biomechanics 32 (1999) 129-134.
5. Lucchetti, L., Cappozzo, A., Cappello, A., Della Croce, U.: Skin movement artefact assessment and compensation in the estimation on knee-joint kinematics. Journal of Biomechanics 31 (1998) 977-984.
6. Andriacchi, T.P., Alexander, E.J., Toney, M.K., Dyrby, C.O., Sum, J.: A point cluster method for in vivo motion analysis: applied to a study of knee kinematics. Journal of Biomechanical Engineering 120 (1998) 743-749.
7. Alexander, E.J., Andriacchi, T.P.: Correcting for deformation in skin-based marker systems. Journal of Biomechanics 34 (2001) 355-361.
8. Benedetti, M.G., Catani, F., Leardini, A., Pignotti, E., and Giannini, S.: Data management in gait analysis for clinical applications. Clinical Biomechanics 13 (1998) 204-215.
9. Gronenschild, E.: The accuracy and reproducibility of a global method to correct for geometric image distortion in the x-ray imaging chain. Medical Physics 24 (1997), 1875-1888.

10. Banks, S.A., Hodge, W.A.: Accurate measurement of three-dimensional knee replacement kinematics using single-plane fluoroscopy. IEEE Transactions on Biomedical Engineering 43 (1996) 638-649.
11. Cappozzo, A., Catani, F., Croce, U.D., Leardini, A.: Position and orientation in space of bones during movement: anatomical frame definition and determination. Clinical Biomechanics 10 (1995) 171-178.
12. Soderkvist,I., Wedin,P.A.: Determining the movements of the skeleton using well-configured markers. Journal of Biomechanics 26 (1993) 1473-1477.
13. Fantozzi, S., Stagni, R., Cappello, A., Leardini, A., Catani, F.: Assessment of skin motion artefact from a combined fluoroscopic and stereophotogrammetric analysis. In Proceedings SIAMOC 2002, 13-15 October 2002, Bologna (Italy). Gait & Posture, 16 (2002) S192-S193.
14. Stagni, R., Fantozzi, S., Cappello, A., Ussia, L., Leardini, A.: Propagation of skin motion artefacts to knee joint kinematics. In Proceedings SIAMOC 2002, 13-15 October 2002, Bologna (Italy). Gait & Posture, 16 (2002) S211-S212.
15. Stagni, R., Fantozzi, S., Cappello, A., Leardini, A.: Quantification of soft tissue artefact in motion analysis by compining 3D fluoroscopy and stereophotogrammetry. Submitted to the Journal of Biomechanics (2002).

An Open Software Framework
for Medical Applications

Erwin Keeve, Thomas Jansen, Bartosz von Rymon-Lipinski, Zbigniew Burgielski,
Nils Hanssen, Lutz Ritter, and Marc Lievin

Surgical Systems Laboratory
research center c a e s a r
Friedensplatz 16, 53111 Bonn, Germany
keeve@caesar.de

Abstract. In this paper we introduce the extendible and cross-platform software
framework *Julius*. Julius combines both pre-operative planning and intraoperative assistance within one single environment. In this paper we discuss
three aspects of Julius: the medical data processing, the visualization pipeline
and the interaction. Each aspect provides interfaces that allow to extend the application with own algorithms and to build complex applications. We believe
that this approach facilitates the development of image guided navigation and
simulation procedures for computer-aided-surgery.

1 Introduction

Image-guided surgery becomes more and more indispensable these days. The vision
of reality is enhanced using information from CT, MR and other medical scans to help
the surgeon in his every-day work. Several software products are available, such as
specialized non-profit systems or commercial applications [1, 2]. We propose to develop a general software framework for a complete medical processing-pipeline,
including data processing, visualization and navigation in pre-operative planning and
intra-operative guidance. It should be possible to use this framework as a test-
platform for research, providing a high-level visualization infra-structure as well as an
architecture with extendible interfaces for easy implementation of professional algorithms, including rendering and simulation techniques. Our system also supports easy
integration of third-party software libraries for medical processing and rendering (e.g.
ITK and VTK) [3, 4].

The remainder of the paper is structured as follows: In section 2 we present the
software architecture of our framework that is the basis for our interaction and navigation model described in section 3. Section 4 gives an overview of some exemplar
software components for visualization and interaction that we have implemented with
Julius. Section 5 discusses the results of our work. Finally, section 6 gives a summery
and addresses future work.

2 Software Architecture

The general software framework *Julius,* together with its graphical user interface
(GUI) and many software components, was presented in [5]. In this paper we will

N. Ayache and H. Delingette (Eds.): IS4TM 2003, LNCS 2673, pp. 302–310, 2003.

focus on the software architecture and components from the visualization and simulation perspective – see Figure 1.

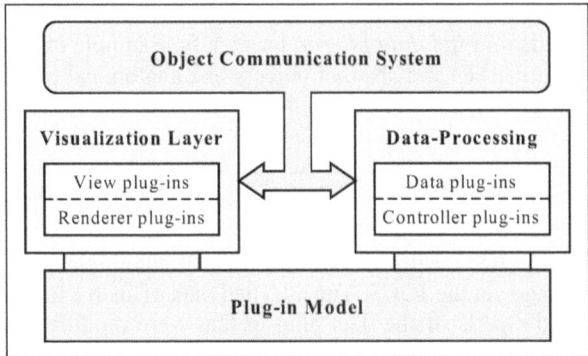

Fig. 1. Overview of the core infra structure of Julius. The plug-in model is the basis for component implementations, in particular for data processing and the visualization layer. The communication system manages the co-operation between both.

The main part of our framework is the object-oriented *JCore* library, based on Trolltech's multi-platform application toolkit Qt [6]. The core library encapsulates fundamental functionality required for an extendible medical processing pipeline. In the following paragraph we present an overview of the core functionality while in the second paragraph we discuss the visualization infrastructure.

2.1 Core Functionality

To achieve extendibility, modularity and scalability the JCore library features a plug-in model that enables third-party developers to connect own software components to the Julius framework at run-time – without recompiling the source code [7]. This allows replacing existing plug-ins with more optimized or introducing dedicated implementations. Available plug-in versions can be easily activated or deactivated, using the plug-in configuration tool of the GUI. This facilitates the testing of new implementations.

The JCore library also addresses the communication between objects of the framework and plug-ins that are currently activated. Our *Julius object communication system* (JOCS) features message-based information exchange between distributed objects in the framework. Complex communication protocols can be created by attaching additional data to the message. JOCS supports different addressing modes to minimize the number of potential listeners, i.e. direct, multicast and broadcast addressing. Therewith, developers have the possibility to implement efficient update schemes, e.g. for rendering of dynamic data such as results of volume deformation during simulation. JOCS also features posting of messages for safe communication in a multi-threaded environment. This greatly simplifies the use of parallelism in multi-processor architectures. Extending the communication system is simple because messages are represented by objects, derived from standard Julius classes.

Data processing is a fundamental part of the JCore library. Our concept is based on the separation between data representation and operations that process that data. By utilizing this pattern we can efficiently build medical workflows.

- Data representation – Information for the medical processing pipeline is stored in plug-ins, derived from the *data plug-in* base class. Example data plug-ins are volumes, polygonal meshes and abstract objects like anatomical properties, finite element nets or scene objects, like the camera. Different data plug-ins can be linked together to build a complex and modular hierarchy, e.g. a 3D scene graph.
- Operations – Algorithms acting on the data or introducing new data are implemented by the so-called *controller plug-ins*. Examples are import, export, segmentation, registration and mesh generation.

Each time a controller creates or modifies data it automatically multicasts a data modification message via the JOCS. The attached data of such a message contains bit-flags, identifying the parts of the data plug-in that were modified by the controller. Other objects that receive this message can efficiently update their internal state. This leads to minimal memory transfer for local dynamics during simulation.

An example is the hardware-accelerated volume renderer (e.g. on SGI graphics hardware) that does not need to upload the whole voxel data to texture memory, if just a small sub-volume has been modified.

2.2 Visualization Infrastructure

An important infrastructure based on the core library of our software architecture is the *Julius visualization layer* (JVL). The JVL is responsible for the visualization of results produced by the data processing pipeline – see Figure 2. It introduces two plug-in types:

- *View plug-ins* specify view ports to the current scene. A scene consists of several data plug-ins encapsulating different data types. The view plug-in is linked to a Qt widget in the user interface, which allows receiving user-input, such as key strokes.
- *Renderer plug-ins* are responsible for the visualization of data plug-ins in a specific view. Their interfaces are very slim and they are not directly linked to the GUI. However, renderer plug-ins have to handle data updates, resulting e.g. from interaction and navigation.

Both visualization plug-in types are based on the context of a rendering library, such as OpenGL or VTK. Julius also provides interfaces for extending the visualization system with other rendering libraries. Different but compatible renderers can be combined dynamically in the same view, resulting in a maximum of configurability and flexibility, e.g. for rendering of a volume that is overlaid by a vector field.

The Active Mode

A view plug-in can be set to the so-called *active mode*: On each incoming data modification message, it automatically updates the rendering and creates new compatible renderer plug-ins for the modified data, if necessary. Such a mechanism is useful especially for displaying updates during interaction and temporary results from a long data-processing pipeline. In addition, successive updates caused by the same manipulation, e.g. interactive dragging of a control point during volume deformation, are gathered before the scene is rendered. This avoids visualization of inconsistent results.

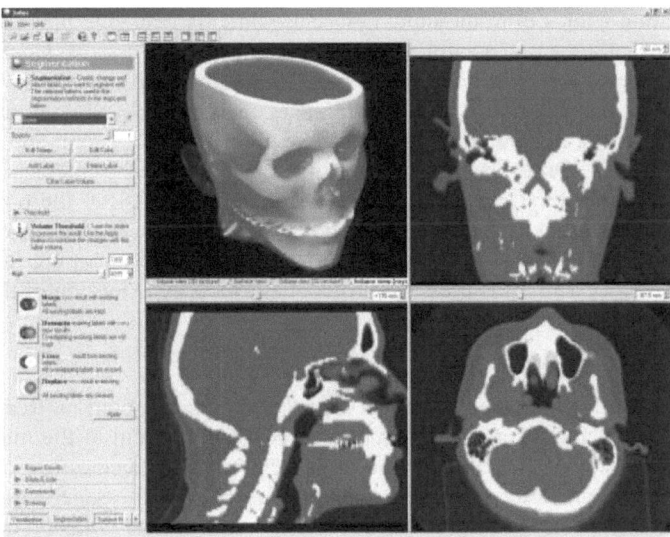

Fig. 2. Visualization of interactive segmentation results using a threshold-based approach.

3 An Infra Structure for Interaction and Navigation

Another essential part for the development of simulation applications, besides data processing and visualization, is our interaction layer. We have developed a general plug-in infrastructure, based on the JCore library that facilitates the development of interaction and navigation applications. We call this infrastructure *DIR model*. It is based on three base plug-ins *device*, *interactor* and *reactor*.

Device
The device plug-in represents input hardware. Examples are mouse, haptic interface and optical tracking. Each device plug-in communicates with the underlying input hardware and stores the information received from the device tool, like transformation (i.e. position and orientation) and button states. Each input device is associated with a list of geometric data plug-ins that are under control of the corresponding device. An example is a polygonal mesh data plug-in, representing a surgical instrument that is navigated through the surgical scene. Note that exploiting the concept of the associated list of data plug-ins allows to manipulate multiple objects with one single input device.

Interactor
The interactor plug-in can be seen as the central of the DIR model and coordinates the complete interaction process. The interactor is triggered by a simulation timer and works on the active input device plug-ins. Its task is to transform the associated data plug-ins using the information read from the input devices. The interactor supports two modes of transformation, *absolute* and *relative*. In the *absolute mode* the position

and orientation values from the input device are directly copied to the associated data plug-ins. This assures best accuracy and performance. In the *relative mode* each transformation is computed relatively to the position and orientation values from the last time step. This mechanism allows to manipulate one single object with multiple input devices at the same time, which can be useful e.g. for manual or semi-automatic registration operations. On each step of the simulation process the interactor sends two types of messages: First, an *interaction message* is invoked after the transformation data from the input device has been read, but not applied yet. Second, a *reaction message* is sent when all position and orientation updates of the data plug-ins have been successfully processed. This message protocol allows easy extension of the interactor plug-in.

Reactor
Finally, the last plug-in type of the DIR model is the reactor plug-in. Reactors implement specific interactive applications just by connecting them to the message interface of the interactor. Typically, a reactor implementation executes one or more controller plug-ins on each incoming interaction/reaction message, resulting in a manipulation of the objects in the current scene. Examples are collision detection and response, force feedback object exploration and object deformation. More detailed examples can be found in the section 4.2 of this paper.

4   Exemplar Plug-Ins

In the following part we will present examples of software components that we have integrated into the Julius framework. First, we will focus on plug-ins for medical visualization. In the second part we will describe exemplar components that implement basic interactive operations, based on the DIR model presented in the previous section.

4.1   Plug-Ins for Visualization

The *3D view plug-in* supports OpenGL and VTK-based renderer plug-ins of polygonal and volumetric datasets. The view plug-in may be seen as a container for the renderer plug-ins (see 2.2). Different renderer plug-ins can be used stand-alone or in combination with others to visualize data plug-ins: *Polygonal data* – A standard renderer plug-in for VTK-based polygonal mesh data objects. *Volumetric data* – A software-based solution using the ray-casting approach is available for high-quality images as well as a hardware accelerated implementation for interactive visualization via the 2D or 3D texture mapping technique. *Cut planes* – A renderer plug-in for visualization of free-oriented cutting planes through a polygonal mesh or volumetric dataset. There are two implementations available, one for software- and one for hardware-based rendering.

Primarily, we use the 3D view plug-in (with various configurations of renderer plug-ins) for execution of long controller sequences, as required during complex medical procedures. The co-operation between the Julius data-processing and the visualization layer provides visual feedback and interfaces for data manipulation in each step of that sequence.

Another view plug-in is responsible for *slice visualization* (see Figure 3), which is fundamental for most medical applications. It supports OpenGL and VTK-based renderer plug-ins showing a planar cut through a volumetric dataset or polygonal mesh similar to the cut-planes for the 3D view plug-in. The slice is represented by a data plug-in that can be shared among different views. By changing the position or orientation of the plane, e.g. via a slider in the GUI, the corresponding cut plane moves synchronously in all related 2D and 3D views. Additionally, the slice view offers interfaces to connect blending plug-ins for the composition of multiple slice images for visualization. This feature is useful particularly for semi-automatic registration.

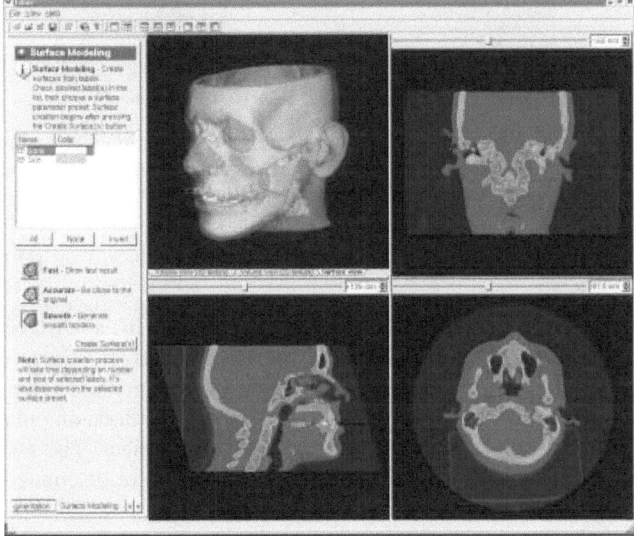

Fig 3. Screen shot with mixed renderer plug-ins for visualization. The 3D view shows overlaid transparent polygon meshes, while the slices display a composition of cuts through the original volume and the corresponding polygon dataset.

All view plug-ins can be linked together by setting the same input data for their renderer plug-ins. Transformations and deformations in one view will cause automatic update in all other views.

4.2 Plug-Ins for Interaction

One interactive application that we have integrated into the Julius software framework is the so-called *operation central*. This application is based on the DIR model and can be configured in a simple graphical user interface. The user has the possibility to setup the following application parameters: *Input device* – Specifies the DIR input device that has to be used for the operation, e.g. mouse, haptic interface. *Operation type* – Specifies the DIR reactor. This parameter determines the type of the interactive procedure, e.g. object deformation. *Tool object* – Specifies the tool that is controlled by the input device. This object can represent a surgical instrument, like a scalpel.

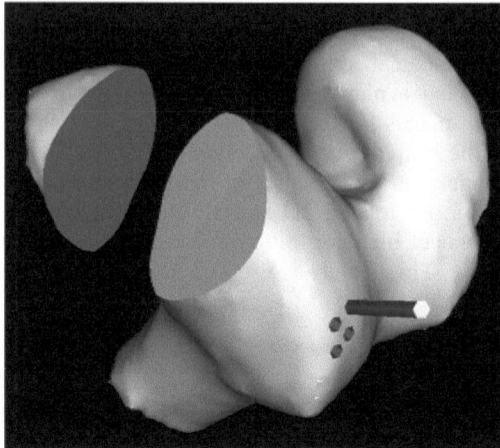

Fig 4. Screenshot, showing an exemplar interactive polygon cutting procedure, implemented with the DIR model. The femur bone was cut and separated into two individual pieces. A cylindrical instrument is used for drilling holes in the object.

We have implemented two exemplar operation types based on polygonal surface meshes:

- *Polygon cutting* – This procedure is implemented with a DIR reactor and two controller plug-ins. Both controller plug-ins are iteratively executed on each incoming reaction message, if an input device button is pressed. The first controller encapsulates a VTK-based algorithm for boolean operations on polygonal meshes, used for subtracting the tool object from the target object. The second controller separates the cut object into individual regions that were disconnected in the first step. Figure 4 shows a screenshot of this procedure.

- *Object exploration* – The object exploration is also implemented with a DIR reactor and is based on the collision detection algorithm of the Proximity Query Package (PQP library) [8]. On every collision between an object in the scene and the currently selected tool in the operation central, the reactor computes response forces that are encapsulated in JOCS messages. Therewith, force feedback devices that receive this message have the possibility to react and pass this information to the haptic hardware. By pressing an input device button, objects can be grabbed and transformed in the scene.

5   Results

We have developed the *JuliusLight* sample application based on our software framework. As described before, the framework can easily be extended and tailored to specific applications in the medical field by supporting a broad range of plug-in types that can be added to the application at run-time. The intention of JuliusLight is to show the versatility and flexibility of this framework.

The JuliusLight sample application (see Figure 5) is dedicated to segmentation and measurement but can easily be extended with simulation algorithms based on previous segmentation results. Accessing per-voxel attributes is simple: As a general rule, volumetric datasets that have been loaded are never modified, instead, all data that is generated during segmentation is stored in a separate labeled volume (also referred as the segmented volume), which has the same dimensions and position as the loaded dataset. With this concept, each voxel in the dataset has an unambiguous assignment to a freely definable structure (including biomechanical parameters) that is in turn represented by the appropriate label.

Both the original and the label volumes can be visualized simultaneously, being in the same location in space, to show which voxels of the dataset belong to which entity. In the slice views, the labels are always translucent in order to show the underlying data. Furthermore, by adjusting the opacity of each segmented entity, the spatial alignment of the structures can be studied in the 3D view.

The described functionality is encapsulated in plug-ins that were independently developed from the rest of the framework. This gives a good impression about the aspects of a medical pipeline that can be extended using the Julius software framework.

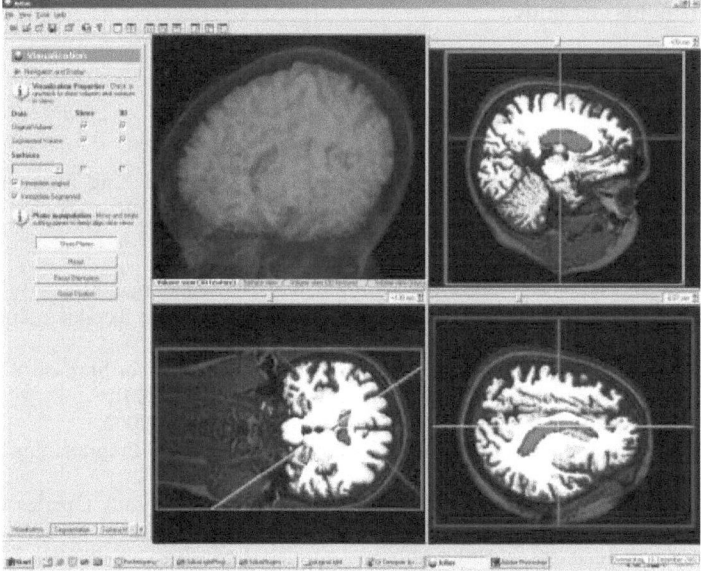

Fig. 5. Brain segmentation results in Julius. The picture shows free oriented slice planes through the volume. Visualization was done using 3D texture mapping functionality.

6   Conclusions

With Julius we have developed a general and flexible software framework for medical applications. The crucial elements of this framework are the cross-platform portability, modularity and application-driven configurability. The plug-in model, the data-processing concept and the visualization layer make Julius easily extendible with new algorithms and rendering techniques especially in the context of visualization and simulation. Our communication system is used to efficiently coordinate medical processing and the rendering components, especially during interaction and navigation.

Therefore, using Julius as a platform for rapid application development benefits to the end-user and the scientific developer.

The DIR model allows easy development and extension of navigation and simulation applications. New input hardware can be easily integrated via device plug-ins. Using the message interface and reactor plug-ins facilitates the incorporation of interactive operations. Additionally, optimizations and features of the interactor plug-in have a global impact on the whole interaction system. One example that we plan to implement is Kalman filtering for the transformation data, received from the input devices. Another example is the adaptive control of the simulation time step, based on the current system workload and the desired simulation accuracy [9].

Our framework is in daily-use and acts as a base for other research. We plan to integrate additional rendering libraries, including scene-graph support (e.g. via OpenSG [10]). The JCore library will provide interfaces for distributed communication, visualization and data-processing.

Julius has been made public available under a non-commercial license at http://www.julius.caesar.de.

References

1. D.T. Gering, A System for Surgical Planning and Guidance using Image Fusion and Interventional MR, Master's Thesis, MIT, 1999.
2. Analyze Software System from Mayo Clinic, http://www.mayo.edu/bir/Software/Software.html, 2002.
3. Insight Segmentation and Registration Toolkit, http://www.itk.org, 2002.
4. W. Schroeder, K. Martin and B. Lorensen, "The Visualization Toolkit – 2nd Edition", Prentice Hall PTR, New Jersey, 1998.
5. E. Keeve et al., "Julius – An Extendable Software Framework for Surgical Planning and Image-Guided Navigation", MICCAI, Utrecht, October 14-17, 2001.
6. K. Dalheimer, "Programming with Qt (2nd Edition)", O'Reilly, 2002.
7. C. Szyperski, "Component Software: Beyond Object-Oriented Programming", Addison-Wesley, 1998.
8. E. Larsen, S. Gottschalk, M. C. Lin, D. Manocha, "Fast Proximity Queries with Swept Sphere Volumes", Technical report TR99-018, Department of Computer Science, University of N. Carolina, Chapel Hill, 1999.
9. A.P. Witkin, D. Baraff, M. Kass, "An Introduction to Physically Based Modeling", SIGGRAPH '97 Course, 1997.
10. D. Reiners, G. Voss and J. Behr, "OpenSG – Basic Concepts", OpenSG Symposium, Darmstadt, January 29, 2002.

Haptic Simulation of a Tool in Contact with a Nonlinear Deformable Body

Mohsen Mahvash[1] and Vincent Hayward[2]

[1] Real Contact Inc.
Lorne Crescent, Montréal, Québec, Canada H2X 2B1
mahvash@cim.mcgill.ca
[2] Center for Intelligent Machines
McGill University, Montréal, Québec, Canada, H3A 2A7
hayward@cim.mcgill.ca

Abstract. This paper presents a method to artificially re-create haptic feedback while moving and sliding an arbitrary virtual tool against a virtual deformable body with nonlinear elastic properties. The computation of the response in such general cases is a task which does not yet admit computational solutions suitable for realtime implementation. To address this, we describe an approach based on the bookkeeping of force-deflections curves stored at the nodes of a triangulated body surface. For realism, normal and lateral deformations at each node are represented in a range of deflection distances. The response everywhere is synthesized via area interpolation of response curves stored at the nodes of the mesh. The mathematical continuity of the synthetic response is the result of both local coordinates interpolation and of response function interpolation, which previous methods did not account for. This guarantees the absence of haptic 'clicks' and 'pops' which are unacceptable artifacts in high fidelity simulations. Sliding contacts are also considered.

1 Introduction

A haptic display is capable of re-creating artificially the forces that simulate the interaction of a tool with a deformable body. This capability can be used for various applications such as in training for surgical tasks performed with a tool.

The computational engine of a haptic display, namely the *haptic engine*, should be able to provide these forces in *real time*. The construction of a haptic engine that can simulate the continuous mechanics of the contact between a tool and a nonlinear deformable body in realtime is a daunting task. Instead, we introduce a pre-calculation approach. It relies on the offline determination of force-deflection curves at the nodes of a triangulated body surface mesh for a range of deflection in a pre-processing step. Online, forces are calculated at arbitrary locations via area interpolation.

The pre-calculation method approach can be viewed as a systems approach. Contact forces are the outputs of a multi-input multi-output system. The inputs are the position and the orientation of the tool tip, and the internal variables

N. Ayache and H. Delingette (Eds.): IS4TM 2003, LNCS 2673, pp. 311–320, 2003.
© Springer-Verlag Berlin Heidelberg 2003

represent the position of the contact point. The behavior of the system is represented by a construction of piecewise polynomial curves designed to approximate the behavior of an actual deformed body. The coefficients of these polynomials are the system parameters. The systems formulation allows us ensure two properties for the simulation which are also properties of the original physical system, namely, continuity and passivity [15,14]. Continuity is provided by the interpolation of the coordinates defined at the surface of virtual body and by the interpolation of the response functions.

The identification of the system parameters is done offline. It is performed by fitting the outputs of the system to actual contact forces obtained through measurement or via accurate offline simulation. The parameters can be stored in standard data file format since at any given time, only a small number of them is needed for simulation.

Section 2 reviews previous work. Section 3 describes a mathematical framework to calculate normal and lateral forces. Section 4 explains how the precalculated responses can be constructed from actual or accurate forces at a set of test points. Section 5 evaluates the interpolation approach through a simple implementation. Section 6 concludes the paper.

2   Related Work

Various techniques were proposed to make the haptic simulation of contact interaction possible. We divide them into two groups.

2.1   Pre-calculation Methods

Cotin et al., James and Pai, and Bro-Nielsen introduced the pre-calculation method [4,11,12,16]. Linear models, obtained either from the finite element method or from the boundary element method, represent the deformation of a body. A set of algebraic linear relations among the nodal quantities at the free boundary are then derived [12]. The deformation responses given by unit forces (or tractions) applied to each node of the free boundary are calculated during a preprocessing phase. The nodal forces in the contact region are calculated in advance from a reduced-order linear system that directly relates the nodal displacements to forces. A small linear system is derived from the displacement responses at runtime. The deformation response of the boundary due to a displacement constraint imposed by the user is calculated as the superposition of responses of each nodal force in a preset contact region.

Delingette et al. as well as Astley and Hayward used pre-calculation techniques to compute the contact between a part of a body (peripheral to the interaction point) and the rest of the body in a hybrid body structure [7,3].

Pre-calculation methods are very effective for reducing the computation time of deformation. However, they are applicable only to linear elastic bodies and small displacements. Due to discretization, even for linear elastic deformation, accurate computation of the force deflection response requires a large number of elements to represent a body well. In the contact region, this number can be arbitrarily large, and hence can yield a prohibitively large computational load.

In contrast, the approach introduced in this paper is capable of handling nonlinear deformation (arising either from large deformation or from nonlinear elasticity) due to massive pre-computation of the forces responses in a pre-processing step. Moreover, the pre-computation burden can be reduced by computing forces for a mesh structure coarser than the mesh structure used during the simulation.

2.2   Time Integration Multi-resolution Methods

High order dynamic deformation models represent the deformation of the body. Explicit or semi-explicit time integration are used to solve these models efficiently in realtime. Multi-resolution calculations in space and time are used to refine the calculation in regions of interest.

Astley and Hayward proposed multi-layer finite element mesh for modeling deformation of a viscoelastic body [2]. Multi-rate integration could be used to compute the model. Zhang and Canny used the explicit time integration method for haptic rendering of large deformation [18]. D'Aulignac et al. used implicit integration in a particle-based method [5]. Wu et al. used explicit integration and adaptive meshing for simulating large deformations of non-Hookean materials [17]. Debunne et al. used adaptive multi-rate integration with multi-spatial resolution for simulating tool contact with viscoelastic materials. These approaches essentially simplify either the finite element method or the contact problem to make high rate computation of forces possible [6].

3   System for Tool Contact Simulation

Figure 1 defines a haptic engine system for a single tool. The inputs to the system are x^1, x^2, x^3, the position components of the virtual tool tip (or the end effector of the device) and $\theta^1, \theta^2, \theta^3$ the orientation components of the tool tip. The internal variables of the system are the position components c^1, c^2, c^3 of a contact point c. Output forces and torques are functions both of the inputs and of the internal variables. This representation of the haptic engine system assumes a quasi-static contact. It also assumes that the contact area between the tool and the body always contains the tool tip.

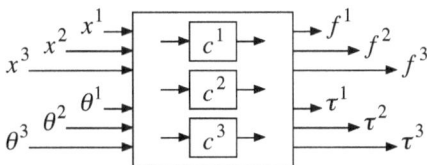

Fig. 1. Haptic engine.

In the following, we describe the calculation of output forces for a haptic engine. A similar approach can be used for the calculation of output torques. We ignore the dependency of contact forces on tool orientation. It could be included following exactly the same approach.

3.1   Normal Contact

The haptic engine receives $x(t)$, the position of tip of a virtual tool. The contact point c represents the position of the initial contact of the tool on the surface of the deformable body *at rest*. The deflection vector at point c is $\delta = x - c$, see Figure 2 and caption.

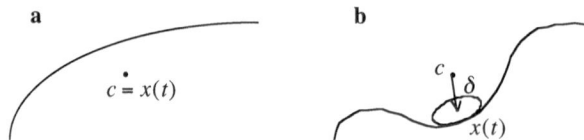

Fig. 2. (a) A tool initially contacts a body at point c. (b) With penetration, a contact surface is formed (shown as a circle). The known tool tip position x gives an approximation for the position of this surface. The deflection δ at point c, a point of initial contact on the surface of an undeformed object (in black outline), is found from the difference between the position of the contact surface and c. When the object deforms (in black outline), a force response is produced.

Nodes distributed on the surface of the body can be organized in a triangulation used to represent the geometry of the body at rest as a set of triangular elements. With the force response pre-calculation approach, the need to determine the nodal displacements online is eliminated.

Response of a Deformable Body. The force response to contact at each node l of element m is represented in local coordinates. Each node is labeled $x_{l,m}$. The unit vectors $u_{l,m}^z$ and $u_{l,m}^r$ with origin at $x_{l,m}$ are such that $\delta_{l,m} = \delta_{l,m}^z u_{l,m}^z + u_{l,m}^r \delta_{l,m}^r$, as shown in Figure 3a. The quantities $f_{l,m}^z(.)$ and $f_{l,m}^r(.)$ are the components of the force responses as in Figure 3b. The responses can be derived from actual force responses (pre-calculated or pre-measured with suitable approximations [14]) using piecewise polynomial interpolation. Given $x_{l,m}$ and x, the force response at node l of element m is:

$$f_{l,m}(\delta) = f_{l,m}^r(\delta_{l,m}^r)\, u_{l,m}^r + f_{l,m}^z(\delta_{l,m}^z)\, u_{l,m}^z. \tag{1}$$

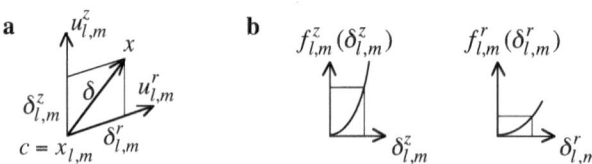

Fig. 3. (a) Deflection in local coordinates. (b) Response vector components.

Interpolated Response. When c is inside an element m, coordinates are defined at point c with a unit vector u_c^z found by interpolation of unit vectors at each node.

The unit vector u_c^r is such that $\delta = \delta_c^z u_c^z + \delta_c^r u_c^r$ as in Figure 4. Call $n_{l,m}(.)$ the interpolation function for node l of element m:

$$u_c^z = \sum_{l=1,2,3} n_{l,m}(c)\, u_{l,m}^z. \tag{2}$$

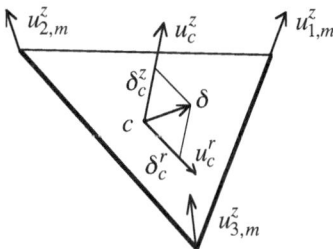

Fig. 4. Local coordinates at c. The normal vector u_c^z points inside the body.

We use a common interpolation scheme used in computational methods for continuous mechanics based on natural coordinates. Within one element of area A_m, these coordinates are given by the areas $A_{l,m}$ of the triangles defined by the contact point and the three vertices as in Figure 5a,

$$n_{l,m}(c) = \frac{A_{l,m}(c)}{A_m}. \tag{3}$$

Note that the vector u_c^z is normally distinct from the vector normal to the surface of the element.

A key property of the interpolation approach is to ensure continuity of coordinate change over the surface of the body. Referring to Figure 5b, the force-deflection response at c is obtained from:

$$f_c(\delta) = f_c^z(\delta_c^z)u_c^z + f_c^r(\delta_c^r)u_c^r, \tag{4}$$

where $f_c^z(.)$ and $f_c^r(.)$ are also interpolated from three pre-calculated force-deflection responses:

$$f_c^z(.) = \sum_{l=1,2,3} n_{l,m}(c)\, f_{l,m}^z(.), \tag{5}$$

$$f_c^r(.) = \sum_{l=1,2,3} n_{l,m}(c)\, f_{l,m}^r(.).$$

The haptic engine outputs are now defined for any initial contact point and tool tip position.

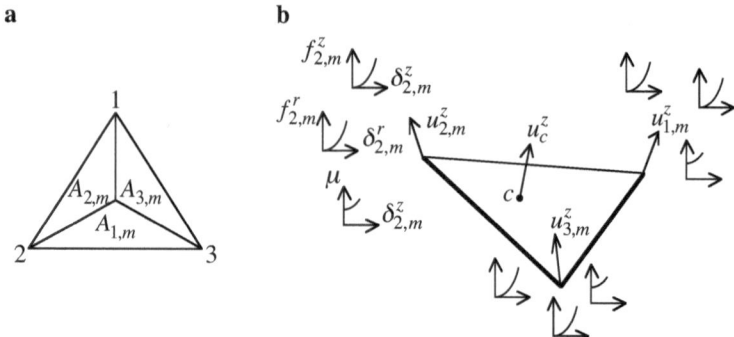

Fig. 5. (a) A point inside a triangle defines three areas labeled as indicated. (b) For each triangular element m, the vertices are associated to $u^z_{l,m}, l = 1, 2, 3$ and to two responses $f^z_{l,m}$ and $f^r_{l,m}$. Similarly, variations in the limiting coefficient of friction as a function of penetration can be encoded over the surface of the body.

3.2   Sliding Contact

Sliding behavior is represented in terms of pre-sliding displacements [10]. During sliding contact, c moves over the body boundary. Let δ^r_s be the limiting sliding deflection such that $f^r_c(\delta^r_s) = \mu(\delta^z_c) f^z_c(\delta^z_c)$, where $\mu(.)$ is the limiting friction coefficient given as function of normal penetration. The pre-sliding distance is not to be confused with the pre-sliding observed with hard materials in contact, which is measured in μm [1]. For soft materials, it can be large and depends mostly on the material properties and on its support. Consider, for example, the case of palpation. The skin can deform over large distances before slip occurs. Applying Coulomb's law is equivalent to assuming that μ is invariant with the contact surface, i.e., is independent from penetration.

In the pre-sliding model, sliding occurs when $\delta^r > \delta^r_s$ [10]. During sliding, c moves over the body surface in such manner that $\delta^r \leq \delta^r_s$ at all times. The movement is in the direction of the projection of u^r_c over the element surface and its magnitude is $|\delta^r_c - \delta^r_s|$, see Figure 6.

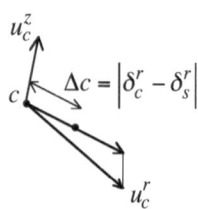

Fig. 6. Vector u^r_c is projected onto the element surface to define the direction of the movement of c. If δ^r_c reaches δ^r_s, c slides by $|\delta^r_c - \delta^r_s|$.

4 Construction of the Pre-calculated Responses

The pre-calculated force-deflections $f^z_{l,m}(\delta^z)$ and $f^r_{l,m}(\delta^r)$, as well as the friction properties described by $\mu(\delta^z)$, can be interpolated by piecewise linear or polynomial interpolators. The parameters of the interpolators are calculated such that the outputs of the haptic system fit either the actual forces obtained by experimental measurement or accurate forces generated by an offline simulator [14,9].

The location and number of testing points can differ from those of the surface nodes used to represent a virtual body. For example, Figure 7 illustrates how a mesh can be refined by obtaining the force response at a new node from pre-calculated forces at three test points, using an interpolation approach similar to the one of the last section. More generally, the number of the test points for a body can be much smaller than the number of virtual body surface nodes used for a simulation. This is useful when it is known that the change of force response over the surface of a body is smoother than, for example, the change in visual aspect or in the geometry of the body surface.

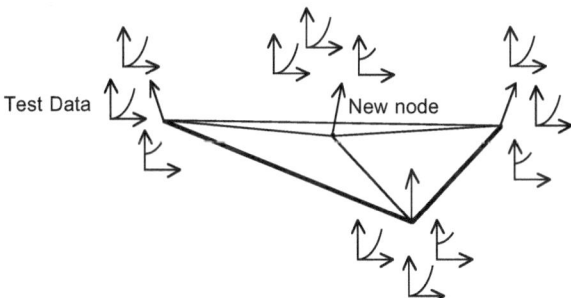

Fig. 7. Force Deflection responses for a new node of a mesh of a body surface is obtained by force-deflection responses of three test points using area interpolation.

5 Implementation

A simple haptic engine system was implemented to evaluate the performance of the interpolation approach. The virtual environment comprised a tool interacting with a deformable cylindrical virtual body made of about 500 patches. The local responses at a node l of element m were:

$$f^z_{l,m}(\delta^z) = \begin{cases} 0 & \delta^z \leq 0 \\ 100\,k^z_{l,m}\,\delta^{z2} & 0 < \delta^z \end{cases} \tag{6}$$

$$f^r_{l,m}(\delta^r) = k^r_{l,m}\,\delta^r \tag{7}$$

The vectors $u^z_{l,m}$ at each mode were in the radial direction.

The test showed that coordinates interpolation, together with force-deflection interpolation provided continuity for the responses. Conversely, eliminating one

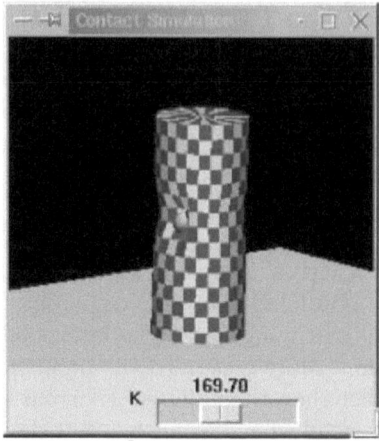

Fig. 8. Graphical user interface and device used in the tool contact simulation.

of the two interpolations caused a variety of artifacts, including unwanted textures or limit cycles resulting from lack of passivity during sliding motions. These could be attributed to discontinuities in deflection or in the produced force at the mesh edges. These unwanted artifacts were called haptic 'clicks' and 'pops' by analogy to the defects in sound production.

Figure 8 shows the user interface used in the test. The simulation program consisted of two independent real-time threads running under RTLinux-3. One thread provided for rendering the forces and the other for finding the active element. The forces were generated by a PenCat/Pro™ haptic device (Immersion Canada Inc.) which is "direct drive" and hence provides good fidelity because of near absence of mechanical damping.

6   Conclusion

This paper described a pre-calculation solution for haptic simulation of contact between a tool and a nonlinear deformable body. The approach relies on pre-calculated forces responses known at nodes of the triangulated body surface mesh. It calculates forces at arbitrary locations through interpolation of both force responses and coordinates. According to the needs and the application, piecewise linear or polynomial interpolators can be used to register normal and lateral forces at each node at each direction for a range of deflection distances. The pre-calculation responses can be derived from actual force responses resulted from measurement or accurate offline simulations. The measurements or simulations can be done only at a set of testing point located over the surface of the body and the results are extended to many nodes of the body surface by using area interpolation.

This approach could be generalized to cases that include damage and plasticity by adding additional internal variables. The case of cutting, a special case of damage, was treated in [13]. The systems approach also allows us to study stability of haptic interactions in a general way [15].

Acknowledgement

This research was funded by the project "Reality-based Modeling and Simulation of Physical Systems in Virtual Environments" of IRIS, the Institute for Robotics and Intelligent Systems (Canada's Network of Centers of Excellence). Additional funding is provided by NSERC, the Natural Sciences and Engineering Council of Canada, in the form of an operating grant for the second author.

References

1. Armstrong-Hélouvry, B., Dupont, P., and Canudas De Wit., C. 1994. A Survey of Models, Analysis Tools and Compensation Methods for the Control of Machines with Friction. *Automatica,* 30(7):1083–1138
2. Astley, O. R., Hayward, V. 1998. Multirate Haptic Simulation Achieved by Coupling Finite Element Meshes Through Norton Equivalents. Proc. *IEEE International Conference on Robotics and Automation,* pp. 989–994.
3. Astley, O., Hayward V. 2000. Design Constraints for Haptic Surgery Simulation. Proc. *IEEE International Conference on Robotics and Automation,* pp. 2446–2451.
4. Cotin, S., Delingette, H., Ayache, N. 1999. Real-time Elastic Deformations of Soft Tissues for Surgery Simulation. *IEEE Transactions on Visualization and Computer Graphics,* Vol. 5:1, pp. 62–73.
5. D'Aulignac, D., Balaniuk, R., Laugier, C. 2000. A Haptic Interface for a Virtual Exam of the Human Thigh. Proc. *IEEE International Conference on Robotics and Automation,* pp. 2452–2457.
6. Debunne, G., Desbrun, M., Cani, M., Barr, A. 2001. Dynamic Real-Time Deformations Using Space and Time Adaptive Sampling. *Computer Graphics and Interactive Techniques,* SIGGRAPH 2001, ACM Press, pp. 31–36.
7. Delingette, H., Cotin, S., Ayache, N. 1999. A Hybrid Elastic Model Allowing Real-time Cutting, Deformations and Force-feedback for Surgery Training and Simulation. *Computer Animation Proceedings,* pp. 70–81.
8. Frank, A. O., Twombly, A. I., Barth, T. J., Smith, J. D. 2001. Finite Element Methods for Real-Time Haptic Feedback of Soft-Tissue Models in Virtual Reality Simulators. Proc. of the *Virtual Reality Conference,* pp. 257–263.
9. Greenish, S., Hayward, V., Chial, V., Okamura, A., Steffen, T. 2002. Measurement, Analysis and Display of Haptic Signals During Surgical Cutting. *Presence: Teleoperators and Virtual Environments,* MIT Press. Vol. 6(11). pp. 626–651.
10. Hayward, V., Armstrong, B. 2000. A New Computational Model of Friction Applied to Haptic Rendering. In *Experimental Robotics VI*, P. I. Corke and J. Trevelyan (Eds.), Lecture Notes in Control and Information Sciences, Vol. 250, Springer-Verlag, pp. 403–412.
11. James, D. L., Pai, D. K. 1999. ArtDefo, Accurate Real Time Deformable Objects. *SIGGRAPH 99 Conference Proceedings,* pp. 65–72.
12. James, D. L., Pai D. K. 2001. A Unified Treatment of Elastostatic and Rigid Contact Simulation for Real Time Haptics. *Haptics-e, the Electronic Journal of Haptics Research,* Vol. 2, No. 1.

13. Mahvash, M., Hayward, V. 2000. Haptic Rendering of Cutting, A Fracture Mechanics Approach. *Haptics-e, the Electronic Journal of Haptics Research*, Vol. 2, No. 3.
14. Mahvash, M., Hayward, V., Lloyd, J. E. 2002. Haptic Rendering of Tool Contact. Proc. *Eurohaptics 2002*. pp. 110–115.
15. Mahvash, M., Hayward, 2003. Passivity-Based High-Fidelity Haptic Rendering of Contact. 2003. Proc. *IEEE International Conference on Robotics and Automation*, in print.
16. Bro-Nielsen, M. 1998. Finite Element Modeling in Surgery Simulation. *Proceedings of the IEEE*, 86:3, pp. 490–503.
17. Wu, X., Downes, M. S., Goktekin, T., Tendick, F. 2001. Adaptive Nonlinear Finite Elements for Deformable Body Simulation Using Dynamic Progressive Meshes. Proc. *Eurographics 2001*, Vol. 20, No. 3, pp. 349–58.
18. Zhuang, Y., Canny, J. 2000. Haptic Interaction with Global Deformations. Proc. *IEEE International Conference on Robotics and Automation*, pp. 2428–2433.

Patient-Specific Biomechanical Model of the Brain: Application to Parkinson's Disease Procedure

Olivier Clatz[1,*], Hervé Delingette[1], Eric Bardinet[1,2],
Didier Dormont[2], and Nicholas Ayache[1]

[1] Epidaure Research Project, INRIA Sophia Antipolis
2004 route des Lucioles, 06902 Sophia Antipolis, France
Olivier.Clatz@inria.fr
[2] Neuroradiology Dept and LENA UPR 640-CNRS La Pitié-Salpêtrière hospital
91, 105 boulevard de l'Hôpital 75013 Paris, France

Abstract. Stereotactic neurosurgery for Parkinson's disease consists of stimulating deep nuclei of the brain. Although target coordinates are calculated with high precision on the pre-operative images, cerebrospinal fluid (CSF) leakage during the procedure can lead to a brain deformation and cause potential error with respect to the surgical planning. In this paper, we propose a patient-specific biomechanical model of the brain able to recover the global deformation of the brain during this type of neurosurgical procedure. Such a model could be used to update the pre-operative planning and balance the mechanical effects of the intra-operative brain shift.

1 Introduction

It is usually the case in neurosurgery that pre-operative planning is based on the assumption that anatomical structures do not move between the image acquisition time and the operation time. In reality, the position of brain tissues changes during the operation and significantly decreases the accuracy of the planning made on the pre-operative image. Two approaches have been proposed to solve this problem. One uses intra-operative images (ultra-sound devices [7], stereo-vision systems [8], open-configuration magnetic resonance scanners [3]) to register the pre-operative images with the intra-operative ones. The second one attempts to predict the brain deformation. This approach can be either based on statistical models [2] or on biomechanical models [5].

The biomechanical approach we propose here is based on a finite element model (FEM) of the brain. It takes into account the patient specificity and anatomical cerebral structures to predict the brain deformation.

---

* Corresponding author

N. Ayache and H. Delingette (Eds.): IS4TM 2003, LNCS 2673, pp. 321–331, 2003.
© Springer-Verlag Berlin Heidelberg 2003

2 Stereotactic Neurosurgery for Parkinson's Disease

2.1 Overview of the Surgical Procedure

The Parkinson's disease procedure consists of the deep implantation of electrodes to stimulate the subthalamic nuclei in the brain. It recently renewed the interest for functional stereotactic surgery.

This method uses a metallic non-magnetic stereotactic frame visible on the different imaging modalities (T1, T2, scanner) as a geometrical referential fixed on the patient's head. The frame serves as a guide for the electrodes paths during the procedure. After locating the target nuclei on pre-operative images, the path of electrodes is planned through the parenchyma, avoiding crossing critical structures like sulci, blood vessels or ventricles. We notice that no consensus quo has been reached yet in the definition of these critical structures. The surgical procedure starts the day after the planning. The surgeon first makes two holes in the skull for

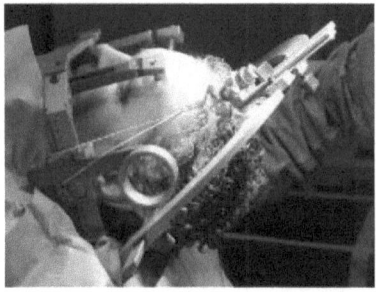

Fig. 1. Stereotactic operation at La Pitié Salpêtrière hospital (courtesy of Medtronic Inc).

accessing the two hemispheres but perform the electrodes implantation one after the other. Because all images in this study come from a single surgeon, the hemispheres are treated in the same order for every patient. After the dura mater opening, the surrounding zone around the target nuclei is explored in a $\pm 5mm$ zone around the pre-operative located position by five different electrodes testing each position to determine the best stimulating location. However, depending on the intervention duration (6 to 10 hours) and on the dura-mater opening size, the CSF leakage induces a brain deformation which might compromise the localization of the targeted nucleus.

2.2 The Data

This study is based on the analysis of 7 cases of Parkinson's disease treated at La Pitié Salpétrière hospital in Paris. The images considered for this study are T1 weighted MRIs (IR-FSPGR). The 7 cases are staggered over two years between 1999 and 2001. All patients have been operated following the same protocol: the pre-operative images and planning are performed the day before the procedure while the procedure starts early in the morning, following the previously described protocol, and in the same position. The post-operative MRI images are acquired in the morning of the day after the procedure. The patient is kept lying down until the post-operative image acquisition. (figure 2).

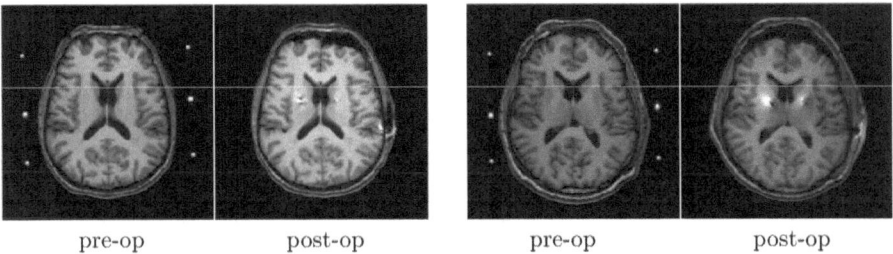

pre-op   post-op   pre-op   post-op

Fig. 2. Two cases extracted from the seven pre and post operative T1 weighted MRI used for this study.

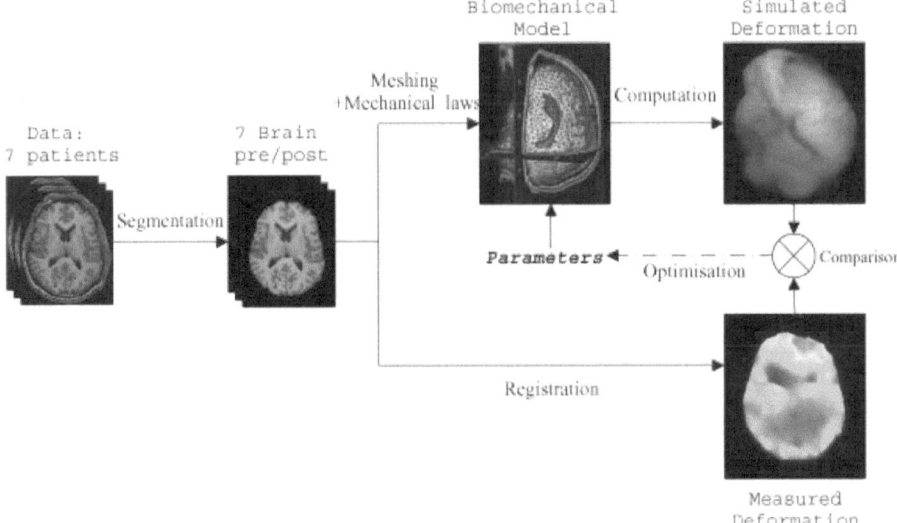

Fig. 3. Overview of the proposed method.

3   Methods

The model we propose is based on a retrospective study of the deformation measured on the 7 patients, rigidly registered on an atlas in a pre-treatment stage (see [6] for the rigid registration algorithm). Figure 3 presents an overview of the proposed method. The model creation can be decomposed in different steps:

- brain segmentation
- patient pre to post-operative deformation generation
- deformation analysis and characteristic features extraction
- patient specific biomechanical model building
- model parameters optimisation

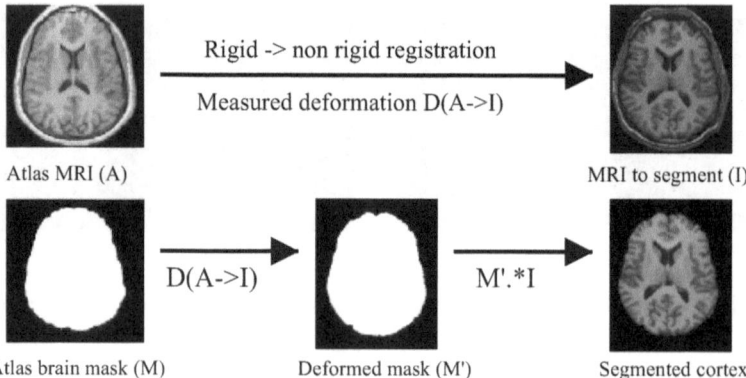

Fig. 4. Brain segmentation method.

3.1   Displacement Field Extraction from the Non-rigid Registration

Brain Parenchyma Segmentation. We developed a simple automatic method based on the registration of a digital atlas[1] on the pre/post operative MRI. The overall method is presented in figure 4.

To optimize the precision of the registration procedure, we split it into different stages, with increasing complexity, initializing each displacement field by the previously computed one:

$$\text{similarity } [6] \implies \text{affine } [6] \implies \text{free-form } [1]$$

Finally, we used the final free-form displacement field obtained from the registration software to deform the brain mask.

Building an Average Displacement Field. In order to analyze the deformations that occur in the brain, we compute the non rigid displacement field between each pre and post-operative MRI from the 7 cases of Parkinsonians, already rigidly registered in the atlas geometry. The average displacement field is then built by averaging these 7 displacement fields (figure 5).

3.2   Characteristic Feature Extraction

Displacement Analysis. One can, from these first displacement fields generated, put forward different phenomena:

- assuming an horizontal position of the patient in the MRI scanner, we computed the average displacement direction. This one is mostly aligned with the gravity (average direction difference = 17°).

---

[1] http://splweb.bwh.harvard.edu:8000/pages/papers/atlas/

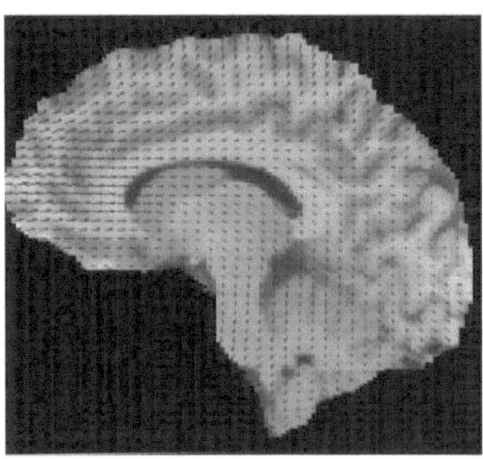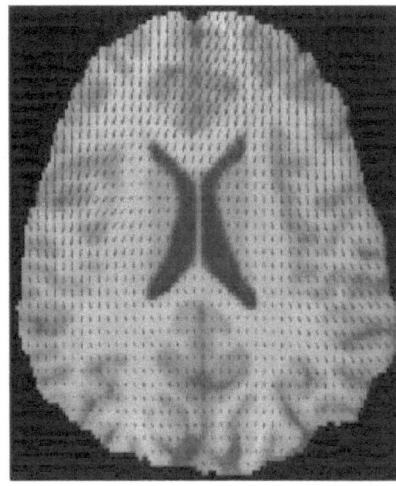

Fig. 5. Average displacement field computed over the 7 cases, affine registered on the atlas brain.

– we observed for almost every cases an important asymmetry in the displacement field of the two hemispheres (also distinguishable on the pre and post-operative MRIs, see figure 2).

The metallic nature of the electrodes introduces very important artifacts on the computed displacement fields, so that we cannot make any deformation analysis on this area.

Deformation Analysis. Let Φ denote the relationship which associates the new position vector of a material point $\mathbf{X}$ to its rest position after transformation in a fixed reference. $\mathbf{U}(\mathbf{X})$ is the displacement function.

$$\Phi(\mathbf{X}) = \mathbf{X} + \mathbf{U}(\mathbf{X}).$$

We can then compute the image (figure 6) which represents the local volume variation (defined by the ratio of the local volume variation by the initial volume):

$$\frac{d\Omega - d\Omega_0}{d\Omega_0} = det(\nabla\Phi(\mathbf{X})) - 1 = det(\mathbf{Id} + \nabla\mathbf{U}(\mathbf{X})) - 1$$

This new image (6) revealed an incompressible behavior for the brain parenchyma, with a compression rate comprised between +5% and -5%, except in the electrode area where artifacts induced large unrealistic compression rate values (over 15%).

4   The Biomechanical Model

To build a physically-based deformable model, one must follow different steps. After the generation of a tetrahedral mesh (4.1), we have to choose a biome-

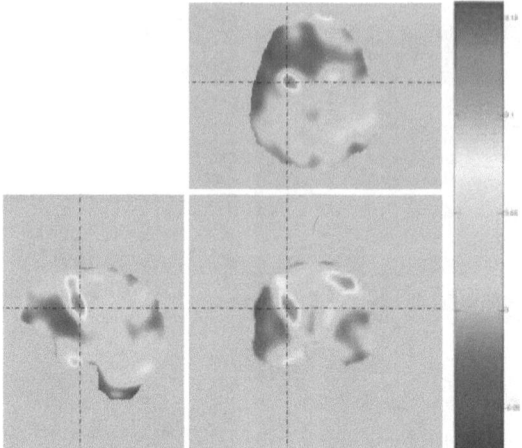

Fig. 6. Local deformation rate: $\frac{d\Omega - d\Omega_0}{d\Omega_0}$.

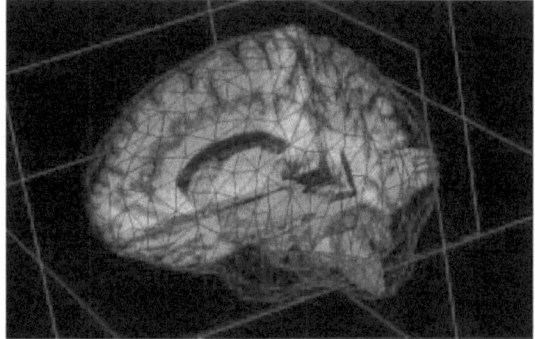

Fig. 7. Final tetrahedral mesh of the brain.

chanical constitutive behavior law for the brain material, based on continuum mechanics laws (4.2). The next step is boundary conditions determination, depending on the observed behavior (4.3). Last step is the computation of the solution (4.4).

4.1 Mesh Generation

We use the masks of the brain presented in section 3.1 to generate the volumetric mesh (figure 7) with the package GHS3D developped at INRIA Rocquencourt[2]. This software minimizes the shape variability between all tetrahedra in the final mesh. We limited the number of tetrahedra in the model to 8000, in order to keep the computation time under two minutes. The proposed method allows us to automatically create a patient-specific mesh of the cortex (figure 7).

---

[2] http://www-rocq1.inria.fr/gamma/index.html

4.2   Material Parameters

We propose a model for brain tissue material which takes into account the observation made on the deformation:

1. We assume a brain tissue material almost incompressible, which is consistent with the known mechanical properties of the brain parenchyma [4] and with the deformation analysis made in section 3.1. We have thus fixed its Poisson ratio to 0.45.
2. With no a-priori information about the brain tissue stiffness, we used the Young Modulus measured in Karol Miller experiments (See [4] for more details) $E = 2000Pa$. We can then compute the Lamé constants:

$$\mu = 689 \quad \lambda = 6200$$

3. In a first stage, we assumed an isotropic material, with linear elasticity behavior law. We wish to insist on the fact that this kind of behavior law does not aim to model a dynamic behavior (like viscoelastic) but only the static effects of an external loading.

4.3   Boundary Conditions

With respect to the displacement field obtained in section 1 we propose a new model including the brain anatomy and the mechanics of static fluids.

1. We consider that a Cerebro Spinal Fluid leak leads to an equivalent liquid level in the skull. The brain part under this liquid level is then subject to the fluid force applied on the brain surface and the gravity:
   - The local volumetric force induced by gravity is:

$$\mathbf{F}_{Gravity} = - \int_{Brain\ Volume} \rho\, \mathbf{g}\, dV$$

   ρ is the mass density of the brain ($\simeq 1Kg.dm^{-3}$)
   $\mathbf{g}$ is the gravity vector ($|\mathbf{g}| = 9.81 N.Kg^{-1}$)
   - Considering no friction between the brain and the skull, the force of the fluid applied o the brain is:

$$\mathbf{F}_{Fluid} = \iint_{Brain\ Surface} P_{fb}\, dS = \iiint_{Brain\ Volume} grad(P_{fb})\, dV$$

   P_{fb} is the pressure of fluid on the brain
   The explicit pressure law in a fluid is given by:

$$P_{fb} = \rho g h + P_0$$

   h is the distance to the liquid level
   P_0 is the air pressure

$$\text{Then:}\quad \mathbf{F}_{Fluid} = \iiint_{Brain\ Volume} \rho g\, dV$$

Therefore, if no friction exists between the brain and the skull, we can consider that the resultant force applied to the part of the brain under the liquid level is null.

2. The part of the brain which is over the liquid level is subject to gravity only ($\int P_0$ is balanced over the total surface of the emerged part). On each tetrahedron's vertex, we have then applied the force:

$$\mathbf{F} = \Sigma_{Adjacent\ Tetrahedron}(-\iiint_{Tetrahedron\ Volume} \frac{\rho\, g\, dV}{4})$$

3. To take into account the asymmetry in the computed displacement field, we propose to separate the hydraulic behavior of each hemisphere. This hypothesis is based on the anatomy (the falx cerebri). It is also based on the fact that during a this type of neuro-surgical procedure, the two hemispheres are treated one after the other. We have then a different CSF liquid level for each hemisphere leading to two different volumetric forces applied to each hemisphere.

4. To balance the force moment due to the difference of forces between both sides, we implicitly modeled the falx cerebri, allowing vertices belonging to tetrahedron intersecting the mid-sagittal plane to slide along this plane.

5. Finally, we fixed vertices at the basis of occipital lobe (on 10 % of the total height) since displacement analysis showed no displacement for this part of the brain.

4.4   Numerical Solving of the Mechanical Problem

Once boundary conditions have been applied, we have to solve the linear system:

$$[\mathbf{K}]\, \mathbf{U} = \mathbf{F}$$

[$\mathbf{K}$] Stiffness matrix
$\mathbf{U}$ Displacement vector
$\mathbf{F}$ External force vector

We do not explicitly compute the stiffness matrix inverse $[\mathbf{K}]^{-1}$. Due to the sparseness of $[\mathbf{K}]$, we precondition the $[\mathbf{K}]$ matrix with incomplete Cholesky method using tools of the Matrix Template Library[3]. The final solution is obtained with the conjugate gradient method of the Iterative Template Library[4].

5   Results

Matching Criterion. To estimate the quality of the prediction, we need to have an error criterion sensitive not only to voxel-value error, but also to displacement error over the brain. We propose to compute the sum of squared displacements over the brain obtained by registering the predicted MRI with the post-operative MRI:

---

[3] http://www.osl.iu.edu/research/mtl/
[4] http://www.osl.iu.edu/research/itl

$$Error = \iiint_{Brain\ Volume} \left| \overrightarrow{u} \right|^2 .dV$$

where $\overrightarrow{u}$ is the measured displacement.

This criterion, more than a precise estimate of the real displacement error, is used as an energy function we minimize to adjust our model's parameters which best correspond to the target post-operative image.

Model Performance Evaluation. We performed our model tests in two stages: first, we optimized an average CSF liquid level in the skull with respect to our matching criterion (the optimum value is 18.5%), then we computed the two different CSF liquid levels, initializing the two levels with the optimal average one.

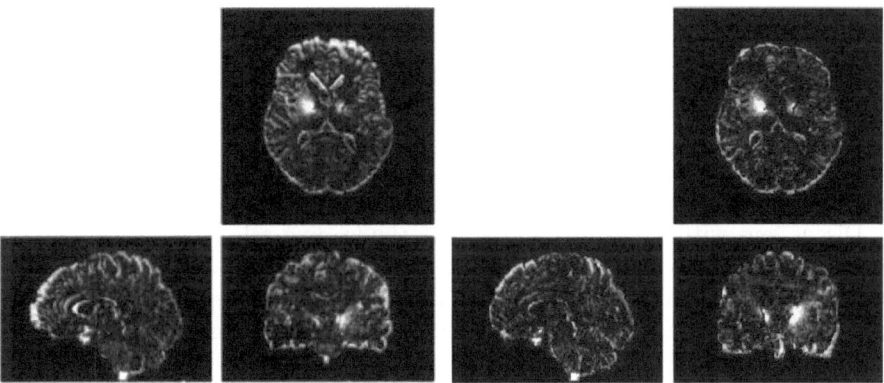

Difference pre / post-operative MRI   Difference predicted / post-operative MRI.

Fig. 8. Comparison of the difference images with and without prediction (patient with the largest brain-shift).

Figure 8 presents the absolute difference of pre and post-operative MRI compared to the absolute difference of predicted and post-operative MRI. The predicted MRI have been computed with the two CSF liquid levels that minimize the global error with the post operative MRI. We can verify that our model is able to balance the brain shift phenomenon as well as the asymmetry. Nevertheless, the model does not seem to reduce the difference residue on the sulcus area. This problem is, in our opinion, due to the homogeneous material model proposed for the full brain and maybe also to displacement boundary conditions imposed at the base of the occipital lobe.

The observation of the image difference allows us to evaluate the quality of the biomechanical prediction:

– The model relates well to asymmetrical deformation.
– The gravity-induced deformation is quantitatively well modeled by the physically-based biomechanical model we propose.

- As previously seen, the measured displacement is very disturbed in the electrode area. We thus have no way to estimate the accuracy of the displacement in this area.
- We observed a general error increase from the center of the brain to its surface, which might originate from the incompressibility constraint.
- Finally, we can see on the sagittal view of the right block of figure 8 that the predicted displacement in the pre-motor area of the frontal lobe is significantly higher than the real one. We therefore propose to prevent the brain to move too far in this direction by modeling the **contact with the skull** in our model.

6   Discussion

This paper demonstrates the ability of a biomechanical model to predict gravity-induced deformations of the brain. The model shows good results for the global displacement and the asymmetrical effects for both brain surface and ventricles. However it still has some weaknesses, especially for the displacement of the frontal lobe. Actually, this displacement observation reveals that the brain collides with the skull in the pre-motor area of the frontal lobe. We thus need to improve our model with a brain/skull contact model.

Discussions with surgeons also revealed the leading role of physiological parameters on the mechanical behavior of the brain. Effects of the mannitol (a drug administered to the patient during the surgery), could explain for instance some of the errors measured on the brain surface. Thus drug effects might balance the mechanical incompressibility constraint. These kind of physiological reaction need to be taken into account in future models.

Using a physics-based model seems to be well suited to explaining the real deformations measured on pre-operative and post-operative MRIs. Although the model cannot predict the entire deformation yet, we believe these are encouraging preliminary results. Further work will include other anatomical structures to simulate more complicated procedures. Combining anatomical and physiological constraints will help understanding and simulating the brain deformation in the future.

References

1. P. Cachier, E. Bardinet, D. Dormont, X. Pennec, and N. Ayache. Iconic Feature Based Nonrigid Registration: The PASHA Algorithm. *CVIU — Special Issue on Nonrigid Registration*, 2003. In Press.
2. C. Davatzikos, D. Shen, A. Mohamed, and S. Kyriacou. A framework for predictive modeling of anatomical deformations. *IEEE Trans. on Med. Imaging*, 20(8):836–843, 2001.
3. Matthieu Ferrant, Arya Nabavi, Benoît Macq, Black P.M., Ferenc A., Jolesz, Ron Kikinis, and Simon K. Warfield. Serial registration of intraoperative mr images of the brain. *Medical Image Analysis*, 6(4):337–360, 2002.

4. Miller K. *Biomechanics of Brain for Computer Integrated Surgery.* Warsaw University of Technology Publishing House, 2002.
5. M. I. Miga, K. D. Paulsen, J. M. Lemry, F. E. Kennedy, S. D. Eisner, A. Hartov, and D. W. Roberts. Model-updated image guidance: Initial clinical experience with gravity-induced brain deformation. *IEEE Trans. on Med. Imaging*, 18(10):866–874, 1999.
6. A. Roche, A. Guimond, N. Ayache, and J. Meunier. Multimodal Elastic Matching of Brain Images. In *Computer Vision - ECCV 2000*, volume 1843 of *LNCS*, pages 511–527, Dublin, Irlande, June 2000. Springer Verlag.
7. A. Roche, X. Pennec, G. Malandain, and N. Ayache. Rigid registration of 3D ultrasound with MR images: a new approach combining intensity and gradient information. *IEEE Transactions on Medical Imaging*, 20(10):1038–1049, October 2001.
8. Oskar Skrinjar, Arya Nabavi, and James Duncan. Model-driven brain shift compensation. *Medical Image Analysis*, 6(4):361–374, 2002.

Volume Modeling of Myocard Deformation with a Spring Mass System

Matthias B. Mohr[1], Leonhard G. Blümcke[2], Gunnar Seemann[1],
Frank B. Sachse[1], and Olaf Dössel[1]

[1] Institut für Biomedizinische Technik, Universität Karlsruhe (TH), Germany
Matthias.Mohr@ibt.uni-karlsruhe.de,
[2] med3D GmbH, Heidelberg, Germany

Abstract. The deformation of myocard takes a vital part in the pumping function of the heart muscle. Knowledge of myocard anatomy and physiology makes it possible to create models of cardiac behavior. These models can be used for surgery planning, educational and research purposes. Simulations can be performed, which are beyond the capability of physical experiments. Volume models of myocard are necessary for realistic simulations. The deformation model presented in this work is based on a mass spring system parameterized by a continuum mechanics deformation model. Electrophysiology and excitation propagation of myocard were simulated and the resulting force development was used as input for the deformation models. Electromechanical coupled simulations with simple myocard geometries were carried out to show the capability of the deformation model in reconstruction of cardiac deformation.

1 Introduction

The heart's pumping function is dependent on its contracting muscle cells. Knowledge about muscle cells leads to a better understanding of functional and pathological states of the heart. With computer aided simulations of the heart, experiments with new treatments can be carried out and the results can be evaluated. Simulations can also be utilized for preparing surgeries and for educational purposes.

In this work an anatomical model of the heart is used as reference for the creation of myocard patches. Anatomical and physical properties are implemented, e.g. fiber orientation and electrical conductivity . Models of cellular electrophysiology and excitation propagation are linked with force development models. The combination of models describing the electrophysiological properties of the heart is base for the mechanical models (Fig. 1). The force development is input for the spring mass system to calculate the deformation. This paper describes the basics of electrophysical and force development models. The paper focuses on the creation of a three dimensional elastomechanic myocard model on the base of a mass spring system and the deformation simulation.

N. Ayache and H. Delingette (Eds.): IS4TM 2003, LNCS 2673, pp. 332–339, 2003.

2   Modeling

Electrophysiological Modeling. A myocard cell consists of various cell components. The membrane is the separator of intra- and extracellular space. Various channels penetrate the membrane for ion transfer and metabolism purposes. Passive ion channels as well as ion pumps and exchangers are found, e.g. for calcium and potassium. An interconnectivity exists between single cells via gap junctions. The ion concentrations inside and outside the cell result in a transmembrane voltage.

Models of the electrophysiology describes e.g. the state of ion channels and concentration of ions by coupled nonlinear differential equations. Electrophysiological cell models were first introduced by Hodgkin et al. [1].

In this work simulations were done with an extended cell model proposed by Noble et al. [2]. The transmembrane potential and the ion concentration change were calculated. Of special interest is the intracellular calcium concentration as the parameter for the electromechanical coupling.

Ion currents through gap junctions and extracellular currents can result in a propagation of excitation to neighboring cells. This phenomenon can be described by the bidomain model [3]. Two domains are specified: One is the extracellular and the other the intracellular space. A Poisson equation describes the cellular potential for each domain. The equations define e.g. the stimulus current for the adjoining cells.

Force Development. A myocyt consists of nucleus, mitochondria, myofibrils, sarcoplasmatic reticulum, cytoskeleton and further components. The myofibrils are composed of myofilaments formed by actin and myosin filaments.

The myofibrils are arranged in the sarcomere and responsible for mechanical force development. The electromechanical coupling is dependent on the intracellular calcium concentration [4]. Calcium initiates the sliding of actin and myosin filaments under consumption of adenosine triphosphate (ATP) and therefore the contraction.

Force development models describe e.g. the binding probabilities of calcium to troponin, the state of tropomyosin and the state of the myofilaments actin and myosin with the transition rates from force to none force generating states. In this work force development was calculated by solving differential equations describing the system [5].

Elastomechanical Modeling. The mechanical properties of myocard, i.e. strain-stress relationship taking anisotropy and isovolumetry into account, have to be modeled. The calculation of deformation of myocard cells can be implemented in two ways: As a continuum mechanical or as a mass spring system. The continuum mechanical model describes the myocard with the equation of motion of a continuous medium in the total Lagrangian formulation [6,7]. The mass spring system applies pointwise discretization of the spatial domain. Both models implement the anisotropy and isovolumetry of myocard. The spring mass system was derived from Bourgouignon et al. [8].

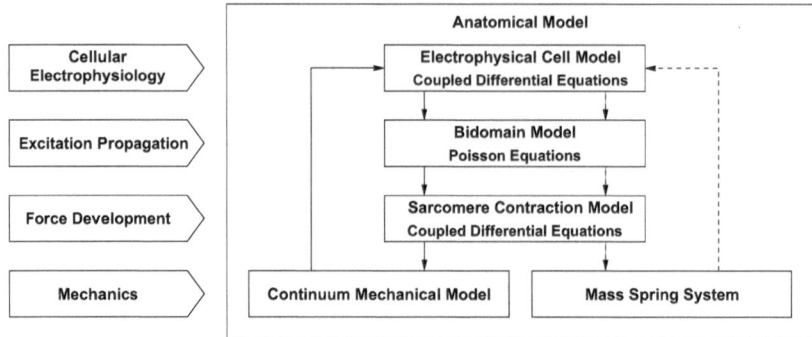

Fig. 1. Overview of myocard deformation simulation. The solid line shows the simulation with the continuum mechanical deformation model. The dashed line represents the simulation with the mass spring system. The parameterization of spring constants was performed by comparing the results of the two deformation models.

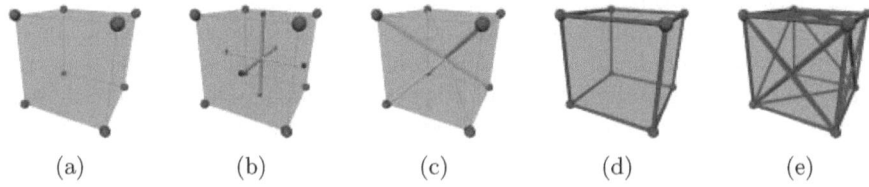

Fig. 2. A cubic voxel of biological tissue is modeled with masses and springs. Masses are denoted as spheres at the corner of a voxel (a). Anisotropy is modeled with 3 springs, displayed as cylinders and are fixed to the center of the voxel. They describe the fiber, sheet and the sheet normal orientation (b). Isovolumetry is modeled by 8 springs running from the center of the voxel to the corner masses (c). The classical structural springs are displayed in (d) and the classical surface springs in (e).

This paper presents an extended mass spring system to simulate the deformation of myocard. The spring constants were parameterized by comparison of deformation simulations with the continuum model.

3   Mass Spring Model

A given geometry was represented by cubic voxels, which are modeled by masses and springs. For simplicity only one cubic voxel is described (Fig. 2).

Masses. The mass of a voxel is split evenly into eight parts, which are assigned to voxels' vertices (Fig. 2(a)). Adjoining voxels contribute to the mass by summation.

Anisotropy. The first set of three springs denote the fiber, the sheet and sheet normal direction (Fig. 2(b)). Their intersection points with the voxel planes $\mathbf{x_P}$ (Fig. 3) are calculated with the bilinear form function:

$$\mathbf{x_P} = \zeta \eta \, \mathbf{x_A} + (1 - \zeta) \, \eta \, \mathbf{x_B} + (1 - \zeta) \, (1 - \eta) \, \mathbf{x_C} + \zeta \, (1 - \eta) \, \mathbf{x_D}$$

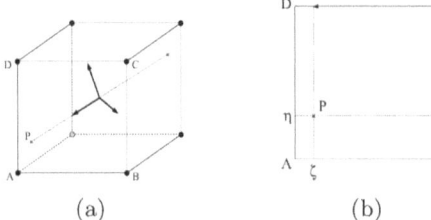

(a)   (b)

Fig. 3. Myocard voxel with 8 corner masses. The directions of fiber, sheet and sheet normal are set up at the barycenter and intersect the voxel planes (a). The position of an intersection P is described by bilinear form factors ζ and η. As an example the front plane of the voxel is extracted (b).

Here, $\mathbf{x_A}$, $\mathbf{x_B}$, $\mathbf{x_C}$ and $\mathbf{x_D}$ denote the vectors describing the position of corner masses of the given voxel and ζ and η the form factors. The force $\mathbf{f_P}$ acting on an intersection point is distributed to the neighboring masses by the following equations:

$$\mathbf{f_A} = \zeta\eta\,\mathbf{f_P}, \; \mathbf{f_B} = (1-\zeta)\,\eta\,\mathbf{f_P},$$
$$\mathbf{f_C} = (1-\zeta)\,(1-\eta)\,\mathbf{f_P}, \; \mathbf{f_D} = \zeta\,(1-\eta)\,\mathbf{f_P}. \tag{1}$$

The forces $\mathbf{f_A}$, $\mathbf{f_B}$, $\mathbf{f_C}$ and $\mathbf{f_D}$ are forces distributed to the corner masses (Fig. 3).

Angular Springs. The angles between anisotropic springs are kept nearly constant by using of three further springs called angular springs [8].

Isovolumetry. Different approaches can be distinguished form model of isovolumetry. An approach described by Bourguignon et al. [8] can be seen in Fig.2(c). A spring is constructed between the voxel center (barycenter) and each corner mass. For simple geometries the volume is approximately conserved during deformations. The determination of spring constants is time consuming and for complex geometries only partly useful. In the modeling process these passive springs were converted to active springs by the following means.

In this work the approach is extended using continuum mechanics foundation. The volume of the deformed voxel is calculated by establishing the deformation gradient tensor $\mathbf{F}$.

$$\mathbf{F} = \frac{\partial \mathbf{x}}{\partial \mathbf{X}}$$

It describes the transformation of a line element $d\mathbf{X}$ from the reference $\mathbf{X}(t)$ to a line element $d\mathbf{x}$ in the momentary configuration $\mathbf{x}(t)$ [9]. The tensor $\mathbf{F}$ can be set up by knowing the initial and the momentary position of the eight corner masses of a voxel. The momentary volume V and the initial volume V_0 is calculated.

$$V = V_0 \det \mathbf{F}, \qquad \Delta V = V - V_0 = V_0 \left(\det \mathbf{F} - 1\right).$$

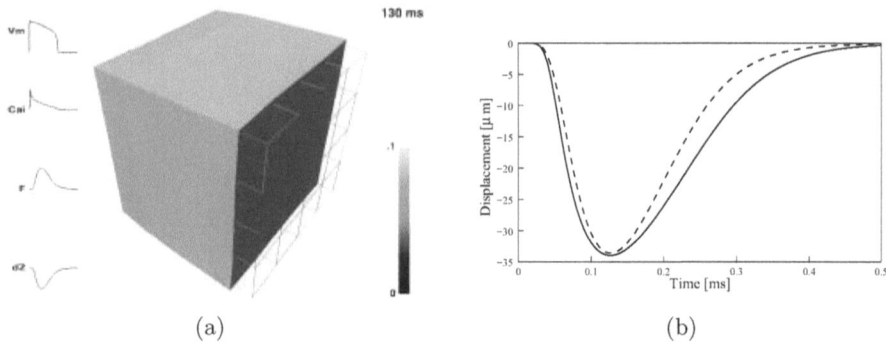

Fig. 4. Myocard patch with 4^3 voxels and fiber orientation in Z direction. At 130 ms the largest deformation is visible (a). The scales describe the transmembrane voltage V_m, the intracellular calcium concentration Ca_i, the force development F and the displacement in Z direction over time. The wireframe shows the relaxed model. The force acting on the model is gray value coded. Displacements of the point with coordinates (4,4,4) were calculated with the continuum model (solid line) and the mass spring system (dashed line) aiming at parameterization of spring constants (b).

The change in volume ΔV is scaled with the factor k_s and issued as additional volume preserving forces $\mathbf{i}$ to the corner masses $\mathbf{x_i}$.

$$\mathbf{f_i} = -k_s \Delta V \frac{\mathbf{l}}{\|\mathbf{l}\|}, \qquad \frac{\mathbf{l}}{\|\mathbf{l}\|} = \frac{\mathbf{x_i} - \mathbf{x_s}}{\|\mathbf{x_i} - \mathbf{x_s}\|}$$

Therefore a positive change in volume results in a scaled volume force $\mathbf{f_i}$ towards the barycenter. The vector of the barycenter is $\mathbf{x_s}$. A negative change in volume results in a force from the barycenter to the corner.

4   Simulation and Parameterization

For dynamic simulations the presented models were combined to an electromechanical model. The simulation was performed starting with the electrophysical models followed by the mechanical models. Therefore the spring mass system was created and the linear spring parameters were set. Then internal iterations were done as follows:

The prior calculated tension for the time step Δt was issued to the end points of the fiber springs. The contribution of each shortened or lengthened spring was calculated and distributed to the masses. The resulting force vector of a corner mass distinguished its displacement direction by Newtons law $\mathbf{F} = m\mathbf{a}$. The mass was displaced according to the internal time step. The change in volume was determined and the volume preserving forces were distributed.

The number of internal iterations was the decisive factor for simulation time. This number had to be sufficiently large to allow for spring relaxation. The deformation at time step Δt was used by the electrophysiologic and force development model.

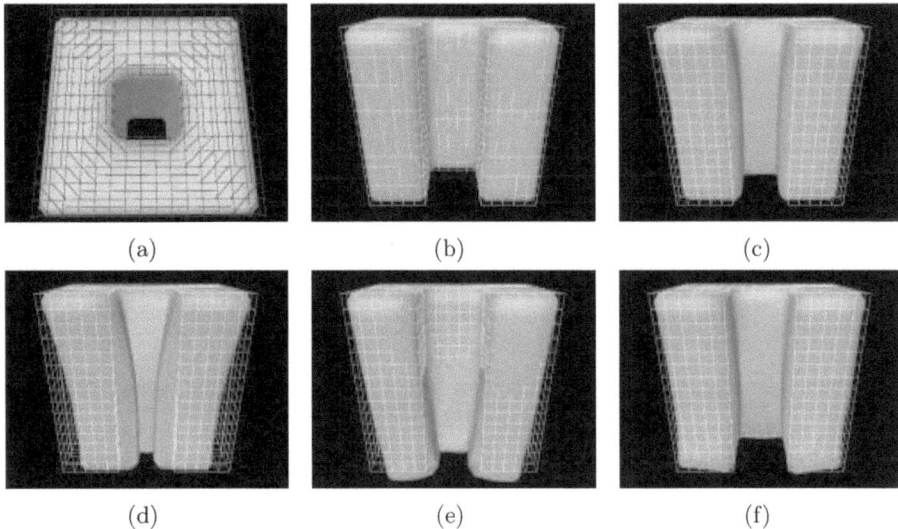

(a)   (b)   (c)

(d)   (e)   (f)

Fig. 5. Simulations with a tube like structure. A circular fiber orientation was con-structed in each plane to test anisotropic behavior. The black lines indicate the fiber orientation (a). A lateral cut of the model is shown in relaxed state with no force act-ing (b). The following figures display the deformation for rising force (c), at maximum force peak (d), (e) and (f) for the fading force.

Determination of Spring Parameters. A myocard patch of $4\times4\times4$ voxels was used. Each voxel had a lenght of 100 μm. The fiber orientation was set to Z direction for all voxels and the plane $Z = 0$ fixed. The simulation of electro-physiology and tension development was performed. The deformation simulation was done with both the continuum mechanical and the mass spring model. The total deformation was compared and the spring parameters adapted accordingly (Fig. 4(b)). A simulation with the mass spring model of the myocard patch at maximal contraction is displayed (Fig. 4(a)).

5   Results

Anisotropic Simulation. The parameterized springs were used to create a my-ocard patch of $15\times15\times15$ voxels with a volume of $15^3 mm^3$ to do simulations with anisotropic fiber orientation. In the middle a free space of $5\times5\times5$ voxels was introduced to visualize effects for a tube like structure (Fig. 5). A circular fiber orientation was constructed in each Z plane (Fig. 5(a)) and the top plane was fixed. A linear ramp function from 0 to 370 kPa was used for applied tension and all voxels were subject to this tension at the same time. Hereby, electrophys-iology was not calculated. The spring mass system was created with all above explained springs. Volume preservation was achieved with an average of 75%.

Anisotropic Simulation with Realistic Fiber Orientation. Simulations were ex-tended to a simplified ventricle model. A cylinder was rendered in a lattice of

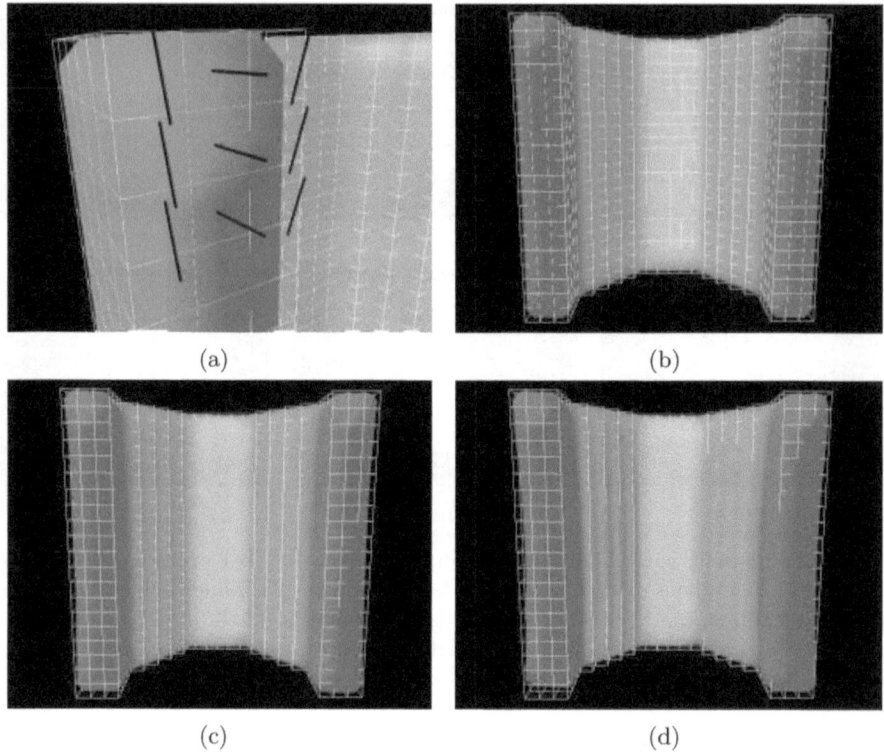

Fig. 6. Simulations with a cylinder model. A closeup of the cylinder wall shows the fiber orientation (a). The figures denote the deformation of the model from rest position (b), through force rising (c) to maximal force peak (d). The white wireframe shows the relaxed model.

$21 \times 21 \times 21$ voxels with a wall thickness of 3 voxels (Fig. 6(b)). The fiber orientation was set according to anatomical studies (Fig. 6(a)).

The top plane was fixed and a stimulus was set at the lowest layer. The active models were calculated with the presented geometry including mechanoelectrical feedback. The resulting force was used with the mass spring model to calculate the deformation (Fig. 6).

6   Discussion

An anisotropic mass spring system for description of cardiac deformation was presented, which was parameterized by continuum mechanics techniques. Simulations of electrophysiology resulted in force development, which provided input for elastomechanical simulations. The process was coupled to feed mechanical deformation back to the electrophysical and force development model. The simulations with the presented mass spring model showed, that a deformation of myocard patches can be realistically described.

The used springs were linear. The results for simple geometries compared with the continuum model were satisfactory. The active spring applying the deformation gradient tensor **F** supported isovolumetry. The additional force made it necessary to implement the classical springs (Fig. 2(d), 2(e)) to be able to preserve continuity of the voxel. This led to an increase of springs per voxels and therefore an increase in calculation time. Furthermore, the evaluation of volume preservation of the spring mass system showed deficiencies.

7 Future Work

Future work will be done concerning reduction of calculation time aiming at deformation simulation of more complex ventricle models. Nonlinear springs were already implemented and their effect in simulations will be tested. The volume preservation will be enhanced and the software will be extended for parallel computing.

References

1. A. L. Hodgkin and A. F. Huxley, "A quantitative description of membrane current and its application to conduction and excitation in nerve," *J. Physiol*, vol. 177, pp. 500–544, 1952.
2. D. Noble, A. Varghese, P. Kohl, and P. Noble, "Improved guinea-pig ventricular cell model incorporating a diadic space, I_{Kr} and I_{Ks}, and length- and tension-dependend processes," *Can. J. Cardiol.*, vol. 14, pp. 123–134, Jan. 1998.
3. C. S. Henriquez, A. L. Muzikant, and C. K. Smoak, "Anisotropy, fiber curvature and bath loading effects on activation in thin and thick cardiac tissue preparations: Simulations in a three-dimensional bidomain model," *J. Cardiovascular Electrophysiology*, vol. 7, pp. 424–444, May 1996.
4. A. F. Huxley, "Muscle structures and theories of contraction," *Prog. Biophys. and biophys. Chemistry*, vol. 7, pp. 255–318, 1957.
5. J. J. Rice, R. L. Winslow, and W. C. Hunter, "Comparison of putative cooperative mechanisms in cardiac muscle: length dependence and dynamic responses," *Am. J. Physiol. Circ. Heart.*, vol. 276, pp. H1734–H1754, 1999.
6. F. B. Sachse, G. Seemann, and C. D. Werner, "Combining the electrical and mechanical functions of the heart," *Int. J. Bioelectromagnetism*, vol. 3, no. 2, 2001.
7. M. Mohr, "Vergleich von mikroskopischen und makroskopischen Modellen der Deformation im Myokard," Diploma Thesis, Universität Karlsruhe, Institut für Biomedizinische Technik, Universität Karlsruhe (TH), Sept. 2001.
8. D. Bourguignon and M.-P. Cani, "Controlling anisotropy in mass-spring systems," in *Computer Animation and Simulation '00, Proc. 11th Eurographics Workshop, Interlaken, Switzerland*, Springer Computer Science, pp. 113–123, Springer, Aug. 2000.
9. K.-J. Bathe, *Finite Element Procedures*. Upper Saddle River, New Jersey: Prentice Hall, 1996.

Towards Patient-Specific Anatomical Model Generation for Finite Element-Based Surgical Simulation

Michel A. Audette, A. Fuchs, Oliver Astley,
Yoshihiko Koseki, and Kiyoyuki Chinzei

National Institute of Advanced Industrial Science and Technology- AIST,
Surgical Assist Technology Group
Namiki 1-2, Tsukuba, 305-8564, Japan
m.audette@aist.go.jp
http://unit.aist.go.jp/humanbiomed/surgical/

Abstract. This paper presents ongoing research on a semi-automatic method for computing, from CT and MR data, patient-specific anatomical models used in surgical simulation. Surgical simulation is a software implementation enabling a user to interact, through virtual surgical tools, with an anatomical model representative of relevant tissues and endowed with realistic constitutive properties. Up to now, surgical simulators have generally been characterized by their reliance on a generic anatomical model, typically obtained at the cost of extensive user interaction, and by biomechanical computations based on mass-spring networks.
We propose a minimally supervised procedure for extracting from a set of CT and MR scans a highly descriptive tissue classification, a set of triangulated surfaces coinciding with relevant tissue boundaries, and volumetric meshes bounded by these surfaces and comprised of tetrahedral elements of homogeneous tissue. In this manner, a series of models could be obtained with little user interaction, allowing surgeons to be trained on a large set of pathologies which are clinically representative of those they are likely to encounter. The application of this procedure to the simulation of pituitary surgery is described. Furthermore, the resolution of the surface and tissue meshes is explicitly controllable with a few simple parameters. In turn, the target mesh resolution can be expressed as a radially varying function from a central point, in this case coinciding with a point on the pituitary gland.
A further objective is to produce anatomical models which can interact with a published finite element-based biomechanical simulation technique which partitions the volume into separate parent and child meshes: the former sparse and linearly elastic; the latter dense, centered on the region of clinical interest and possibly nonlinearly elastic.

1 Introduction

Surgical simulation is a software implementation enabling a user to interact, through virtual surgical tools, with an anatomical model representative of relevant tissues and endowed with realistic constitutive properties. The anatomical

N. Ayache and H. Delingette (Eds.): IS4TM 2003, LNCS 2673, pp. 340–352, 2003.

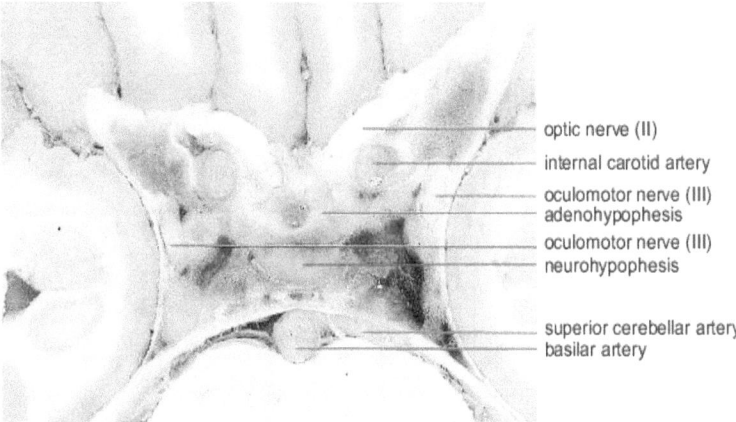

optic nerve (II)

internal carotid artery

oculomotor nerve (III)
adenohypophesis

oculomotor nerve (III)
neurohypophesis

superior cerebellar artery
basilar artery

Fig. 1. Illustration of anatomy relevant to pituitary surgery: pituitary gland and surrounding critical tissues (reproduced with permission from [8]).

model must feature tissues whose consideration is essential to the clinical situation, such as critical vasculature and nerves, so as to appropriately simulate and penalize any damage to them. This is particularly true in our clinical application, the simulation of *transnasal pituitary surgery* [8][10], where the target is surrounded by the optic and oculomotor nerves and by cranial arteries (see figure 1).

Up to now, virtual anatomical models for surgical simulation have been highly task-specific and obtained by elaborate computer-user interaction, including segmentation and meshing. This paper presents on-going research on a minimally supervised procedure for extracting from a set of MR and CT scans a highly descriptive anatomical model, leading not merely to one generic model, but to a family of patient-specific models. This procedure integrates a distortion-tolerant mutual information MR-CT registration, a new tissue classification exploiting global spatial cues, a simplex-based surface meshing model to identify and triangulate the relevant anatomical boundaries, and an automatic almost-regular tetrahedralization of tissue volumes surrounded by these boundaries. This procedure is designed to be extensible to other surgical applications.

Lastly, a further objective of our research is to apply a new finite element (FE) software architecture, proposed by Astley [1][2] and specifically designed for surgical simulation, to a real clinical problem. As shown on figure 2 (a)-(c), this architecture partitions the underlying volume into one or more relatively dense child meshes and a sparse parent mesh, and decouples them in a manner analogous to the Norton equivalent in circuit analysis. The static FE equation is represented as follows:

$$\mathbf{Ka} - \mathbf{f} = 0 \ , \tag{1}$$

where $\mathbf{K}$ is the *stiffness* matrix, $\mathbf{a}$ is the vector of *node displacements*, and $\mathbf{f}$ is the set of *node forces*. Each represents the assemblage of *elemental* stiffness matrices

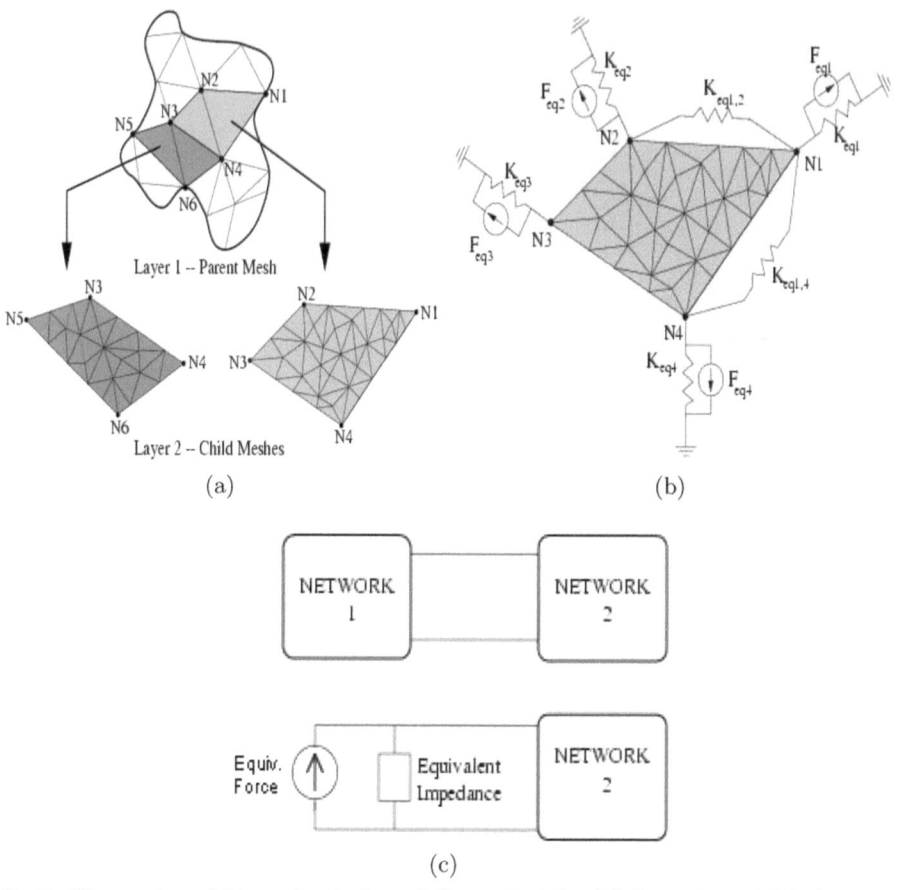

Layer 1 -- Parent Mesh

Layer 2 -- Child Meshes

(a)

(b)

(c)

Fig. 2. Illustration of biomechanical modeling principle. (a) Partition of FE domain into parent and one or more child meshes. Subsequently, either subregion, parent or child, can be represented in the stiffness matrix of the other as an equivalent based on the Norton equivalent in circuit analysis. Each node shared by parent and child is considered (b), and, in computing the decoupled stiffness matrix of the child mesh, the parent mesh as seen by this node can be expressed as (c) force and stiffness equivalents (reproduced with permission from [2]).

$\mathbf{K}^e$, displacements $\mathbf{a}^e$ or forces $\mathbf{f}^e$ [28]. The set of forces $\mathbf{f}^e$ includes a concentrated loads term which accounts for user-controlled virtual cutting forces [18] [9].

This FE architecture considers each node shared by the parent and child meshes: for each subregion, parent or child, it expresses the other subregion(s) encountered at each node as *one* equivalent impedance and force. A system characterized by a large stiffness matrix $\mathbf{K}$ is then reduced to $n + 1$ decoupled systems with significantly smaller stiffness matrices, where n is the number of child meshes. Each system can be resolved independently and at different rates over time, but the child system(s) surrounding the surgical tool(s) must be solved at haptic rates, typically of the order of 500 Hz [7]. Decoupling the problem naturally leads to parallelization on $n + 1$ processors, which can make optimal use of even a dual-processor Pentium computer, particularly if the haptic [19] and visual [21] rendering can be handled by peripheral hardware. Finally, the FE method can be extended to nonlinearly elastic models [22] [5]; this architecture allows for nonlinearly elastic children and has been demonstrated at haptic rates on non-anatomical geometries, justifying its selection over methods that are constitutively limited or that require extensive precomputation [3].

This perspective imposes on our surface and volumetric meshing stages a requirement of explicit control over mesh resolution. Control is exercised with a smooth, radially varying mesh scale function defined from a user-provided central point (e.g.: on the pituitary gland), which naturally leads to a conformal mesh composed of a dense child and a sparse parent. It should also be emphasized that our procedure identifies anatomical surfaces prior to volumetric meshing, rather than proceed directly from the classification to tissue-guided volumetric meshing, because in general the latter approach will not produce a mesh boundary that is smooth and that closely agrees with the anatomical boundary. From a haptic and visual rendering standpoint, an anatomical model with jagged or badly localized boundaries would detract from the realism and clinical relevance of a surgical simulator as well as confuse the user.

2   Materials and Methods

2.1   Locally Weighted Mutual Information Registration

Pituitary surgery imposes requirements on the surgeon of a highly accurate trajectory through both bone and soft tissue, which entails their accurate resolution near the pituitary gland [8]. To do so, we must co-register CT and MR volumes, and resample MR data in a manner compatible with CT sampling. However, this registration is complicated by MR distortion caused by magnetic susceptibility variations near the pituitary gland and sinus cavities [25]. The prevalent co-registration technique for MR and CT is the global mutual information (MI) procedure [27]. It registers two volumes in manner which iteratively minimizes the dispersion of their joint intensity distribution, is widely used and provides globally accurate results. However, this method suffers from misregistration precisely where the distortion is most pronounced, and where the greatest accuracy is required by our clinical application. Our registration technique, which addresses this shortcoming, is comprised of the two following stages:

1. a standard *global 7-parameter MI procedure* (rigid + scale), followed by
2. a *locally weighted linear MI procedure* based exclusively on information "near" the anatomy of interest, using a spherical mask easily specified by the user.

2.2   Global Structure-Preserving Voxel Classification

Voxel Classification is a mapping of feature vectors, typically comprised of tomographic modalities such as CT or MRI, to a discrete set of tissue classes. Up to now, virtual anatomical models for surgical simulation have been obtained by highly elaborate manual segmentation. Ultimately, the tissue classification on which the anatomical model is based must account for clinically relevant tissues, including critical tissues such as vasculature and nerves, while also fulfilling our objective to produce a series of such models with little supervision. These conflicting requirements lead to the necessary consideration of a priori anatomical information in the classification, as feature space alone does not provide enough information to discriminate between all classes relevant to the simulation. Recent techniques [16] [20] exploit essentially local spatial constraints on tissue classification. In constrast, we exploit the *global spatial structure of tissues*, whose recruitment in disambiguating two or more clinically relevant tissue types, overlapping in feature space, has been neglected up to now in the literature.

Our classification method, somewhat inspired from the semi-supervised Fuzzy C-Means technique [6], begins with a Minimum Distance (MD) classification from a small training set, and then tries either to consolidate, or to invalidate and recompute, membership on the basis of global spatial constraints. We integrate the Fast Marching (FM) [24] method with our minimally supervised, iterative classifier. Constraints can easily be placed on the FM front propagation to implement assumptions about global structure. These assumptions include (for more details see [4]):

1. *Contiguity with training points*[1]: given an initial classification, the contiguity of voxels $\mathbf{x}_k$ with the training points of class C_i can be enforced by an outward front from these points, propagating only on voxels of class C_i.
2. *Prior assumptions about spatial extent*: the FM method described in 1 can be restricted by a scale factor, preventing the front propagation to continue beyond this limit; e.g.: pathology, cranial nerve and vasculature voxels can be assumed to be close to their training points, if the latter are well chosen.
3. *Embedded structure*: a very useful cue for classification relates to how contiguous tissues fit within each other, particularly in relation to a tissue class which can be identified reliably such as bone; e.g.: grey and white matter occur inside, while muscle and fat are found outside, the cranium.
4. *Similarity and proximity to confidently classified voxels*: if there is still ambiguity, feature similarity and spatial proximity to the boundary Γ_i of a set of contiguous blobs of confidently classified voxels $\tilde{\mathbf{x}}_{k,i}$ can be exploited.

The philosophy of this method is not to favour any single class over all others, but merely to discount from consideration any class which at a given position has essentially zero likelihood, based on prior information.

In our tissue classification, there is currently no special treatment of critical tissues other than the global constraints discussed so far, but given their importance to the simulation, we recognize that our framework can be refined to consider them as a separate case. In research currently underway, user-provided

---

[1] For classes where this constraint is useful, training points should be chosen not only for their intensity, but for their spatial relation to tissues. We consider this added burden on the user to be manageable, since the training set remains small.

"training points" from blood vessels and cranial nerves are being used to anchor a minimal path (MP) through them [11][14], and a tissue classification which proceeds outwards from these minimal paths. In contrast with existing MP techniques, which exploit only image *gradient magnitude* information and may be undermined by low-contrast areas, our MP computation makes use of a robust image feature for detecting tubular structures. This feature, which takes *gradient direction* into consideration, is the *image gradient flux* [26], which distinguishes between *sources* and *sinks* of a gradient vector field, coinciding with points where the *outward flux* of this field is *positive* and *negative* respectively. For visible vessels and nerves, tubular structures of higher intensity than their surroundings, points inside their boundary coincide with strong gradient sinks, particularly along their central axis, whereas points beyond coincide with gradient sources. Consequently, a robust minimal path along this central axis can be identified based on a flux-weighted potential. The subsequent critical tissue classification can make use of the sign of image gradient flux, as well as the proximity of a voxel to a particular minimal path.

2.3   Simplex-Based Tissue-Guided Surface Meshing with Resolution Control

Once a descriptive tissue classification is computed, our next step is to establish the triangulated tissue boundaries relevant to the simulation by exploiting this classification, a discrete surface model, and the position of the training points of each class. The *n-simplex* mesh is a discrete active model [12], characterized by each vertex being linked to each of $n+1$ neighbours by an edge. A balloon force can act on this mesh to cause it to expand until some image-based force halts this expansion. A surface model in 3D is realized as a 2-simplex, characterized by each vertex having 3 neighbours, and this representation is the dual of a triangulation, with each simplex vertex coinciding with a center, and each simplex edge being bisected by an edge, of a triangle. Furthermore, this surface model also features other internal forces [12] which nudge each simplex face, and consequently each dual triangle, towards having edges of equal length, and towards $\mathcal{C}_0$, $\mathcal{C}_1$ or $\mathcal{C}_2$ continuity, for example.

We apply the 2-simplex surface model to our anatomical meshing problem because of the explicit control on mesh characteristics which can be achieved, many of which are already implemented [12][13]. Each simplex is initialized with a small spherical mesh centered on a tissue class training point, and expanded until halted by image information, in the form of a set of tissue classes assumed "outside" the volume of interest. A few adaptations to the simplex model are proposed here for automatic FE mesh generation. First, we can endow the surface model with absolute mesh size thresholds, expressed as a target simplex area scale A_s and a percent tolerance ε_{A_s}: a simplex face falling below the area minimum $A_s - \varepsilon_{A_s}/100$ results in the fusion of two contiguous faces into one by eliminating a shared edge [12], while a face above the maximum area $A_s + \varepsilon_{A_s}/100$ is subdivided into two by a new edge. This *constant* target area can be replaced with a *radially varying function* $A_s(\mathbf{x})$, determined by the distance $R(\mathbf{x}) =$

$\|\mathbf{x} - \mathbf{x}_c\|$ between the centroid $\mathbf{x}$ of each simplex face and a user-defined central point $\mathbf{x}_c$, e.g.: inside the pituitary gland:

$$A_s(\mathbf{x}) = \begin{cases} A_{s,min} & \text{if } R(\mathbf{x}) <= R_{child} \\ A_{s,min} + (A_{s,max} - A_{s,min}) \left\{ 1 - \exp\left[\frac{-(R(\mathbf{x}) - R_{child})}{R_{scale}}\right] \right\} & \text{otherwise,} \end{cases}$$

(2)

where $A_{s,min}$ and $A_{s,max}$ specify smallest and largest target areas, and R_{child} and R_{scale} determine the behaviour of the function bridging the two values: a small, constant scale for a radius less than R_{child} and an exponential function tending towards $A_{s,max}$ as the distance of the simplex face to the central point increases. This scale function thereby produces small mesh faces near the pituitary gland and larger faces away from it. The triangulated surface coinciding with the anatomical boundary, obtained by duality with the final simplex mesh, is the objective of this stage. More than one children are possible, at the cost of selecting other "central" points of interest $\mathbf{x}_{c,i}$, in which case we define $R(\mathbf{x}) = \min_i \|\mathbf{x} - \mathbf{x}_{c,i}\|$.

Finally, a number of other adaptations are underway, including:

- A *conformality-preserving force*, which would cause two contiguous boundaries to share vertices wherever desirable, and
- *Topological adaptivity* applied to the 3D simplex model (previously demonstrated for 2D simplex models [13]), which would allow two or more spheres emerging from neighbouring training points to fuse together upon contact, or conversely would allow a simplex boundary to break apart over large gaps. In particular, a promising approach is that of Lachaud [17], which for a triangular mesh model sets an inner bound on edge length and detects topological changes on the basis of an inter-vertex distance falling under this bound. In our case we could substitute our simplex target area for a target edge length, and proceed analogously. This quality may turn out to be essential to identifying the boundaries of various sinus and air passages, for example, whose topology can vary considerably from patient to patient.

2.4   Almost-Regular Volumetric Meshing with Radial Resolution Control

The last stage in our procedure partitions each volume bounded by a triangulated mesh, coinciding with a relevant tissue class, into tetrahedral elements consistent with the FE method. The volumetric meshing stage is essentially a published technique [15] which automatically produces an optimal tetrahedralization from a given polygonal boundary, such as a triangulated surface. In this case, optimality is defined as near-equal length of the tetrahedral edges, along with a sharing of each inner vertex by a nearly consistent number of edges and tetrahedra. This method features the optimal positioning of inner vertices, expressed as a minimization of a penalty functional, followed by a Delaunay tetrahedralization. The resulting near-regularity is important for FE stability and efficiency [23].

We modify this technique by integrating into the penalty functional the now-familiar radially varying scale function, which is specified as a target edge length $L_t(\mathbf{x})$ for each tetrahedron. Based on the relationship between the number of simplex and triangle vertices $V_t \approx V_s/2$ [12], a target simplex mesh size of A_s works out to a triangular area of $A_t \approx A_s/2$, and to the following triangular and tetrahedral target edge length (assuming edges of near-equal lengths):

$$L_t(\mathbf{x}) \approx \sqrt{[2A_s(\mathbf{x})/\sqrt{3}]} \,. \tag{3}$$

The separation of the resulting tetrahedral mesh into child and parent is as follows: contiguous tetrahedra whose edge lengths approach

$$L_{t,min} = \sqrt{[2A_{s,min}/\sqrt{3}]}$$

comprise the child mesh, while the other elements constitute the parent.

3   Results

Ongoing validation for each stage is based on the application of the procedure to real patient data obtained in collaboration with the Tokyo Women's Hospital. Validation of the registration exploits fiducial registration error (FRE) statistics, based on manually identified surgical fiducials in both spaces, and is documented in [4]. While we could have exploited the fiducials in the registration, we proceed otherwise for the sake of extensibility, as they may not be present in other surgical applications which eschew image guidance, to which we would extend this method. A study evaluating the relative merits of different linear transformations is currently under way and will be published shortly. We use either a small set of homologous anatomical landmark pairs or a principal axis transform to provide the global MI stage with an initial transformation. The size of the spherical mask is typically 100 mm. An illustration of typical results is shown in figure 3.

Classification validation, while currently qualitative, provides a stark justification for incorporating global spatial constraints, given the number of classes which can be discriminated, in comparison to what is achievable otherwise, as illustrated in figure 4. Future validation will make use of a digital anthropomorphic phantom as well as CT and MR simulators.

The validation of both the surface and volumetric meshing are also based on qualitative studies with patient data, as shown in figure 5, and in the future will exploit synthetic anthropomorphic data.

4   Conclusion and Future Directions

This paper presented results of ongoing research on a semi-automatic procedure for computing anatomical models for patient-specific surgical simulation. The procedure features distortion-tolerant MI registration of MR and CT, classification which exploits the global structure of tissues, simplex-based tissue-guided

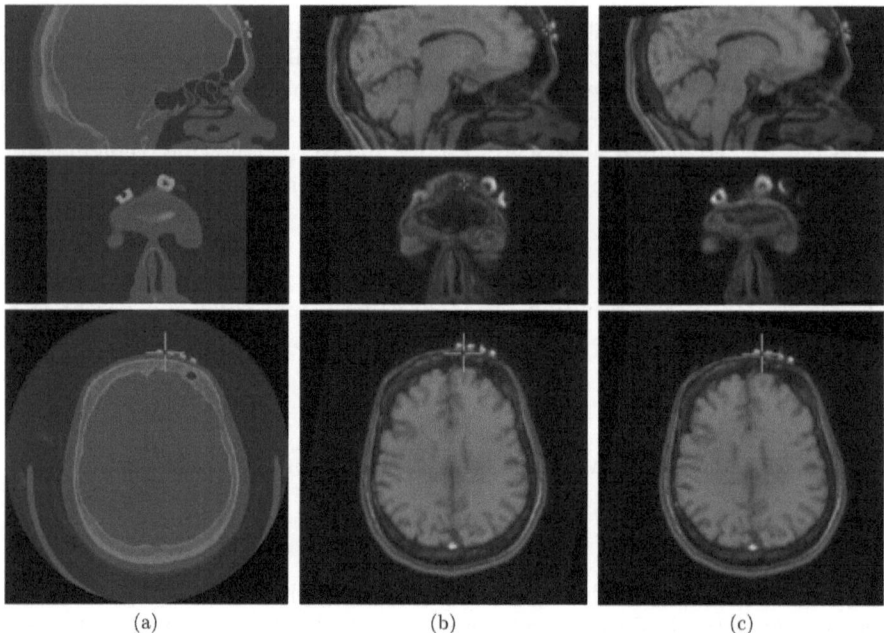

Fig. 3. Illustration of local weighting: (a) CT scan with manually identified fiducial; (b) MR scan transformed to CT space based on global MI registration, with CT fiducial overlaid; (c) transformed MR data as in (b), using the locally weighted method.

surface meshing, and automatic almost-regular volumetric meshing, with explicit resolution control on both meshing techniques.

It should be emphasized that this meshing strategy will tend to produce a mesh of radially varying density and of heterogeneous material properties, resulting from the consideration of triangulated boundaries of different tissues and necessary for constitutive realism. In the event that a particular tissue is essential to the simulation but impractical to tessellate at the computed target size, it may become imperative for this constraint to be relaxed somewhat. For example, a multiresolution surface meshing strategy [17] may be beneficial to first identify a complex surface, and might then be followed by a few iterations where the simplex model tends towards the spatially varying target area. Moreover, some tubular structures may benefit from a biomechanical modeling as curvilinear, rather than volumetric, elements, for the sake of computational performance. Ultimately, forthcoming experimentation with the multirate FE architecture of [2] will validate our meshing strategy, or suggest modifications to it.

As for the tissue classification, we do not claim that the resulting description is the equivalent of a manual classification everywhere, from a histological standpoint, but we believe that *for the sake of the simulation*, a sufficiently descriptive tissue map is achievable with limited user interaction. Another issue worth exploring in the future is whether our tissue map can serve as a starting point for a high-quality description based on warping a manual classification on a patient-by-patient basis.

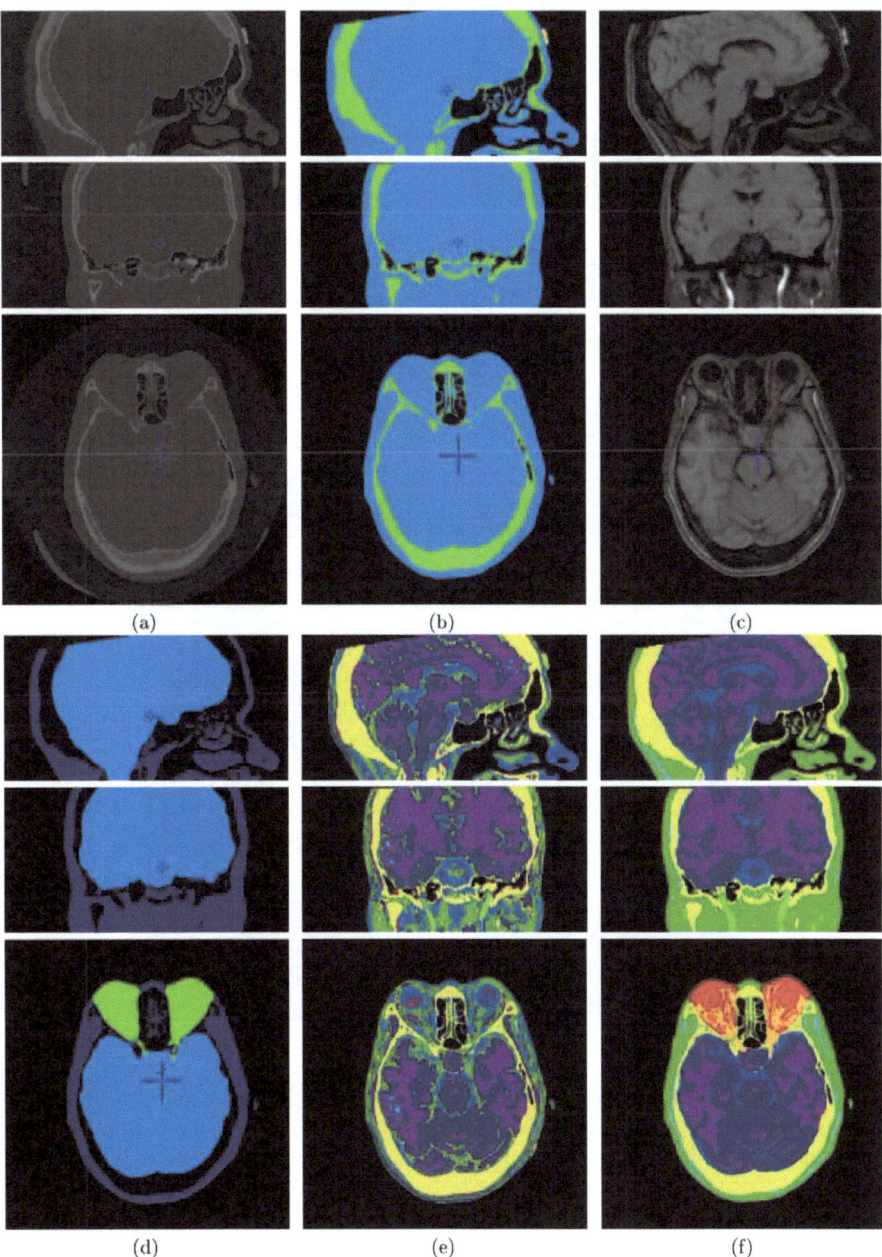

(a)   (b)   (c)

(d)   (e)   (f)

Fig. 4. Illustration of improvement due to global spatial constraints: (a) CT data; (b) CT-based classification of hard and soft tissues, via constraints 1, 2 and 3; (c) MR data; (d) embedded intracranial and intraorbital regions (constraint 3); classification of 9 soft tissues (e) without, and (f) with, constraints 1, 2, 3 and 4. In (e) there are many corticospinal fluid (royal blue) and muscle (green) false positives, as well as many vasculature (bright green) and ocular tissue (orange and red) false negatives.

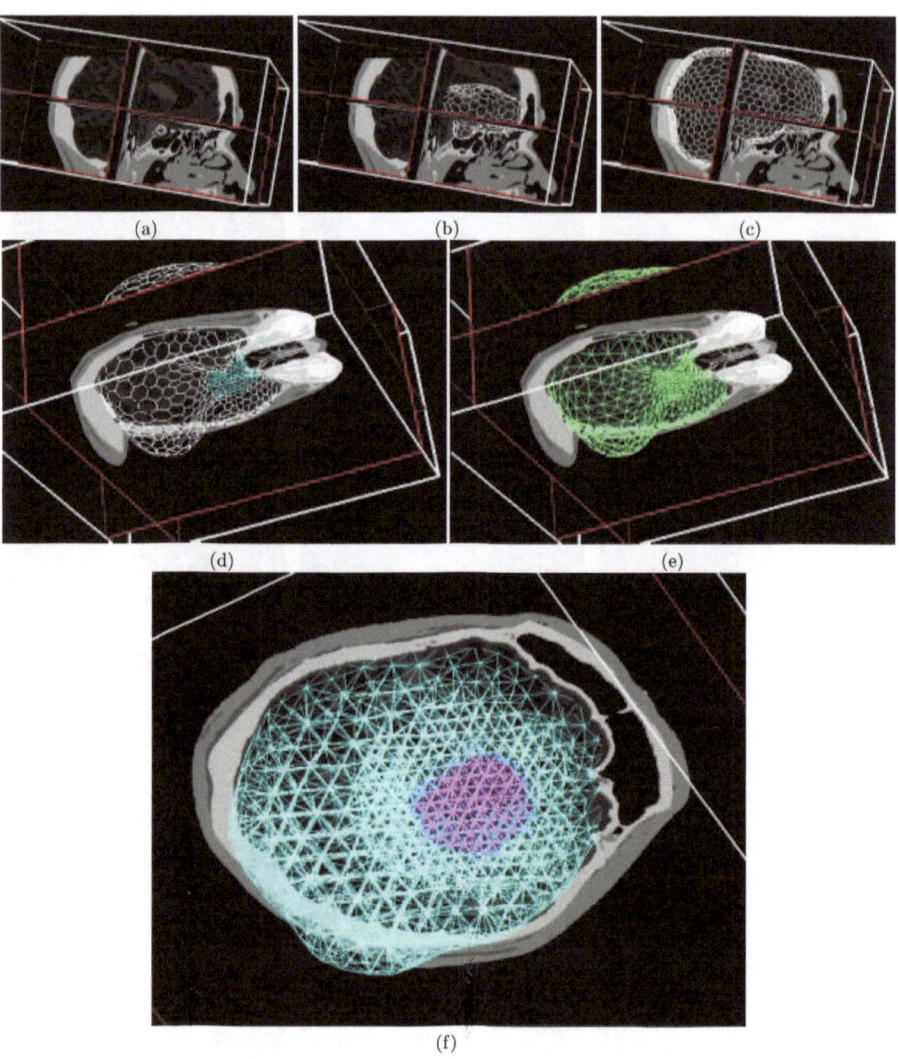

Fig. 5. Illustration of surface and volumetric meshing: (a)-(d) evolution of simplex mesh, at (a) 10, (b) 300 and (c) 1300 iterations at constant target mesh size; (d) radial target mesh size, with faces with minimal target area shown in turquoise; (e) final dual triangulated surface; (f) volumetric mesh, bounded by the surface in (e) (with child mesh shown in pink), featuring visibly near-regular structure.

Acknowledgements

The authors thank Dr. H. Delingette (INRIA) for generously providing simplex software, as well as Dr. H. Iseki and Mr. M. Sugiura (Tokyo Women's Hospital).

References

1. O.R. Astley & V. Hayward, Multirate Haptic Simulation Achieved by Coupling Finite Element Meshes Through Norton Equivalents, *IEEE Int. Conf. Rob. & Auto.*, 1998.
2. O.R. Astley, *A Software Architecture for Surgical Simulation Using Haptics*, Ph.D. thesis, McGill University, 1999.
3. O.R. Astley & V. Hayward, Design Constraints for Haptic Surgery Simulation, *Proc. IEEE Int. Conf. Rob. & Auto*, pp. 2446-2451, 2000.
4. M.A. Audette & K. Chinzei, Global Structure-preserving Voxel Classification for Patient-specific Surgical Simulation, *Proc. IEEE EMBS-BMES Conf.*, 2002.
5. M.A. Audette et al., A Review of Biomechanical Modeling of the Brain for Intrasurgical Displacement Estimation and Medical Simulation, submitted for publication, *Annals of Biomed. Eng*, 2003.
6. A.M. Bensaid et al., Partially Supervised Clustering for Image Segmentation, *Pattern Recognition*, Vol. 29, No. 5, pp. 859-871, 1996.
7. G.C. Burdea, *Force and Touch Feedback for Virtual Reality*, John Wiley & Sons, 1996.
8. P. Cappabianca et al., *Atlas of Endoscopic Anatomy for Endonasal Intracranial Surgery*, Springer, 2001.
9. V.B. Chial, S. Greenish & A.M. Okamura, On the Display of Haptic Recordings for Cutting Biological Tissues, *Haptics 2002 - IEEE Virtual Reality Conference*, 2002.
10. I. Ciric et al., Complications of Transsphenoidal Surgery: Results of a National Survey, Review of the Literature, and Personal Experience, *Neurosurg.*, Vol. 40, No. 2., pp. 225-236, Feb. 1997.
11. L.D. Cohen & R. Kimmel, Global Minimum for Active Contour Models: A Minimal Path Approach, *Int. J. Comp. Vis.*, Vol. 24, No. 1, pp. 57-78, 1997.
12. H. Delingette, General Object Reconstruction Based on Simplex Meshes, *Int. J. Comp. Vis.*, Vol. 32, No. 2, pp. 111-146, 1999.
13. H. Delingette and J. Montagnat, Shape and Topology Constraints on Parametric Active Contours *Comp. Vis. and Image Under.*, Vol. 83, pp. 140-171, 2001.
14. T. Deschamps & L.D. Cohen, Fast Extraction of Minimal Paths in 3D Images and Applications to Virtual Endoscopy, *Med. Imag. Anal.*, Vol. 5, pp. 281-299, 2001.
15. A. Fuchs, Almost Regular Triangulations of Trimmed NURBS-Solids, *Eng. w. Comput.*, Vol. 17, pp. 55-65, 2001.
16. K. Held et al., Markov Random Field Segmentation of Brain MR Images, *IEEE Trans. Med. Imag.*, Vol. 16, No. 6, pp. 878-886, 1997.
17. J.-O. Lachaud & A. Montanvert, Deformable Meshes with Automated Topology Changes for Coarse-to-fine Three-dimensional Surface Extration, *Med. Imag. Anal.*, Vol. 3, No. 2, pp. 187-207, 1998.
18. M. Mahvash & V. Hayward, Haptics Rendering of Cutting: A Fracture Mechanics Approach, *Haptics-e - The Electronic Journal of Haptics Research*, Vol. 2, No. 3, Nov. 20, 2001.
19. Microstar Laboratories, *www.mstarlabs.com*.
20. N.A. Mohamed et al., Modified Fuzzy C-Mean in Medical Image Segmentation, *Proc. IEEE ICASSP*, pp. 3429-3432, 1999.
21. NVidia Corp., *www.nvidia.com*.
22. J.T. Oden, Finite Elements of Nonlinear Continua, McGraw-Hill, 1972.

23. V.N. Parthasarathy et al., A Comparison of Tetrahedron Quality Measures, *Fin. Elem. in Anal. & Des.*, Vol. 15, pp. 255-261, 1993.
24. J.A. Sethian, *Level Set Methods and Fast Marching Methods: Evolving interfaces in computational geometry, fluid mechanics, computer vision, and materials science*, 2nd ed., Cambridge University Press, 1999.
25. T.S. Sumanaweera et al., MR Susceptibility Misregistration Correction, *IEEE Trans. Med. Imag.*, Vol. 12, No. 2, June 1993.
26. A. Vasilevskiy & K. Siddiqi, Flux Maximizing Geometric Flows, *IEEE Trans. Patt. Anal. & Mach. Intel.*, Vol. 24, No. 2, pp. 1565-1578, 2002.
27. P. Viola & W.M. Wells, Alignment by Maximization of Mutual Information, *Proc. 5th Int. Conf. Computer Vision*, pp. 15-23, 1995.
28. O.C. Zienkiewicz, *The Finite Element Method*, McGraw-Hill, 1977.

Model Predictive Control for Cancellation of Repetitive Organ Motions in Robotized Laparoscopic Surgery

Romuald Ginhoux[1], Jacques A. Gangloff[1], Michel F. de Mathelin[1], Luc Soler[2],
Joël Leroy[2], and Jacques Marescaux[2]

[1] LSIIT UMR 7005 CNRS — Strasbourg I University, Bd. Sébastien Brant,
BP 10413, F-67412 Illkirch Cédex, France, http://eavr.u-strasbg.fr
[2] IRCAD/EITS — Hôpitaux Universitaires de Strasbourg, 1, Place de l'Hôpital,
F-67000 Strasbourg, France

Abstract. Periodic deformations of organs which are due to respiratory movements may be critical disturbances for surgeons manipulating robotic control systems during laparoscopic interventions or tele-surgery. Indeed, the surgeon has to manually compensate for these motions if accurate gestures are needed, like, *e.g.*, during suturing. This work presents a model predictive control scheme that is applied to the problem of maintaining a constant distance in the endoscopic images from a surgical tool's tip to the organ's surface. A new optimization criterion is developed for an unconstrained generalized predictive controller based on a repetitive input-output model, where contributions of the control input to reference tracking and to disturbance rejection are split and computed separately. Thanks to this approach, mechanical filtering of the repetitive disturbances and teleoperation by the surgeon can run simultaneously and independently on the robot arm. The system is tested on both an endotraining box with a surgical robot and in *in vivo* conditions on a living pig. Results are shown to validate the control scheme and its application.

1 Introduction

Robotic systems appeared a few years ago in the field of laparoscopic surgery. Commercial systems like ZEUS (Computer Motion, Inc.) or DaVinci (Intuitive Surgical, Inc.) make use of robot arms to manipulate surgical tools and the endoscopic camera. The robot is actually tele-operated through master arms by the surgeon looking at the visual feedback from the endoscope (see Fig. 1). Advantages of these systems are numerous: the surgeon's tiredness is reduced as he can comfortably seat during the operation; natural tremor is eliminated by translating the surgeon's hand motions to robotic movements; the use of a high master/slave motion ratio can increase the surgeon's motion accuracy. Furthermore, teleoperation allows long distance surgical procedures to be performed (see, *e.g.*, the Lindbergh Operation (Marescaux *et al.*, 2001)).

Research works recently reported new possible developments for providing the surgeon with further assistance from the robot or even for having the robot

N. Ayache and H. Delingette (Eds.): IS4TM 2003, LNCS 2673, pp. 353–365, 2003.

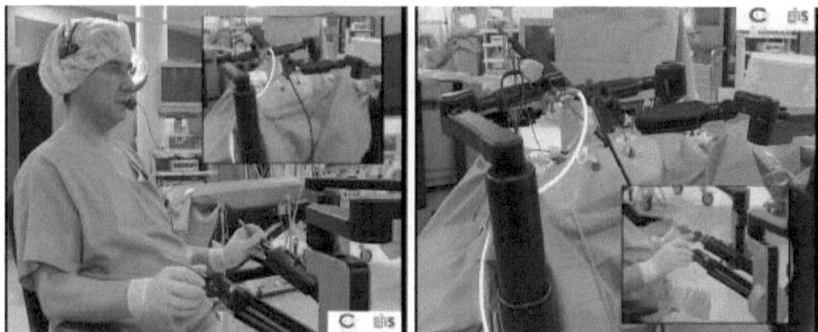

Fig. 1. Laparoscopic robotic system at work (IRCAD, Strasbourg). *The master arms are driven by the surgeon (left) to control the movements of the instruments into the patient's body (right).*

itself realize semi-autonomous tasks. For instance, systems appeared that use visual servoing techniques and patterned (Casals *et al.*, 1995) or coloured (Wei *et al.*, 1999) marks to make the endoscope autonomously track instrument's motions. A specific lightweight endoscope manipulator was shown in (Berkelman *et al.*, 2002) with the use of an external optical localizer that allows the surgeon to automatically position the endoscope. The system proposed by (Krupa *et al.*, 2002) is made of optical markers and tiny laser pointers; a visual servoing control scheme is developed to automatically bring the surgical tool at the center of the endoscopic image; servoing of the insertion depth of a standard instrument is also proposed.

Nevertheless, systems above do not explicitly consider motions of the organs in tele-operated laparoscopic surgery; these motions and deformations are induced by the patient's breathing or heart beating and have to be manually compensated for by the surgeon through the manipulation interface when accurate tasks are required, such as suturing. Using standard surgical robot controllers, accurate servoing of the distance between an instrument and organ surfaces is not possible, since PID-type controllers are unable to reject repetitive disturbances due to breathing; curves shown in (Krupa *et al.*, 2002) and in Fig. 2 actually exhibit large residual errors. A possible solution would be to do high-gain control to reduce these errors, but this is not appropriate since surgical robots are mechanically designed to be lightweight and enable only slow displacements for safety reasons.

In our research, we focus on the design of new controllers for standard surgical robots that explicitly take into account repetitive disturbances. We address the problem of keeping constant the distance from the instrument to the moving organ, as it is viewed from a fixed endoscope. Basic idea is to use a model of the robotic servo loop and a model of the perturbation to predict future disturbances so as to get the robot anticipate the movement thereby cancelling its effect. Several applications of Model Predictive Control (MPC) to the rejection of periodic disturbances can be found in the literature: a dynamical model describ-

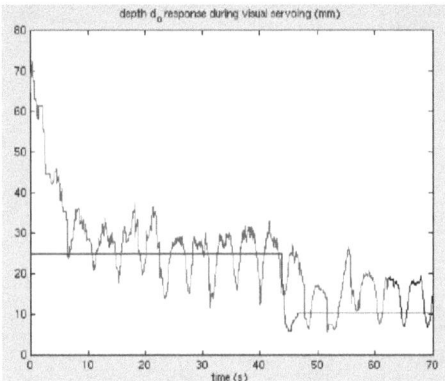

Fig. 2. Example of perturbations induced by respiratory movements during servoing of the tool insertion depth in in-vivo conditions, with a classical PID-type controller (courtesy of Krupa *et al.*, 2002).

ing the oscillatory behavior of a chemical process is shown in (Zhu *et al.*, 2000), a mixture of repetitive control and predictive control is found in (Natarajan and Lee, 2000) (for a chemical application too). These controllers are developed using the state-space formulation of MPC and linear time-invariant or even time-varying (as in (Lee *et al.*, 2001)) descriptions of the plants; applications shown consider steady-state control where no distinction is made between a periodic output disturbance or a periodically varying reference.

In this paper, we formulate an input-output model in the ARIMAX form that includes a repetitive noise model and derive a new Repetitive Generalized Predictive Control (R-GPC) scheme. A new cost function is formulated where the control commands are split into two independent components, the first one depending only on the reference trajectory tracking, the second one depending only on disturbance rejection.

The remainder of the paper is organized as follows. The next section presents the laparoscopic robotic setup and visual servoing system used. Section 3 details the R-GPC controller and its two control outputs and discusses the impact of varying or unknown breathing periods as well as a suddenly stopped breathing. Results are finally given in the last two sections; experimental results obtained on an endo-trainer box with a simulated breathing motion and a laparoscopic surgery robotic arm are first shown in section 4 ; results from *in vivo* tests on a anaesthetized pig in the operating room of IRCAD are presented in sec. 5.

2   Laparoscopic Robotic Setup

The robotic setup we consider in this work is shown in Fig. 3. The robot is an AESOP laparoscopic arm (Computer Motion, Inc.) which is modelled by the identified transfer function of its translational joint. The endoscope is a monochrome PAL camera mounted on a fixed rigid stand; images from the patient's abdomen

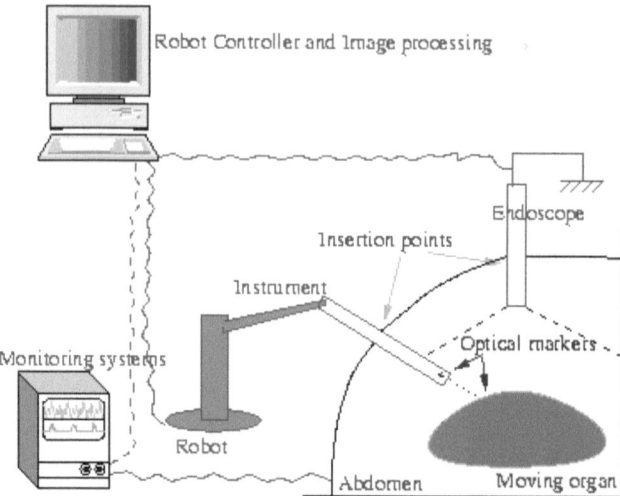

Fig. 3. System setup. *The robot is an* AESOP *surgical arm holding a laser-pointing instrument. The breathing period is controlled by external systems during the procedure.*

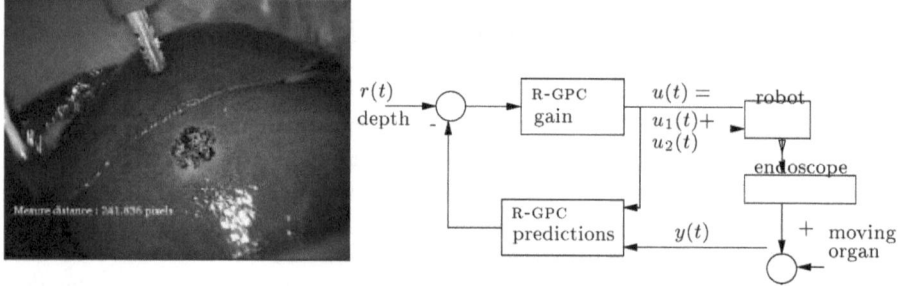

Fig. 4. Left: distance measurements are performed using a laser-pointing instrument with optical markers. Right: Block-diagram of the predictive control system. *Instrument insertion depth $y(t)$ is measured in the endoscope images from the instrument's tip to the organ surface; it is controlled by visual servoing, using references $r(t)$.*

are updated every $T_e = 40\,ms$ through a incision point. The robot arm is holding a laser pointing instrument equipped with optical marks (Fig. 4). Distance from the instrument's tip to the moving organ is estimated through the observed images using the technique described in (Krupa *et al.*, 2002). Measurements are fed into the predictive controller, which, in turn, returns the optimal speed to be applied along the instrument's axis. The vision thread (see (Krupa *et al.*, 2002) for more detail) and the predictive control thread are hosted by a 800 MHz dual-processor Linux PC computer that communicates with the surgical robot via a serial link. Controller computations are synchronized with the image acquisition.

The system we want to control is a single input / single output plant. Its input (hereinafter $u(t)$) makes the robot translate the instrument along its axis, and the output (hereinafter $y(t)$) is the distance from the tip of the instrument to the organ surface that is measured in the endoscope. The visual servoing loop is accordingly summarized in Fig. 4: predictions are made using the robot model and current and past outputs and control inputs; command $u_1(t)$ is computed according to the finite receding horizon strategy of MPC so as to ensure an optimal tracking of the reference distance $r(t)$ in the future; command $u_2(t)$ is computed according to the disturbance period T^* so as to cancel the effect of the moving organ; command $u(t) = u_1(t) + u_2(t)$ is sent to the robot and controls the insertion depth of the instrument.

This setup will be applied to a laboratory experiment in sec. (4) and for *in vivo* tests in sec. (5).

3   Generalized Predictive Control

This section introduces a new unconstrained Generalized Predictive Control (GPC) scheme based on a repetitive input-output model of the system to be controlled. Separate contributions of the control input to reference trajectory tracking and disturbance rejection are computed by means of a new cost function that ensures no interaction between both components.

3.1   Repetitive ARIMAX Model

Unconstrained GPC was originally introduced by (Clarke *et al.*, 1987), where the system model is represented by an ARIMAX equation,

$$A(q^{-1})y(t) = B(q^{-1})u(t-1) + \frac{C(q^{-1})}{\Delta(q^{-1})} \xi(t) \qquad (1)$$

where q^{-1} is the backward operator, $T_e = 1\,s$ is the (normalized) sampling period, A and B are two polynomials modelling the system dynamics (B may also include pure delays), and polynomial C is used to colour the zero-mean white noise $\xi(t)$. Polynomial Δ is used to make noise ξ/Δ be non-stationary, which is suitable to model any perturbation in a control loop (Camacho and Bordons, 1999). For instance, Δ is set to a pure integrator,

$$\Delta(q^{-1}) = \delta(q^{-1}) \triangleq (1 - q^{-1}),$$

when disturbances are only supposed to be constant steps (Clarke *et al.*, 1987; Camacho and Bordons, 1999). However, this classical setting is not appropriate when the actual disturbances vary periodically over time; we therefore propose to modify the ARIMAX model by including repetitive features of disturbances and write Δ as:

$$\Delta(q^{-1}) = \delta(q^{-1}) \Delta_R(q^{-1}) \qquad (2)$$
$$\text{with } \Delta_R(q^{-1}) \triangleq 1 - q^{-T} \qquad (3)$$

and $T \in \mathbb{N}$, $T \geq 2$, is the number of sampling periods in one period T^* of the disturbance. The perturbation model ξ/Δ is actually made periodic with a period equal to T.

Earlier results were shown in (Ginhoux et al., 2002) where the standard unconstrained GPC cost function was considered with $\Delta_R = 1 - \alpha\, q^{-T}$. The float α was chosen in $]0;1]$ so as to act as a forgetting factor, whose main impact was to filter control commands in order to increase the robustness against errors in the system model or noise in the visual measurements. In the next section, we propose to split the control input into two terms so that they can be filtered independently.

3.2   Separation of Control Input

Writing $y(t)$ in equation (1) as $y(t) = y_{th}(t) + \epsilon(t)$ is equivalent to the two following equations:

$$A\, y_{th}(t) = B\, u_1(t-1) \tag{4}$$

$$A\, \epsilon(t) = B\, u_2(t-1) + \frac{C}{\Delta}\, \xi(t) \tag{5}$$

where $u(t)$ is now written as $u(t) = u_1(t) + u_2(t)$. Command $u_1(t)$ is input to the theoretical system model (4), leading to the theoretical output measurement $y_{th}(t)$. Command $u_2(t)$ is responsible for the actual plant to exhibit measurement error $\epsilon(t)$ when subject to noise and disturbances in the measurement signal.

Then, following and adapting the method from (Clarke et al., 1987; Camacho and Bordons, 1999) to this formulation of the controller, the expression of the the cost function for the unconstrained R-GPC is derived from equations (4) and (5) as:

$$J(u = u_1 + u_2, t) = \sum_{j=N_1}^{N_2} \|\hat{y}_{th}(t+j) - r(t+j)\|^2 + \sum_{j=N_1}^{N_2} \|\hat{\epsilon}(t+j)\|^2$$

$$+\lambda \sum_{j=1}^{N_u} \|\delta u_1\,(t+j-1)\,\|^2 + \mu \sum_{j=1}^{N_u} \|\delta u_2\,(t+j-1)\,\|^2 \tag{6}$$

N_1, N_2 are respectively, the lower and upper bound of the cost horizon, and N_u is the length of the control cost horizon; $N_u < N_2$ and $\delta u_i(t+j-1) = 0$ for $j > N_u$, $i = 1$ or 2; λ and μ weight the relative importance of both control energies. The reference trajectory is denoted by $r(t)$. The aim is to compute the N_u future control increments $\delta u_1(t+j-1)$ so that the error between the predictions of the theoretical model outputs and the future references $r(t+j)$ is minimized; and the control increments $\delta u_2(t+j-1)$ so as to drive the actual system outputs towards the theoretical ones, or, equivalently, to compensate for the measurement disturbances. Note that the two sets of control increments separately contribute to the minimization of the cost function. As the control law is receding, only the first control increment $\delta u(t) = \delta u_1(t) + \delta u_2(t)$ is sent to the system, and the overall minimization is performed at each time step.

Advantage of this decomposition is manifold: first, the control command u_2 that acts on the disturbance rejection is equivalent to an autonomous mode for the robot, as it is independent of command u_1 that is directly driven by the surgeon. Secondly, the commands responsible for the cancellation of the perturbation can be filtered separately from the commands acting on the reference tracking; this means that low-pass filters or even nonlinear filters can replace and increase the performance of the single forgetting factor used so far (Ginhoux *et al.*, 2002), thereby increasing the robustness against errors in the system model or noise in the depth estimation; monitoring component $u_2(t)$ ensures a reliable estimation of the perturbation period since it remains periodic whatever changes are found in the reference trajectory; different levels of saturation can be put on both control inputs in order to prevent teleoperated component $u_1(t)$ from saturating with no influence on the perturbation cancellation in case of large changes in the reference signal.

3.3   Comments on the Breathing Period

This section discusses the problem of unknown or time-varying breathing periods and the effect of a suddenly stopped respiration in the control system.

For effective cancelling of the breathing disturbance, the actual respiratory frequency is required to be known in Δ_R (see eq. (2)); this is not a limitation in real operating conditions since the breathing frequency can be precisely known from the external medical systems that control the anaesthetics (Fig. 3, sec. 5). One can nevertheless take advantage of the control input decomposition in the R-GPC controller to apply algorithms for period estimation, as it is ensured that component $u_2(t)$ remains periodic over time and reflects the effect of the disturbance in the feedback loop; the use of the recursive procedure proposed by (Tsao and Qian, 1993) was shown in (Ginhoux *et al.*, 2002) ; the application is to account for temporal variations in the disturbance signal's period by providing online updates to polynomial Δ_R using the best estimate $\hat{T}$ of the true motion period T^* (results are shown in the next section).

In the case of a respiration that would suddenly stop, the measurement disturbance would actually disappear from the control loop in Fig. 4 ; this means that the predictive controller would then stop rejecting the disturbance, thereby concentrating only on reference tracking.

4   Experimental Results

In this section we provide results from the laboratory experiment shown in Fig. 5. The system is an endo-training box where the organ motion is simulated thanks to an additive semi-rotating motor that makes the plane shown in Fig. 5 oscillate periodically. Movement of this oscillating plane is set so that $T^* = 2.4\,s$ and the oscillations make measurements of the instrument insertion depth vary of about $+/- 6\,mm$. Controller parameters are set to $N_1 = 8$, $N_2 = 45$, $N_u = 30$, $\lambda = 0.55$ and $\mu = 0.70$.

A classical GPC controller with the same parameters as above (except that λ is not considered) is first used in figure (6) where the reference is kept to a

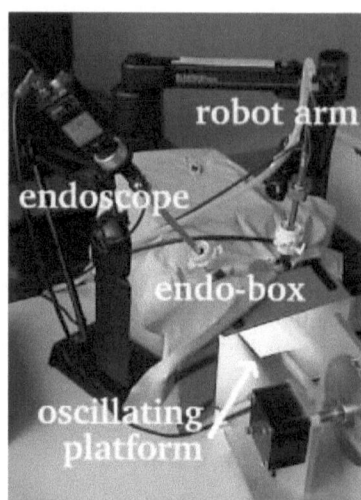

Fig. 5. Laboratory experiment with an endo-training box. Respiratory motions are simulated using a periodically oscillating plane.

constant distance. The perturbation is badly attenuated (residual amplitude is $+/- 4\,mm$) and makes the system output vary periodically with time. Figure (7) shows the system response as it is driven by the R-GPC controller with varying references. Effect of the disturbance is clearly reduced and residual error has an amplitude of about $2\,mm$. The corresponding command $u(t)$ and its two components $u_1(t)$ and $u_2(t)$ are shown in Fig. 8. Curve for $u_1(t)$ is shown to be in accordance with reference changes whereas $u_2(t)$ is made oscillating for reducing the disturbance amplitude.

In figure (9), the period of the R-GPC controller was originally set to $2.16\,s$ and switched to $2.4\,s$ once the period estimate has been recovered by the gradient descent algorithm. The period switch is made online thereby making the controller adaptive with respect to T and improving the disturbance cancellation.

Fig. 10 depicts the gradient descent and the evolution of the period estimate before the R-GPC switches in Fig. 9.

5   Validation in *in vivo* Conditions

In this section, the R-GPC controller is again compared to a classical GPC, but the laser-pointing instrument and the endoscope are now inserted into the abdomen of an anaesthetized pig at the operating room of IRCAD (see Fig. 11). Respiration is set to 20 movements a minute. Curves in fig (12) show that the residual error in steady state with the R-GPC controller is twice as small as the error with the GPC controller. Note that the transient for R-GPC reflects the learning behaviour of repetitive controllers. Curves in Fig. 13 show that changes in the reference

System output (mm)

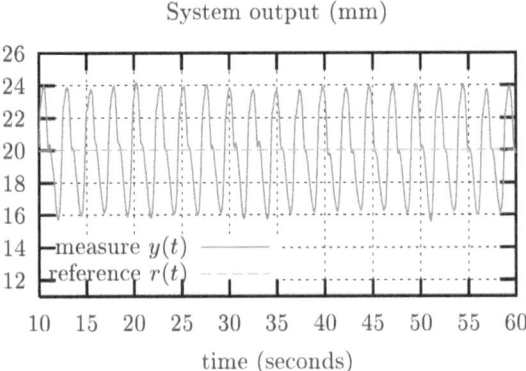

Fig. 6. System output with a classical GPC and repetitive disturbances.

System output (mm)

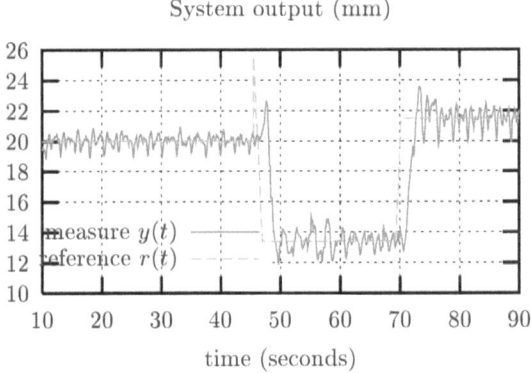

Fig. 7. System output with a varying reference.

signal (which are driven by the surgeon) do not affect the performance of the cancellation of the breathing-induced disturbances.

6   Conclusion

This paper presented a repetitive model predictive controller approach for the active mechanical filtering of periodic motions induced by respiration or even heart beating in laparoscopic robotized interventions. Periodic property of the perturbation has been included into the input-output model of the controlled system so as to have the robotic system anticipate perturbating motions. A new cost function has been presented for the unconstrained generalized predictive controller where reference tracking is decoupled from the rejection of predictable oscillatory motions. Experimental results have been shown for visually servoing the insertion depth of a surgical tool in an endoscopy-training box and in real conditions with an AESOP surgical robotic arm. The surgeon can interact with the

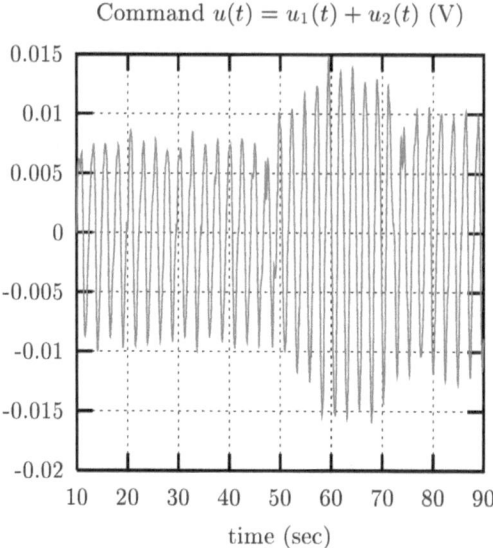

Command $u(t) = u_1(t) + u_2(t)$ (V)

time (sec)

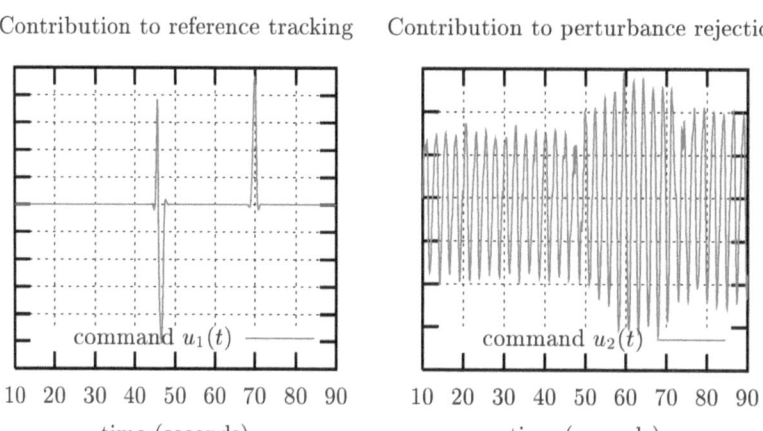

Contribution to reference tracking   Contribution to perturbance rejection

command $u_1(t)$

command $u_2(t)$

time (seconds)   time (seconds)

Fig. 8. System command for Fig. 7. *The system is driven to the varying reference as shown by command $u_1(t)$ (bottom left). Term $u_2(t)$ is periodic and reflects the controller behaviour for disturbance rejection (bottom right).*

control system to change the distance reference, while the robot is autonomously cancelling the breathing motions.

Acknowledgements

The authors thank Computer Motion Inc. that has graciously provided the AE-SOP medical robot.

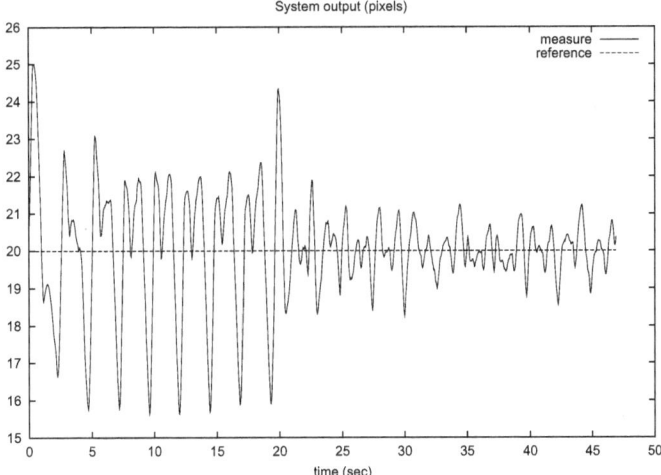

Fig. 9. Period switching in the R-GPC controller. *The period switch was initiated at* $t = 20\,s$ *by the estimation algorithm of (Tsao et al., 1993). Disturbance rejection is improved after the switching instant.*

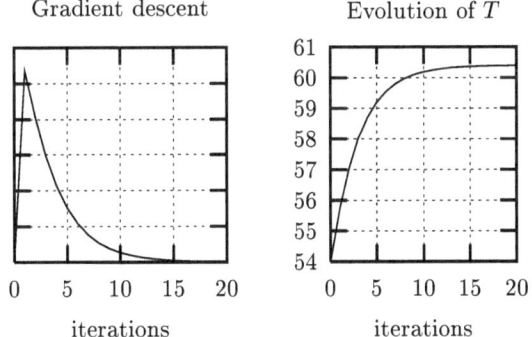

Fig. 10. Gradient descent (left) and evolution of the period estimate (right) for Fig. 9. *The initial period of the controller is set to* $2.16\,s$ *and the true period is* $2.4\,s$ *(* $T_e = 40\,ms$ *).*

References

Berkelman, P. J., P. Cinquin, J. Troccaz et al. (2002). Development of a compact cable-driven laparoscopic endoscope manipulator. In: *Proc. of the fifth International Conference on Medical Image Computing and Computer-Assisted Intervention (MICCAI)*. Lecture Notes in Computer Science, 2488. Springer. Tokyo, Japan. pp. 17–24.

Camacho, E. F. and C. Bordons (1999). *Model Predictive Control*. Springer-Verlag. London.

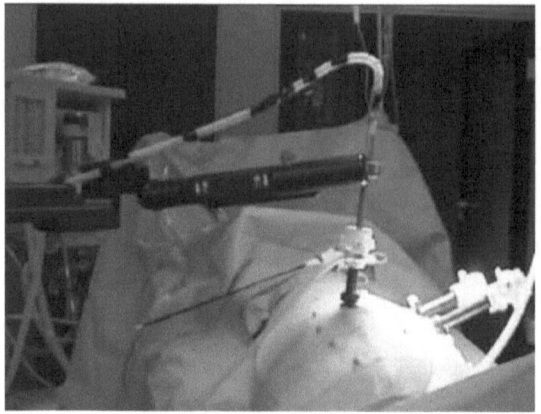

Fig. 11. Setup of sec. (2) in *in vivo* conditions. *Experiments are carried out on a anaesthetized pig.*

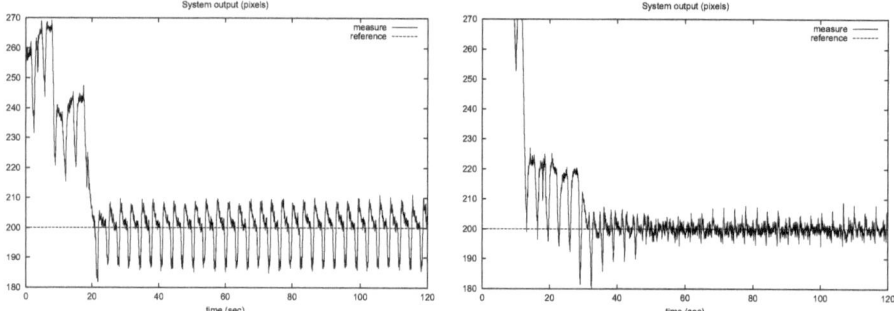

Fig. 12. Transient and steady state after the controller is switched on. *(a) left, with a* GPC *controller, initial instrument/organ distance is 255 pixels ; (b) right,* R-GPC *controller, initial distance of 320 pixels.*

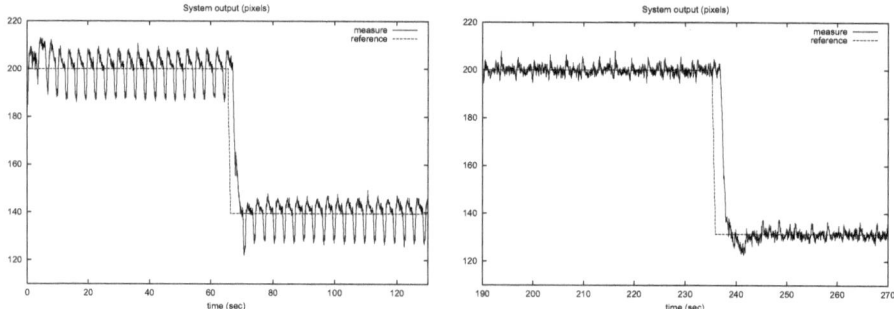

Fig. 13. System output when the distance reference is changed by the surgeon. *(a) left,* GPC *controller; (b) right,* R-GPC *controller.*

Casals, A., Amat J. et al. (1995). Vision-guided robotic system for laparoscopic surgery. In: *Proc. of the 1995 IFAC International Congress on Advanced Robotics.* Barcelona, Spain. pp. 33–36.

Clarke, D. W., C. Mohtadi and P. S. Tuffs (1987). Generalized predictive control - part. 1: The basic algorithm. *Automatica* **23**, 137–160.

Ginhoux, R., A. Krupa, J. Gangloff, M. de Mathelin and Soler L. (2002). Active mechanical filtering of breathing-induced motion in robotized laparoscopy. In: *Surgetica-CAS, 1st European Conference on Computer Assisted Surgery* (Sauramps Medical, Ed.). Grenoble, France.

Krupa, A., C. Doignon, J. Gangloff and M. de Mathelin (2002). Combined image-based and depth visual servoing applied to robotized laparoscopic surgery. In: *Proc. of the 2002 IEEE/RSJ International Conference on Intelligent Robots and Systems.* Lausanne, Switzerland.

Lee, J. H., S. Natarajan and K. S. Lee (2001). A model-based predictive control approach to repetitive control of continuous processes with periodic operations. *Journal of Process Control* (11), 195–207.

Marescaux, J., J. Leroy and M. Gagner (2001). Transatlantic robot-assisted telesurgery. *Nature* (413), 379–380.

Natarajan, S. and J. H. Lee (2000). Repetitive model predictive control applied to a simulated moving bed chromatography system. *Computers and Chemical Engineering* (24), 1127–1133.

Tsao, T.-C. and Y.-X. Qian (1993). An adaptive repetitive control scheme for tracking periodic signals with unknown period. In: *Proc. of the American Control Conference.* San Francisco, California. pp. 1736–1740.

Wei, G.-Q., K. Arbter and G. Hirzinger (1999). Real-time visual-servoing for laparoscopic surgery. *IEEE Engineering in Medicine and Biology* **16**(1), 40–45.

Zhu, G.-Y., A. Zamamiri, M. A. Henson and M. A. Hjortsø (2000). Model predictive control of continuous yeast bioreactors using cell population balance models. *Chemical Engineering Science* (55), 6155–6167.

Virtual Radiofrequency Ablation
of Liver Tumors

Caroline Villard, Luc Soler, Nicolas Papier, Vincent Agnus, Sylvain Thery,
Afshin Gangi, Didier Mutter, and Jacques Marescaux

IRCAD, 1, place de l'Hôpital, F-67091 Strasbourg, France,
caroline.villard@ircad.u-strasbg.fr
www.ircad.org

Abstract. In the last few years, radiofrequency ablation has become
one of the most promising techniques to treat liver tumors. But radiologists have to face the difficulty of planning their treatment while only
relying on 2D slices. We present here a realistic radiofrequency ablation simulation tool, coupled with a 3D reconstruction and visualization
project. They help radiologists to have a better visualization of patients
anatomic structures and pathologies, and allow them to easily find an
adequate treatment.

1 Introduction

Recent advancements in medical imaging have allowed to develop several kinds
of minimally invasive techniques for the ablation of liver tumors that used to be
considered as being non-resectable. Among them, percutaneous thermal ablation
has been studied in different forms, such as microwave, laser, high intensity
focused ultrasound, cryotherapy, and radiofrequency (RF) that appears to be
the easiest, safest and most predictable [1].

Radiofrequency is a tumor denaturation method using heat created by ionic
agitation generated by the principle of a microwave located at the tip of a needle-
like probe. It leads to cell death and coagulative necrosis when enough heated.
The probe may be positioned several times in order to treat a larger zone. Radiologists burn the whole tumor volume with a 1 cm security margin [2], which is
mandatory to prevent local recurrence of a tumor after treatment, and to reduce
the effects of a possible inaccuracy of needle placement.

The three important criteria for the success of such a treatment are the choice
of secure trajectories for the probe, the destruction of a maximum number of
cancerous cells, and a minimum amount of affected healthy tissues. Unfortunately, planning such a treatment according to these factors is quite difficult for
a radiologist who can only rely on 2D scanner slices.

Scanner image reconstruction allows a more intuitive 3D visualization of the
patient's anatomy [3], that makes the simulation of needle placement easier. The
expected follows up of this functionality are both the visualization of the necrosis
of treated zones, and the automatic planning of needle trajectories that would
optimize the three criteria.

N. Ayache and H. Delingette (Eds.): IS4TM 2003, LNCS 2673, pp. 366–374, 2003.

In this paper, after a summary of recent studies on lesion size and shape, and on treatment planning, we explain how we simulate the necrosis of the treated area. Then, we show how we plan to automatically compute optimal needle positions, and which future improvements this work will bring in RFA treatment.

2   State of the Art

2.1   Lesion Size and Shape

3D simulation of radiofrequency necrosis depends directly on recent studies on lesion size and shape. Indeed, to accurately simulate lesions, it is necessary to know all the factors that have an influence on them. Literature shows that several kinds of factors are important to predict how lesions will look like [4].

- *Device-dependent factors*:
  The shape of the necrosis zone depends on the type of probe used for the treatment. Different kinds of probes are used in radiofrequency ablation. Expandable multi-array needles, which are insulated needles containing hook-shaped inner electrodes that can be deployed once the target is reached, produce a more or less spherical lesion. Other types, single or clustered cooled needles, produce an ellipsoidal lesion shape. A comparative study of these systems can be found in [5].
  Technological advancements in needle design tend to increase lesion size. Needle design (electrode's diameter and tip length) also modifies lesion size and shape. Moreover, the different kinds of provided generators use various ablation algorithms [1]. According to the power they can supply, the lesion size may be greater. It is possible to control lesion size and shape by changing intensity and duration of the supplied current.

- *strategy-dependent factors:*
  Since larger lesions may avoid multiple needle insertions, different strategies have been studied to increase lesion size. Radiofrequency lesions can be enlarged by a temporary occlusion of the tumor blood supply, because blood flow in large vessels has a "heat-sink" effect that cools the thermal process [6]. Infusion of a saline solution into the target tissue via a cannulated radiofrequency probe also amplifies the ablative volume [7].

- *anatomic factors:*
  As we said previously, blood flow in large vessels causes thermal process cooling. If the needle tip is placed close to large vessels, they will not be affected by the treatment, and the lesion shape will be deformed, and model itself to the shape of the vessels. As for smaller vessels ($\varnothing$ < 2 to 3 mm), they seem to be obstructed by clots after radiofrequency treatment, and have no consequence on lesion shape or size [5].

− *pathologic factors:*
  Some studies showed that in the case of cirrhotic livers, the shape of the thermal lesion was more extended and almost modelled to the tumor shape, whereas less surrounding cirrhotic tissue was burnt. Livraghi et al. attribute this difference to a phenomenon called "oven effect" [8].

All these factors, directly influencing size and shape of the necrosis zone, obviously have to be taken into account for an accurate RFA simulation. Indeed, a computer-aided treatment planning has to offer a maximal guarantee of success and safety for the patient, for a better survival rate.

2.2   Treatment Simulation and Planning

Previous studies on treatment simulation and planning have already been carried out, but they were mainly focused on other minimally invasive treatments such as cryotherapy [9,10], or they were centered on finite element modelling [11], but very few works have been done on radiofrequency real-time 3D simulation and treatment planning.

Cryotherapy is with RF, one of the most often used systems to treat liver tumors. So far, it has been more widely studied in computer science than radiofrequency, and some teams carried out softwares able to simulate iceball growth. Most are based on finite element methods and compute the propagation of freezing inside tissues [9], others approximate lesions by ellipsoids [10].

Both also tried to initiate a treatment planning for cryotherapy [10,12], but with some limitations. For instance, T. Butz proposed a very interesting cryotherapy simulator and planner, included in *3D-Slicer*, that can also be extended to one type of radiofrequency probe. However, it can only compute the best positioning of cryoprobes within a predefined window of the body, and doesn't take into account the presence of surrounding organs. In [11], a finite element model for radiofrequency is presented, based on the resolution of bioheat equation, but it doesn't seem to be real-time.

3   Radiofrequency Ablation Simulation

The researches we carried out, and that we will detail here, were justified by the fact that many radiologists expressed a need of information about their patients that would be easier to visualize, and of having the ability to simulate accurately and realistically the radiofrequency treatment before operating.

We present here a tool, called *RF-Sim*, that links 3D reconstruction and visualization of slices, and a virtual RF probe placement simulator. The whole project allows to easily visualize the patient's anatomy in 3D and to localize his tumors, and has several useful functionalities such as security margin display, volume calculation, and inner navigation. *RF-Sim* adds the ability to append RF probes in the scene, and to place them by taking into account security margins, and surrounding organs and factors.

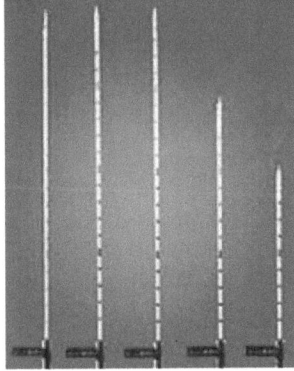

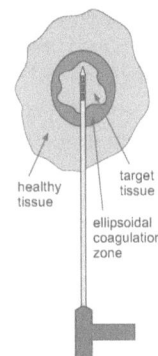

Fig. 1. Berchtold® electrodes

Fig. 2. Expandable needle

Fig. 3. Ellipsoidal lesion zone

3.1 Chosen Framework

First of all, we perform the 3D reconstruction of a patient's liver and tumors from an enhanced spiral CT scan using an automated algorithm described in [3]. All the experiments that were made with this simulator were based on real patients' data that came from Strasbourg's Civil Hospital, and that were performed within a preoperative framework.

For this simulator, aimed at being first tested in this hospital, we chose to represent the electrode model that is mainly used by our radiologists, that is the Berchtold® HITT system (see Fig.1). This kind of needle produces a quite ellipsoidal lesion zone around the tip, as shown on Fig.3. Therefore, we first chose to represent the necrosis zone by a meshed ellipsoid (see Fig.4). However, we could easily extend our simulator to other types of needles, like expandable (see Fig.2) or clustered systems, by simply changing the appearance of the needle and the general shape of the lesion.

Fig. 4. Virtual needle electrode and its associated ellipsoidal lesion around tip

3.2 Enhancement of Realism

Once the basic shape of the lesion zone had been chosen, we tried to make it even more realistic by taking into account surrounding factors. One of the most important of them is the proximity of large vessels, that cools the surrounding tissue zone, and avoids a complete burning of the ellipsoidal zone if the blood flow is not occluded. When done close to a large vessel, a RF treatment causes a deformed necrosis zone. However, we also have to consider that, on the contrary, small vessels don't affect the treatment and are completely burnt.

Deformation of the Ellipsoid: We consider that the heat produced by the electrode and propagating within the tissue is coming from the tip and is directed towards the boundary of the ellipsoid, following the heat flow. We approximate the "heat-sink" phenomenon by saying that if the heat flow comes near a vessel, it is stopped. That way, we can represent the necrosis zone by a quite simple deformation of the ellipsoid, following the shape of the vessel.

Technically talking, for each vertex of the theoric ellipsoid, we test if it is inside the deformation zone. If so, we replace it by the vertex placed on the same vector and that is as far as possible from the center of the ellipsoid, while being outside the deformation zone. Intuitively, we "repulse" the vertex away from the vessel, but keep it on the same radius of the ellipsoid (see Fig.5).

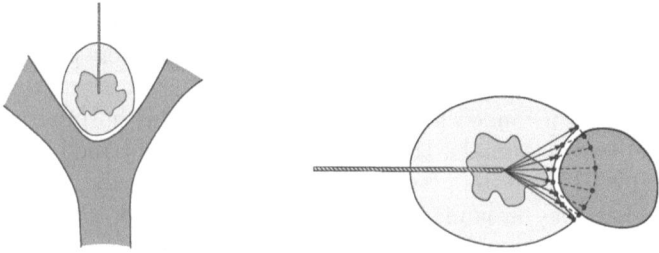

Fig. 5. Deformed shape of the necrosis zone close to large vessels

Computing Deformation Zone from Vessels: As we said earlier, large vessels cool their surrounding area and remain unchanged after the treatment, whereas small vessels are considered as being burnt. To reproduce this effect, we have to compute a deformation zone that includes large vessels but excludes the smallest ones. We chose to use the voxel version of the whole vessel network in order to perform an opening on it. An opening is a composition of one or more erosions and of the same number of dilations [13].

We have to perform a sufficient number of erosions on voxel shape to make small vessels disappear, only thinning large ones. The number of erosions is determined by the size of small vessels to eliminate ($\varnothing < 2$ to 3 mm) and the resolution of the voxel mask. The average resolution of the masks we currently use is $0.6 \times 0.6 \times 0.6$ mm, so we perform 2 erosions in order to eliminate vessels having a radius < 1.2 mm, *i.e.* a diameter < 2.4 mm. Performing 3 erosions would eliminate too many vessels ($\varnothing < 3.6$) and performing only 1 erosion would miss out on some vessels ($1.2 < \varnothing < 2.4$). Then, the same number of dilations bring large vessels to their initial thickness.

In a second step, we perform some more dilations, in order to increase thickness of large vessels. Indeed, the "heat-sink" effect also cools the area surrounding the vessel, so the zone has to be extended to include this area. We obtain a deformation zone that incorporates large vessels and their neighborhood, and that excludes small vessels. An example for a portal vein is shown on Fig.6, where the voxels are represented by their center point.

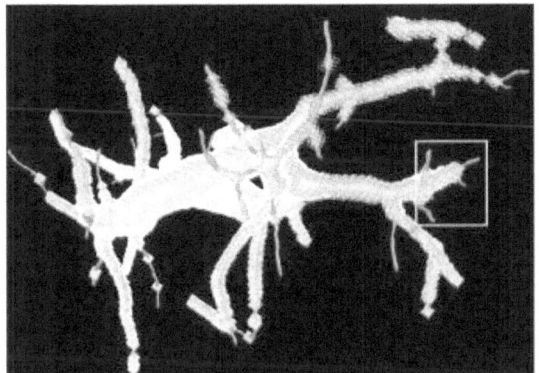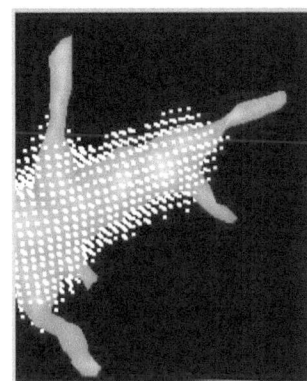

Fig. 6. Voxel deformation zone of a portal vein

4   Results and Perspectives

We present on Fig.7 the results of 2 deformations of a virtual necrosis zone based on the voxel deformation shape. In these snapshots, we can easily see the deformations induced by large vessels. For each one of them, we present 2 views: in the second one, veins are not drawn in order to see the deformation more easily. The manipulation of the needles and of their associated scalable lesion zone is a real-time operation.

Finally, we present on Fig.8 a snapshot of a typical *RF-Sim* scene, showing the simulation of a RF treatment. Several electrodes are placed on tumors, some of them produce overlapping lesions to treat a large tumor. To have a better visualization, skin and liver are drawn in a transparent mode, and lesion zones in a semi-transparent mode.

This example is taken from a set of 18 real study cases on which we made our experiments so far. These patient data were taken from the Strasbourg Civil Hospital radiology service's database, among patients that were candidate for a RF treatment. All these experiments will soon be subjected to post-treatment medical validation by experts.

These experiments consisted, for each case, in opening the set of reconstructed organs and tumors of a patient, and then trying to place manually
as many needles as necessary to cover all tumors plus margin volumes, while
also trying to minimize healthy tissue burning, and to avoid inserting needles
through vital organs. In other words, it consisted in thinking about a placement
plan for each case, trying to apply it by placing virtual needles, checking if the
predicted locations were appropriate by comparing margin and lesion shapes and
volumes, and if necessary correcting the plan until a satisfactory treatment was
found.

The experiments clearly showed that when a tumor is located too close to
vessels, it is often impossible to find a needle configuration that would totally
burn the additional margin volume, because the cooling process interferes. The
direct manipulation of the needles and the real-time deformation of the lesion
shape significantly improved the visualization of this phenomenon, and seem to
be helpful in the prediction of treatment efficiency.

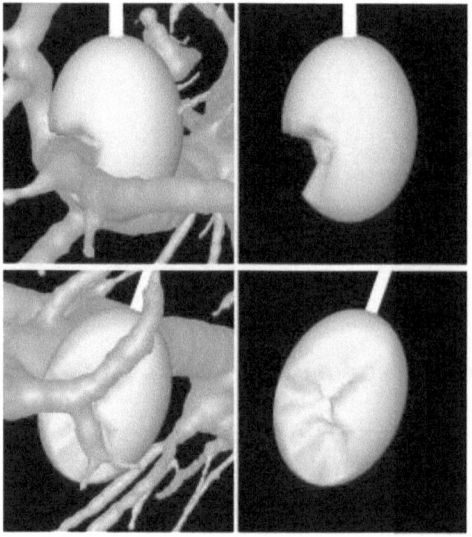

Fig. 7. Deformed virtual necrosis zones

The direct extension of this work is obviously RF treatment planning. Let
us explain here what will be the main points of this study, and how they will be
carried out.

First, we have to define what is an optimal treatment plan. It has to fulfill the
3 criteria of an efficient treatment: burn all the cancerous cells, burn a minimum
of healthy tissue, and be safe. But it also has to take into account accessibility
(bones), and treatment simplicity (minimum number of needle insertions) in
order to be more comfortable for the patient and easier for the radiologist.

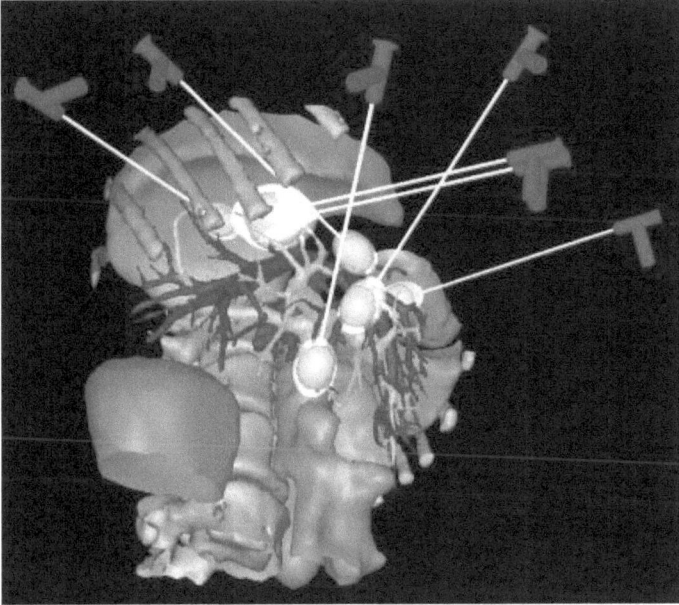

Fig. 8. Snapshot of *RF-Sim*

We plan to compute electrode placements in 3 phases:

– definition of a range of possible paths for needle placements, in order to eliminate unaccessible or unsafe directions,
– computation of a set of accurate overlapping ellipsoid configurations, that produce compound lesions allowing to burn cancerous tissues efficiently,
– choice of the best configuration among those previously computed, minimizing the burning of healthy tissue, and maximizing the simplicity criteria.

The first phase is mainly a collision detection operation, defining a set of 3D shapes inside of which needles can manoeuvre to reach their goals. The second phase is probably an extension of the work of T. Butz [10], that allowed to compute the best overlapping ellipsoid configuration to cover the tumor plus margin shape, the needle being inserted within a predefined window. For us, the window will be extended to the previously computed 3D shapes of possible directions. Then, the third step is a heuristics-based selection of the best solution among many according to several fixed constraints.

5   Conclusion

The simulator *RF-Sim* we presented here represents a first step to a complete treatment planning and simulation tool that will assist radiologists in their work. It allows to see easily the patient's anatomy in 3D, to append virtual RF probes and their associated lesion zone, that are realistically rendered according to surrounding parameters. With the adjunction of the planning tool, radiologists will be able not only to simulate the treatment they plan to perform, but also to ask the computer for advice. This study on RF ablation simulation may lead in the future to considerable improvements in the treatment of cancers.

References

1. McGahan J.F., Dodd III G.D.: Radiofrequency ablation of the liver: Current status. American Journal of Roentgenology **176**(1) (2001) 3–16
2. Cady B., Jenkins R.L., Steele G.D. Jr, et al.: Surgical margin in hepatic resection for colorectal metastasis: a critical and improvable determinant of outcome. Annals of Surgery **227** (1998) 566–571
3. Soler L., Delingette H., Malandain G., Montagnat J., Ayache N., Koehl C., Dourthe O., Malassagne B., Smith M., Mutter D. and Marescaux J.: Fully automatic anatomical, pathological, and functional segmentation from CT scans for hepatic surgery. Computer Aided Surgery **6**(3) (2001) 131–142
4. Goldberg S.N.: Radiofrequency tumor ablation: principles and techniques. European Journal of Ultrasound **13**(2) (2001) 129–147
5. De Baere T., Denys A., Wood B.J., Lassau N., Kardache M., Vilgrain V., Menu Y., Roche A.: Radiofrequency liver ablation: Experimental comparative study of water-cooled versus expandable systems. American Journal of Roentgenology **176**(1) (2001) 187–192
6. Rossi S., Garbagnati F., Lencinoni R.: Unresectable hepatocellular carcinoma: percutaneous radiofrequency thermal ablation after occlusion of tumor blood supply. Radiology **217** (2000) 119–126
7. Livraghi T., Goldberg S.N., Monti F., et al.: Saline-enhanced radiofrequency tissue ablation in the treatment of liver metastasis. Radiology **202** (1997) 205–210
8. Livraghi T., Goldberg S.N., Lazzaroni S., Meloni F., Solbiati L. and G. Scott Gazelle: Small Hepatocellular Carcinoma: Treatment with Radio-frequency Ablation versus Ethanol Injection. Radiology **210**(3) (1999) 655–661
9. Rewcastle J.C., Sandison G.A., Kennedy P., Donnelly B.J., Saliken J.C.: 3-Dimensional visualization of Simulated Prostate Cryosurgery. Society of Cardiovascular and Interventional Radiology 25th Annual Meeting, San Diego, California (March 25-30, 2001)
10. Butz T., Warfield S.K., Tuncali K., Silverman S.G., van Sonnenberg E., Jolesz F.A. and Kikinis R.: Pre- and Intra-operative Planning and Simulation of Percutaneous Tumor Ablation. Medical Image Computing and Computer Assisted Intervention, Pittsburgh, USA. (Oct. 2000) 317–326
11. Tungjitkusolmun S., Staelin S., Haemmerich D., Cao H., Tsai J.-Z., Cao H., Webster J.G., Lee F.T., Mahvi D.M., Vorperian V.R.: Three-dimensional finite element analyses for radio-frequency hepatic tumor ablation. IEEE Trans. Biomed. Eng. **49**(2) (2002) 3–9
12. Sandison G.A. and Rewcastle J.C.: Optimized Treatment Planning for Cryotherapy. Medical and Biological Engineering and Computing **35** Supplement Part 2 (1997) p 708
13. Serra J.: Image Analysis and Mathematical Morphology. Ac. Press, Tome 1 (1982)

Pathology Design
for Surgical Training Simulators

Raimundo Sierra[1], Michael Bajka[2], and Gábor Székely[1]

[1] Computer Vision Group, ETH Zürich, Switzerland
{rsierra,szekely}@vision.ee.ethz.ch
[2] Clinic of Gynecology, Dept. OB/GYN, University Hospital of Zürich, Switzerland

Abstract. Realistic generation of variable anatomical organ models and pathologies are crucial for a sophisticated surgical training simulator. A training scene needs to be different in every session in order to exhaust the full potential of virtual reality based training. We previously reported on a cellular automaton able to generate leiomyomas found in the uterine cavity. This paper presents an alternative approach for the design of macroscopic findings of pathologies and describes the incorporation of these models into a healthy virtual organ. The pathologies implemented are leiomyomas and polyps protruding to different extents into the uterine cavity. The results presented are part of a virtual reality based hysteroscopy simulator that is under development.

1 Introduction

In the last few years many surgical training simulators have been developed [6,8,9]. There is consensus that one of the main advantages of a training simulator is the ability to offer training on many different surgical scenes in a compressed period of time, thus increasing the experience of the trainee. Nevertheless, the generation of variable anatomical models of the healthy organ and the incorporation of pathologies have not been treated as a specific issue. Instead, today's simulators use single static organ models, usually derived from an exemplary anatomy such as an MRI dataset of a volunteer, a specially acquired high resolution dataset, or artificially created with CAD.

Our current research aims at the development of a hysteroscopy simulator. Hysteroscopy is the visualization of the inner surface of the uterus by performing a distension of the cavum uteri, realized through a single hull for manipulation and visualization. It makes minimal invasive surgery on the uterus possible and allows the physician to perform a specific treatment under organ saving conditions. Hysteroscopy is the second most often performed endoscopic procedure after laparoscopy in gynecology. Because of lack of alternatives, training is usually performed during actual interventions, with an experienced gynecologist.

A single organ model is inherently unable to represent an everyday situation of the operating site, thus obstructing the training effect of the simulator. The organs of any two patients will never be alike. Statistical anatomical models, for example those used for the incorporation of prior anatomical knowledge into the

N. Ayache and H. Delingette (Eds.): IS4TM 2003, LNCS 2673, pp. 375–384, 2003.

segmentation process [3,7] offer an appealing way to handle the variability of healthy human anatomy within the organ models used for simulation.

The other requirement for a reasonably realistic surgical simulator is the ability to offer training on a wide variety of pathological cases. The large number of possible pathologies as well as the enormous range of their manifestations makes a similar statistical approach unreasonable if not impossible. We are therefore exploring different approaches to model the pathologies.

We previously reported on a cellular automaton that is able to model different types of leiomyomas. A minimal set of rules was presented that simulates the growth process of a myoma. Thus, by successive and probabilistic application of the rules new myomas are generated [15]. The limitations of the cellular automaton become obvious when investigating pathologies like polyps. The interaction between the surface of the pathology and the healthy organ is an integral part of the growing process. Modeling of collisions in a cellular automaton is far more complex than in any surface based system.

In this work, a different strategy for the generation of pathologies found in the uterine cavity is described and their incorporation into the model of the healthy organ is presented. Instead of growing a pathology model every time a different manifestation is needed, the desired findings are directly designed. In a second step, the pathology is embedded in the healthy organ. Hysteroscopy being the target application, the most frequent pathologies visible from within the uterine cavity have been implemented, namely submucosal leiomyomas and polyps. The implementation of the first one allows for a direct comparison with the previously published approach. In addition, polyps are excellent candidates to discover the suitability and limitations of the different approaches, as they have significantly different properties than myomas.

2   Pathologies in the Uterine Cavity

Leiomyomas and polyps are both the cause of abnormal bleeding and the reason for the majority of hysteroscopic interventions. While the symptoms and the appearance of these pathologies might to some extent be similar, their treatment during hysteroscopy is completely different. Resecting a pedunculated myoma the same way as a polyp is removed can lead to serious problems during intervention.

2.1   Myoma Formation

Uterine leiomyomas are well-circumscribed, non-cancerous tumors arising from the myometrium of the uterus. They are found in 25-40% of women in their child-bearing years and are classified by their location in the uterus. In hysteroscopy submucosal leiomyomas, both sessile and pedunculated, protruding to different extents into the cavity, are visible and treatable. As a myoma is composed of very dense fibrotic tissue, it has a much stronger tendency to keep its shape than any of the surrounding tissues. Therefore the myoma will be able to grow almost independently from its surroundings and keep a spherical shape. The size of a myoma can range from a few millimeters up to several decimeters.

Removing a myoma is usually done by electric resection in small parts. A monopolar loop electrode is used to carve small strips that can subsequently be grasped and extracted from the cavity through the cervix. In doing so, it is crucial to keep the peduncle. Cutting off the peduncle of a myoma can lead to serious problems in hysteroscopy: the rigid sphere that forms the myoma will be floating like a ball in the cavity, making any further resection impossible.

2.2 Polyp Formation

Endometrial polyps are - like myomas - benign tumors. They are encountered most commonly in women between 40 and 50 years of age and occur relatively frequently after the menopause. The prevalence of polyps in the general population is about 24% [10].

As most polyps are small, they do not cause symptoms. The most common clinical presentation is abnormal bleeding, less frequently they are the cause of infertility. Two types of polyps are distinguished, one type originating from the corpus, the other from the cervix [14]. There is no classification based on the size or shape of a polyp.

Polyps originate as focal hyperplasias of the basalis and develop into benign, localized overgrowth of endometrial tissue covered by epithelium and containing a variable amount of glands, stroma, and blood vessels. These components make them - in contrast to myomas - very soft. Polyps may be broad-based and sessile, pedunculated, or attached to the endometrium by a slender stalk. Furthermore, they vary in size from 1.0 mm to a mass that fills and expands the entire uterine cavity. They rarely exceed 1.5 cm in diameter. Large polyps may extend down the endocervical canal or may even protrude into the vagina, being visible on physical examination. The surface is tan and glistening, but occasionally the tip or the entire polyp may be hemorrhagic or characterized by sufusions and petechiae due to irritation of infraction. Polyps are generally solitary, but about 20% are multiple. They may originate anywhere in the uterine cavity, but most occur in the fundus, usually in the cornual region.

The occurrence of carcinoma in benign polyps has been reported to be no more than 0.5%; however, polyps have been found in 12-34% of uteri with endometrial carcinoma. Polyps found in postmenopausal women are therefore always subject to histological investigation, as they are often a side effect of a carcinoma [1]. It is important to remove any polyp completely from the basis, otherwise there will always be a chance of relapse.

Hysteroscopy offers the accepted advantages of direct visualization, targeted biopsy and simultaneous complete surgical resection of polyps and myomas. Ultrasound often fails to accurately differentiate between polyps, hyperplasia and endometrial carcinoma. Hysteroscopy is more specific as a diagnostic tool in cases of post-menopausal bleeding and the combined use of hysteroscopy and biopsy leads to nearly 100% accuracy [16].

The hysteroscopic resection of a polyp - as opposed to the resection of a myoma - is best performed by picking the stem of the polyp and extruding it. Once the polyp is separated from the uterus, it can easily be pulled out of the cavity, as the soft tissue of the polyp can be squeezed through the cervix.

Table 1 summarizes the basic differences between myomas and polyps in the context of surgical training simulation. If polyps and myomas cannot be visually distinguished, haptic feedback can indicate the pathology.

Table 1. Fundamental differences between myomas and polyps

Property	Myoma	Polyp
Origin	Myometrium	Endometrium
Tissue	Stiff	Soft
Shape	Spherical	Elongated
Resection	Carving	Extruding

3   Skeleton Based Pathology Design

Almost a century ago, D'Arcy Thompson already identified the decisive role of axial growth processes in forming the shape of living objects [4]. The concept of local symmetries has later been shown to capture axial growth efficiently and can be handled by a coherent mathematical theory [2]. Local symmetries are described by skeletons, which are therefore a good tool for designing anatomical objects such as the pathologies under scrutiny. For axial growth, a skeleton can be approximated by of a set of connected axes that describe the geometry of the object, and a width r associated with every position p of every axis of the skeleton [13].

The design process is initiated by generating a skeleton S. Even though relatively complex growth patterns can theoretically emerge, only one main axis of growth can be identified in the presented pathologies. However, more complex geometrical structures are sometimes observed in polyps, e.g. in cases where the tip of the polyp seems to split. Yet, extreme tree-like branching does not occur. These small geometrical changes can be handled differently by specific application of the surface distortion, as shown below.

In an unconstrained environment, the growth process can be idealized to follow one straight axis. The restrictions imposed by the organ's shape are considered later in the post-processing steps of the design. The beginning of the skeleton is aligned with the origin of the coordinate system and the tip placed somewhere on the z-axis. The z-value represents the size of the pathology specified by the physician and scaled to the relative size of the organ. Using this orientation, the socket of the pathology lies on an virtual plane Σ which coincides with the xy-plane.

The width function $r(p)$ is determined by a characteristic profile of the desired pathology. This profile is generated by a two dimensional non-uniform rational

B-spline (NURBS) curve C parameterized by $t \in [0..1]$ [5]. The control points of the curve C have been empirically selected to produce profiles that match the descriptions and images of real pathologies. As one basic requirement for the generation of models for surgical simulators is the specification in medical terminology [15], the physician will select the desired pathology (myoma or polyp) as well as the size and width of the body and the peduncle. Consistency tests have to ensure meaningful values before transforming the specifications into the control points.

The profile is discretized at regular intervals of the parameter t. In the case of the polyp, two different radii can be specified to generate elliptical axial shapes. The profile is then scaled by $r_1 \cos(\varphi) + r_2 \sin(\varphi)$. This extension is not needed in the case of myomas, due to their spherical shapes. The surface Ω of the pathology is a 4-connected grid where its vertices are defined by the discrete angles φ_i and the discrete positions t_i of the profile. The resulting structure is open at the socket on the plane Σ.

In addition to the grid connections, every point $p \in \Omega$ is connected to its closest skeleton point, resulting in a segmentation of the skeleton at z-values corresponding to the projections of the t_i positions of the profile. Additionally, a triangulation of the surface is defined for rendering purposes. All connections are modeled by springs. Three different stiffness factors are assigned to the following spring elements in increasing order: surface to surface, surface to skeleton and skeleton to skeleton.

Randomness is introduced by modulating the generated surface Ω. The points p on the surface are displaced in normal directions $\boldsymbol{n}_p$. The amount $d(p)$ of displacement is chosen such as to achieve a predefined global amplitude a_0 and frequency. The values $d(p)$ can be stored in a matrix D since the surface Ω is a 4-connected grid. In principle, the values of every element D_{ij} in the matrix can be randomly selected and the resulting matrix smoothed with a circular Gauss filter. The amplitude a_0 of the perturbation is additionally modulated with the radius $r(p)$ to avoid too large distortions at the peduncle:

$$d(p) = a_0 r(p) x$$

where x is a uniformly distributed random variable in the range $[-1..1]$. The repeated application of this process with different amplitudes and filter-lengths leads to different perturbation patterns at different scales, allowing a simple interactive modification of the shape. In the actual implementation, the modulating and smoothing steps are unified for efficiency. Figure 1 illustrates the results of the process described for a polyp and a myoma.

4   Fusion of Pathology with Healthy Anatomy

Alternative tumor growth models place an initial seed for the pathology directly in the organ's model. Finite element analysis has been used to model the deformation of the organ induced by the growth process [11]. The application of such concepts is coupled with the modeling of the growth processes inside the

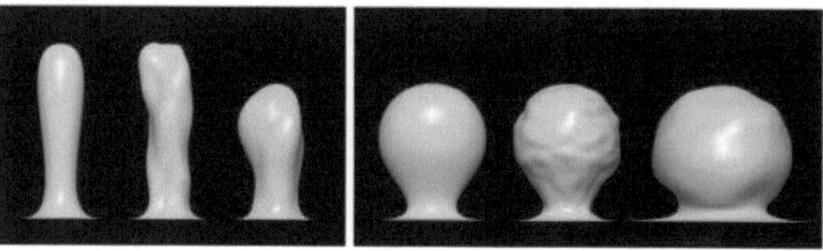

Fig. 1.   The first image exhibits an undistorted and its corresponding distorted polyp profile, as well as a variation of the shape. The second image illustrates the corresponding profiles for myomas.

organ and thus not applicable in the current approaches. In both the cellular automaton and the skeleton method, the pathology models $M_{pathology}$ have to be combined with the organ model M_{organ} after they have been generated.

A model M is an entity associated with a certain representation. In the current context, M_{organ} is an uterus represented by a triangulated, inner and outer surface oriented according to a patient specific coordinate system. In the case of the cellular automaton, M_{CA} is a voxel based representation of different concentrations on which the rules are applied. The skeleton model M_S of a pathology incorporates a surface representation with a local coordinate system at the origin of the pathology. The set $\mathcal{L}$ denotes a representation type, by convention labeled with the original model instance. Therefore $\mathcal{L}_{organ}$ is the set of models parameterized by a surface triangulation, whereas $\mathcal{L}_{pathology}$ is the set of models represented by voxels in the case of the cellular automaton and by surface triangulation in the case of the skeleton-based approach.

The combination of the organ and the pathology corresponds to a transformation $T : \mathcal{L}_a \to \mathcal{L}_b$ and additional operations on the models M. For the case of the cellular automaton this leads to

$$\mathcal{L}_{organ} \xrightarrow{T_1} \mathcal{L}_{pathology} \xrightarrow{T_2} \mathcal{L}_{organ}$$

T_1 being the voxelization of the organs surface, T_2 a voxel to surface conversion algorithm, e.g. Marching Cubes [12]. In the ideal case $T_2 = T_1^{-1}$, so that parts not involved in the tumor gestation process are not altered. As $T_2 \approx T_1^{-1}$, it is better to select the region of growth in $\mathcal{L}_{organ}$ and only transform this portion of the surface. T_2 will then produce a surface fragment that has to be merged with the remaining organ's surface. Possible collisions have to be handled appropriately. The improved procedure can be written as

$$M'_{organ} \quad \subset \quad M_{organ} \tag{1}$$

$$\mathcal{L}_{organ} \xrightarrow{T_1} \mathcal{L}_{pathology} \xrightarrow{T_2} \mathcal{L}_{organ} \tag{2}$$

$$M_{scene} \quad = \quad f_{collision}(M_{organ} \cup M_{pathology}) \tag{3}$$

where $\subset$ defines the selection of a region of interest, $\cup$ the surface merging operation, and $f_{collision}$ the handling of any collisions. This procedure explains, why the

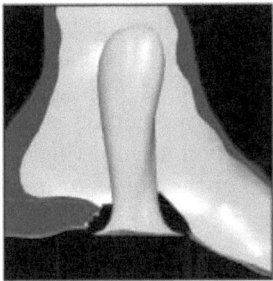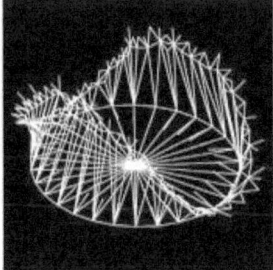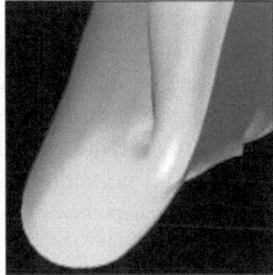

Fig. 2.   Image a) is an example of a polyp inserted in the organ before merging. Image b) shows the resulting connections and image c) the result after relaxation of the springs.

cellular automaton is not suited for the generation of polyps. A cellular automaton cannot handle collisions, since two adjacent cells of the same tissue have no information whether they belong to the same or different surface fragments. To handle collisions, equations 1, 2 and 3 would have to be applied on every growth iteration. The repeated change of representation would lead to an unacceptable accumulation of errors.

For the skeleton approach, the combination is simplified by the fact that the model already entails a surface representation. The only transformation required is

$$\mathcal{L}_{pathology} \quad \xrightarrow{T} \quad \mathcal{L}_{organ} \qquad (4)$$

where T represents a simple rotation and translation of the model M_S to position the local coordinate system in the selected point p of origin of the pathology in the organ. Equation 3 also holds for the transformed model M_S to obtain a coherent surgical scene.

Merging M_S and M_{organ} in $\mathcal{L}_{organ}$ is more involved than in the case of the cellular automaton and will be described in more details. For simplicity of the description, equation 4 is inverted. Thus, merging is accomplished in $\mathcal{L}_{pathology}$. The final result needs to be transformed with T to obtain a scene in the correct coordinate system.

The triangulation surrounding the point $p \in M_{organ}$ is refined to roughly match the resolution of the pathology's rim $\mathcal{R}_{pathology}$. The points closer to the point p than the pathology's radius are removed, thereby creating a hole in M_{organ} of approximately the size of the pathology's base. The resulting rim in M_{organ} is denoted $\mathcal{R}_{organ}$. Image a) in Figure 2 shows an exemplary case.

The points of $\mathcal{R}_{pathology}$ and $\mathcal{R}_{organ}$ have to be connected, so that the surface triangulation is closed and no two points of the same rim are connected. That is, m points of $\mathcal{R}_{organ}$ and n points of $\mathcal{R}_{pathology}$ have to be mapped with the additional constraints of no overlapping connections and smallest possible distances between the connected points. The later restriction is required during the relaxation process of the springs derived from the newly created connections.

First the points on $\mathcal{R}_{organ}$ are projected on the plane Σ by rotating them around the closest tangential segment of $\mathcal{R}_{pathology}$, so that they lie outside of the ellipse described by $\mathcal{R}_{pathology}$. The points $p_i \in \mathcal{R}_{organ}$ and $q_j \in \mathcal{R}_{pathology}$ next to each other are selected as starting points p_0 and q_0. The remaining points of each rim are sorted along the rim with the same orientation around the origin. A matrix of all possible angles between the normals $\boldsymbol{n}_j$ of the points q_j and the vectors $\boldsymbol{m}_{ij} = \overrightarrow{q_j p_i}$ is constructed

$$A_{ij} = \sphericalangle (\boldsymbol{n}_j, \boldsymbol{m}_{ij}) \qquad A \in \Re^{m \times n}$$

A triangle between the two rims is now represented by two adjacent values in A. The global optimal triangulation in the sense stated above corresponds to the cheapest path from A_{00} to A_{mn}. This path can be found by building a new matrix B that accumulates the values in A starting from A_{00}

$$B_{ij} = \min(A_{i,j} + B_{i-1,j}, A_{i,j} + B_{i,j-1}) \qquad B \in \Re^{m \times n}$$

and building the triangles by starting at B_{00} and always selecting the lower of the adjacent values. Finally the points q_j are projected back to their original positions.

The new connections are represented by springs with equilibrium length zero. These springs and the pathology's springs are relaxed to attract the pathology's surface Ω to the uterine cavity. Image b) in Figure 2 illustrates all springs connected to $\mathcal{R}_{organ}$ and $\mathcal{R}_{pathology}$ before relaxation. Image c) in Figure 2 is the result after relaxation of the springs. The boundary between the pathology's rim and the uterine cavity is no longer $\mathcal{C}^1$-continuous, as was the case with the original virtual plane Σ. Nevertheless, as can be seen in image c) in Figure 2, this inconsistency is barely visible on the scale of interest.

The inserted pathology will very likely overlap at some places with the uterine cavity. Depending on the type of pathology, this collisions have to be dealt with differently. As argued in Section 2, myomas have a very strong tendency to keep a spherical shape. The edges of the triangles building the surface of the organ are equally modeled by springs. A collision of a myoma and the cavity will be handled by correcting the cavity's surface while fixing any point of the pathology.

In contrast to the myoma the polyp is a very soft tissue mass, which will be deformed according to the uterine cavity's shape. Only springs belonging to the pathology, both peduncle and polyp, are relaxed during collision handling. The final results of the skeleton-based approach are compared to a real polyp in Figure 3.

5   Conclusion and Future Research

The presented approaches have all been discussed with a experienced gynecologists, who attested them a high resemblance with real cases. As we are aware that such validation will never overcome the limitations of subjective judgment, we are looking for long term validation strategies. This will entail the comparison of different modeling approaches as they emerge and the comparison of the

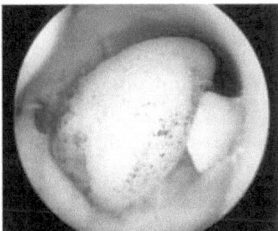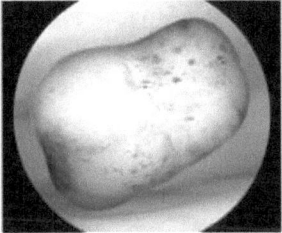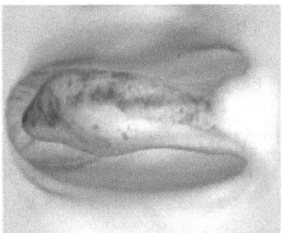

Fig. 3.   Real polyp (left) and two different views of an artificial polyp in the uterine cavity. The cavity has been inflated to allow for a better view, as is done during surgery. The last image is an unusual perspective during actual surgery but gives an excellent overview of the scene.

resulting models during actual training with real cases, giving not only the visual, but also the haptic feedback. Future research will concentrate on the two pathologies presented.

The described approach of designing a pathology for surgical training simulators has both advantages and disadvantages over the cellular automaton. The main advantages are, that both polyps and myomas can be generated and that no growing time is required, making the generation of a pathology model possible in the order of one second. While completely autonomous generation was one of the requirements stated for the generation of pathologies for surgical training simulators, a physician might want to modify the resulting pathologies to more specifically match his expectations. The presented model is well suited for this task, as mass-spring models of this size can be modified at interactive rates.

However, some features might be missing. As an example the inner core of the myomas cannot be discerned from the surface. Therefore the generated myomas might not supply sufficient information for an accurate vascularization model.

Polyps are also a common finding in other organs, e.g. the colon (colon-rectal polyps). The described approach is completely independent of the organ chosen, allowing a direct application to any other surface-based organ model.

Acknowledgments

This work has been performed within the frame of the Swiss National Center of Competence in Research on Computer Aided and Image Guided Medical Interventions (NCCR CO-ME) supported by the Swiss National Science Foundation.

References

1. M. Bajka. *Empfehlungen zur Gynäkologischen Sonographie.* Schweizerische Gesellschaft für Ultraschall in der Medizin, 2001.
2. H. Blum. A transformation for extracting new descriptors of shape. *Models for the Perception of Speech and Visual Form. MIT Press*, pages 362–380, 1967.

3. Cootes et al. Active Shape Models - Their Training and Application. *Computer Vision and Image Understanding*, 61(1):38–59, 1995.
4. D'Arcy W. Thompson. *On Growth and Form, Volume I,II.* Cambridge University Press, 1952.
5. G.Farin. *Curves and Surfaces for CAGD. A practical guide.* Academic Press, San Diego, USA, fourth edition, 1997.
6. G.Székely et al. Virtual Reality-Based Simulation of Endoscopic Surgery. *Presence*, 9(3):310–333, 2000.
7. Kelemen et al. Elastic Model-Based Segmentation of 3-D Neororadiological Data Sets. *IEEE Transactions on Medical Imaging*, 18(10):828–839, 1999.
8. K.Montgomery et al. Surgical Simulator for Hysteroscopy: A Case Study of Visualization in Surgical Training. *IEEE Visualization 2001*, 2001.
9. C. Kuhn. *Modellbildung und Echtzeitsimulation deformierbarer Objekte zur Entwicklung einer interaktiven Trainingsumgebung für Minimal-Invasive Chirurgie.* Forschungszentrum Karlsruhe GmbH, Karlsruhe, 1997.
10. R. Kurman and M. Mazur. *Blaustein's Pathology of the Female Genital Tract.* Springer, Berlin, New York, fourth edition, 1990.
11. S. Kyriacou et al. Nonlinear Elastic Registration of Brain Images with Tumor Pathology Using a Biomechanical Model. *IEEE Transactions on Medical Imaging*, 18(7):580–592, 1999.
12. W. Lorensen and H. Cline. Marching Cubes: A High Resolution 3D Surface Construction Algorithm. *Computer Graphics*, 21(4):163–170, 7.1987.
13. S. Pizer et al. Multiscale Medial Loci and Their Properties. *Int. J. Comp. Vision*, accepted 2002.
14. Pschyrembel, Strauss, and Petri. *Praktische Gynäkologie für Studium, Klinik und Praxis.* de Gruyter, Berlin, New York, fifth edition, 1990.
15. R.Sierra, G.Székely, and M.Bajka. Generation of Pathologies for Surgical Training Simulators. *MICCAI*, Proceedings, Part II:202–210, 2002.
16. *Journal of Obstetrics and Gynecology of India.* http://www.journal-obgyn-india.com/issue_march_april2001/g_papers_115.htm, 2002.

Author Index

Lecture Notes in Computer Science

For information about Vols. 1–2603

please contact your bookseller or Springer-Verlag

Vol. 2645: M.A. Wimmer (Ed.), Knowledge Management in Electronic Government. Proceedings, 2003. XI, 320 pages. 2003. (Subseries LNAI).

Vol. 2646: H. Geuvers, F, Wiedijk (Eds.), Types for Proofs and Programs. Proceedings, 2002. VIII, 331 pages. 2003.

Vol. 2647: K.Jansen, M. Margraf, M. Mastrolli, J.D.P. Rolim (Eds.), Experimental and Efficient Algorithms. Proceedings, 2003. VIII, 267 pages. 2003.

Vol. 2648: T. Ball, S.K. Rajamani (Eds.), Model Checking Software. Proceedings, 2003. VIII, 241 pages. 2003.

Vol. 2649: B. Westfechtel, A. van der Hoek (Eds.), Software Configuration Management. Proceedings, 2003. VIII, 241 pages. 2003.

Vol. 2651: D. Bert, J.P. Bowen, S. King, M, Waldén (Eds.), ZB 2003: Formal Specification and Development in Z and B. Proceedings, 2003. XIII, 547 pages. 2003.

Vol. 2652: F.J. Perales, A.J.C. Campilho, N. Pérez de la Blanca, A. Sanfeliu (Eds.), Pattern Recognition and Image Analysis. Proceedings, 2003. XIX, 1142 pages. 2003.

Vol. 2653: R. Petreschi, Giuseppe Persiano, R. Silvestri (Eds.), Algorithms and Complexity. Proceedings, 2003. XI, 289 pages. 2003.

Vol. 2655: J.-P. Rosen, A. Strohmeier (Eds.), Reliable Software Technologies – Ada-Europe 2003. Proceedings, 2003. XIII, 489 pages. 2003.

Vol. 2656: E. Biham (Ed.), Advances in Cryptology – EUROCRPYT 2003. Proceedings, 2003. XIV, 429 pages. 2003.

Vol. 2657: P.M.A. Sloot, D. Abramson, A.V. Bogdanov, J.J. Dongarra, A.Y. Zomaya, Y.E. Gorbachev (Eds.), Computational Science – ICCS 2003. Proceedings, Part I. 2003. LV, 1095 pages. 2003.

Vol. 2658: P.M.A. Sloot, D. Abramson, A.V. Bogdanov, J.J. Dongarra, A.Y. Zomaya, Y.E. Gorbachev (Eds.), Computational Science – ICCS 2003. Proceedings, Part II. 2003. LV, 1129 pages. 2003.

Vol. 2659: P.M.A. Sloot, D. Abramson, A.V. Bogdanov, J.J. Dongarra, A.Y. Zomaya, Y.E. Gorbachev (Eds.), Computational Science – ICCS 2003. Proceedings, Part III. 2003. LV, 1165 pages. 2003.

Vol. 2660: P.M.A. Sloot, D. Abramson, A.V. Bogdanov, J.J. Dongarra, A.Y. Zomaya, Y.E. Gorbachev (Eds.), Computational Science – ICCS 2003. Proceedings, Part IV. 2003. LVI, 1161 pages. 2003.

Vol. 2663: E. Menasalvas, J. Segovia, P.S. Szczepaniak (Eds.), Advances in Web Intelligence. Proceedings, 2003. XII, 350 pages. 2003. (Subseries LNAI).

Vol. 2665: H. Chen, R. Miranda, D.D. Zeng, C. Demchak, J. Schroeder, T. Madhusudan (Eds.), Intelligence and Security Informatics. Proceedings, 2003. XIV, 392 pages. 2003.

Vol. 2667: V. Kumar, M.L. Gavrilova, C.J.K. Tan, P. L'Ecuyer (Eds.), Computational Science and Its Applications – ICCSA 2003. Proceedings, Part I. 2003. XXXIV, 1060 pages. 2003.

Vol. 2668: V. Kumar, M.L. Gavrilova, C.J.K. Tan, P. L'Ecuyer (Eds.), Computational Science and Its Applications – ICCSA 2003. Proceedings, Part II. 2003. XXXIV, 942 pages. 2003.

Vol. 2669: V. Kumar, M.L. Gavrilova, C.J.K. Tan, P. L'Ecuyer (Eds.), Computational Science and Its Applications – ICCSA 2003. Proceedings, Part III. 2003. XXXIV, 948 pages. 2003.

Vol. 2670: R. Peña, T. Arts (Eds.), Implementation of Functional Languages. Proceedings, 2002. X, 249 pages. 2003.

Vol. 2671: Y. Xiang, B. Chaib-draa (Eds.), Advances in Artificial Intelligence. Proceedings, 2003. XIV, 642 pages. 2003. (Subseries LNAI).

Vol. 2672: M. Endler, D. Schmidt (Eds.), Middleware 2003. Proceedings, 2003. XIII, 513 pages. 2003.

Vol. 2673: N. Ayache, H. Delingette (Eds.), Surgery Simulation and Soft Tissue Modeling. Proceedings, 2003. XII, 386 pages. 2003.

Vol. 2674: I.E. Magnin, J. Montagnat, P. Clarysse, J. Nenonen, T. Katila (Eds.), Functional Imaging and Modeling of the Heart. Proceedings, 2003. XI, 308 pages. 2003.

Vol. 2675: M. Marchesi, G. Succi (Eds.), Extreme Programming and Agile Processes in Software Engineering. Proceedings, 2003. XV, 464 pages. 2003.

Vol. 2676: R. Baeza-Yates, E. Chávez, M. Crochemore (Eds.), Combinatorial Pattern Matching. Proceedings, 2003. XI, 403 pages. 2003.

Vol. 2678: W. van der Aalst, A. ter Hofstede, M. Weske (Eds.), Business Process Management. Proceedings, 2003. XI, 391 pages. 2003.

Vol. 2679: W. van der Aalst, E. Best (Eds.), Applications and Theory of Petri Nets 2003. Proceedings, 2003. XI, 508 pages. 2003.

Vol. 2686: J. Mira, J.R. Álvarez (Eds.), Computational Methods in Neural Modeling. Proceedings, Part I. 2003. XXVII, 764 pages. 2003.

Vol. 2687: J. Mira, J.R. Álvarez (Eds.), Artificial Neural Nets Problem Solving Methods. Proceedings, Part II. 2003. XXVII, 820 pages. 2003.

Vol. 2688: J. Kittler, M.S. Nixon (Eds.), Audio- and Video-Based Biometric Person Authentication. Proceedings, 2003. XVII, 978 pages. 2003.

Vol. 2689: K.D. Ashley, D.G. Bridge (Eds.), Case-Based Reasoning Research and Development. Proceedings, 2003. XV, 734 pages. 2003. (Subseries LNAI).

Vol. 2692: P. Nixon, S. Terzis (Eds.), Trust Management. Proceedings, 2003. X, 349 pages. 2003.

Vol. 2694: R. Cousot (Ed.), Static Analysis. Proceedings, 2003. XIV, 505 pages. 2003.

Vol. 2695: L.D. Griffin, M. Lillholm (Eds.), Scale Space Methods in Computer Vision. Proceedings, 2003. XII, 816 pages. 2003.

Vol. 2701: M. Hofmann (Ed.), Typed Lambda Calculi and Applications. Proceedings, 2003. VIII, 317 pages. 2003.

Vol. 2702: P. Brusilovsky, A. Corbett, F. de Rosis (Eds.), User Modeling 2003. Proceedings, 2003. XIV, 436 pages. 2003. (Subseries LNAI).

Vol. 2706: R. Nieuwenhuis (Ed.), Rewriting Techniques and Applications. Proceedings, 2003. XI, 515 pages. 2003.

Vol. 2707: K. Jeffay, I. Stoica, K. Wehrle (Eds.), Quality of Service – IWQoS 2003. Proceedings, 2003. XI, 517 pages. 2003.

Vol. 2709: T. Windeatt, F. Roli (Eds.), Multiple Classifier Systems. Proceedings, 2003. X, 406 pages. 2003.

Vol. 2716: M.J. Voss (Ed.), OpenMP Shared Memory Parallel Programming. Proceedings, 2003. VIII, 271 pages. 2003.